EVALUATION AND CONTROL OF MEAT QUALITY IN PIGS

CURRENT TOPICS IN VETERINARY MEDICINE AND ANIMAL SCIENCE

EVALUATION AND CONTROL OF MEAT QUALITY IN PIGS

A Seminar in the CEC Agricultural Research Programme, held in Dublin, Ireland, 21–22 November 1985

Sponsored by the Commission of the European Communities, Directorate-General for Agriculture, Division for the Coordination of Agricultural Research

Edited by

P.V. Tarrant
An Foras Talúntais, The Agricultural Institute, Grange / Dunsinea Research Centre Castleknock, Dublin, Ireland

G. Eikelenboom
Research Institute for Animal Production "Schoonoord", Zeist, The Netherlands

G. Monin
Station for Meat Research, INRA-CRZV de Theix, Ceyrat, France

1987 **MARTINUS NIJHOFF PUBLISHERS**
a member of the KLUWER ACADEMIC PUBLISHERS GROUP
DORDRECHT / BOSTON / LANCASTER
for
THE COMMISSION OF THE EUROPEAN COMMUNITIES

Distributors

for the United States and Canada: Kluwer Academic Publishers, P.O. Box 358, Accord Station, Hingham, MA 02018-0358, USA
for the UK and Ireland: Kluwer Academic Publishers, MTP Press Limited, Falcon House, Queen Square, Lancaster LA1 1RN, UK
for all other countries: Kluwer Academic Publishers Group, Distribution Center, P.O. Box 322, 3300 AH Dordrecht, The Netherlands

Library of Congress Cataloging in Publication Data

```
Evaluation and control of meat quality in pigs.

   (Current topics in veterinary medicine and animal
science)
   1. Swine--Congresses.  2. Pork--Quality--Congresses.
I. Tarrant, P.V.  II. Eikelenboom, G.  III. Monin, G.
IV. Commission of the European Communities.  Division
for the Coordination of Agricultural Research.
V. Series.
SF391.3.E93  1987    636.4'0883       86-31175
ISBN 0-89838-854-6 (U.S.)
```

ISBN 0-89838-854-6
EUR 10421 EN

Book information

Publication arranged by: Commission of the European Communities, Directorate-General Telecommunications, Information Industries and Innovation, Luxembourg

Copyright/legal notice

PRINTED IN THE NETHERLANDS

CONTENTS

PREFACE

These are the proceedings of a European Community seminar held at the Agricultural Institute, Dublin 4, Ireland on the 21 and 22 November 1985.

The subject of this seminar, the evaluation and control of meat quality in pigs, is the concentrated on two important and closely related aspects of the pigmeat industry, pig welfare and product quality.

There is great interest in the welfare of pigs during the preslaughter period. The welfare of animals with inherited susceptibility to the stress syndrome is of particular concern. Meat quality is determined by breeding and husbandry practices and in particular by the handling of animals and carcasses immediately before and after slaughter. The major challenge facing the pig industry today is to satisfy the consumers' demand for a competitively priced and high quality food that has been produced under acceptable and humane conditions. The work of this seminar will help the industry to meet this challenge.

In the preparation and running of the seminar, we acknowledge the direction and support of the Commission of the European Communities, Directorate General for Agriculture, Division for the Coordination of Agricultural Research. We thank the scientists who contributed papers on basic aspects and meat industry applications, those who acted as Chairman and Rapporteurs and those who assisted in the numerous aspects of organisation and in the preparation of the published proceedings. A number of important conclusions were reached on pig welfare and pigmeat quality.

P V Tarrant, G Eikelenboom, G Monin

April 1986

SESSION I

AETIOLOGY OF THE PORCINE STRESS SYNDROME

Chairman: D. Lister

THE PHYSIOLOGY AND BIOCHEMISTRY OF THE PORCINE STRESS SYNDROME

David Lister

Animal and Grassland Research Institute
Shinfield, Reading RG2 9AQ, U.K.

ABSTRACT

Much of the pertinent biochemistry and physiology of the Porcine Stress Syndrome has been defined but a number of key problems have yet to be solved. For example, we do not know why individuals differ in their threshold of sensitivity to stressors, exactly how muscle activity is triggered under these circumstances and why some animals may quickly recover from such stimuli and others may not. Neither is it completly clear how sensitivity to stressors and body composition are linked. There is, however, good evidence that the sympathetic nervous system and the catecholamines of the adrenal medulla are essential elements in all of this.

INTRODUCTION

The Porcine Stress Syndrome (PSS) has been an important focus in research on pigs and pigmeat for at least 20 years. The research has taken on many guises to include topics as different as genetic polymorphisms, animal behaviour, mitochondrial function and the classical approaches of stress physiology. What began as enquiries into the nature of aberrations in meat quality, quickly recognised in the aetiology of the problem the important contribution of genetic type and the ways in which animals were handled, and drew in expertise from animal and meat science. The pig was recognised as a model for certain comparable conditions in man and opened the field for contributions from human medicine. It is fair to say, however, that many of the early prognostications of researchers, especially in regard to associations between the leanness of carcases and the quality of meat, fell on deaf ears and only relatively recently have the pig and meat industries begun to ask whether the search for ever leaner carcases should not be tempered to preserve the vigour of the live animal and the quality of its meat.

There have been many reviews of PSS in terms of the associated physiology and biochemistry (Briskey, 1964; Cassens et al., 1975; Lister et al., 1981) and little will be served by rehearsing them. Instead I shall attempt to identify the physiological and biochemical issues behind the following questions and raise more questions which require further research.

1. What is PSS?
2. How can PSS be described in objective terms?
3. How is meat quality affected?
4. What makes some pigs stress prone?

WHAT IS PSS?

All animals (and men) will react characteristically to physical, mental and environmental stressors and, if those stressors are particularly severe, individuals may die. Within species, stress prone lines have been identified and perpetuated by breeding. It may be considered that PSS represents the perpetuation of such proneness amongst pigs which have been selected and bred primarily for, for example, rapidity and efficiency of growth, or body composition and conformation, but which also contain undesirable characteristics which may be called 'stress proneness'.

Typically Pietrain, Poland China and strains of Landrace pigs have been noted for their dramatic and often fatal reactions to such stimuli as transportation, raised environmental temperature and physical exercise, restraint or general handling (Topel et al., 1968). This porcine stress syndrome is characterised by muscle tremor, an increased respiratory rate, a systemic acidosis and a rise in body temperature. After the animal dies, or if it is slaughtered in the reacting state, rigor mortis develops almost immediately to give rise to the characteristic pale, soft and exudative (PSE) condition of the meat. Fatal responses to environmental stress are to be found amongst several species of animals. Capture myopathy is a condition which develops in wild animals such as antelopes and buffaloes being pursued and captured for whatever reason and many will die (Harthoorn et al., 1974). Clinical symptoms and post mortem findings are similar to those of PSS.

Malignant Hyperthermia (MH) which develops in certain pigs (and people) when they are exposed especially to the halogenated gaseous anaesthetics, but also spontaneously during stressful encounters, is another reaction which has been incorporated into the general condition referred to as PSS (Nelson et al., 1974; Lucke et al., 1979). MH has been particularly important in defining the physiological and biochemical events in PSS for it is possible to precipitate MH under laboratory conditions and thereby allow detailed monitoring of blood and tissues throughout the reaction.

HOW CAN PSS BE DESCRIBED IN OBJECTIVE TERMS?

The first steps towards characterising PSS were made in the late 1960s when attempts were made to identify the metabolic events in muscle or more generally associated with the exposure of animals to various and extreme environmental conditions (Kastenschmidt et al., 1965; Forrest et al., 1968; Kallweit, 1969). In some studies (Judge 1969) a period of exercise was included in addition to thermal stress. Results from these experiments paved the way for the more sophisticated approaches of later studies which for the most part only confirmed the earlier findings. Temperature stress, it seems, is likely to lead to respiratory alkalosis which becomes a metabolic acidosis if exercise is included as a treatment.

The extent of the reaction to these stressors, and particularly the ability of animals to maintain or recover homeostasis, was used to identify so-called stress-susceptibility or -resistance amongst the experimental pigs. This terminology has become commonplace in the literature but generally refers to breeds e.g. Poland China, stress-susceptible; Chester White, stress-resistant. In recent times this nomenclature has been refined with the introduction of the terms Halothane-positive or -negative which may be used to describe animals within as well as between breeds.

It is generally accepted that the pattern of metabolic, physiological and biochemical developments in MH is very similar to, if not the same as, that which occurs as a response to environmental stress. The MH reaction has been extensively characterised in the literature (Berman et al.,1970; Hall et al.,1980; Gronert,1980) and only a short description is necessary here.

Essentially, the characteristic signs of MH (and stressor induced reactions generally) are increased oxygen consumption and carbon dioxide production. The developing respiratory and metabolic acidoses are associated with high plasma lactate levels and falling pH. Glucose and fat mobilisation (as identified by plasma glycerol concentration) occur, though plasma free fatty acids may fall. There are substantial electrolyte changes and some haemoconcentration.

Contrary to the early views, muscle stimulation which is the primary source of heat production in MH (Hall et al., 1976) is fuelled almost if not wholly via aerobic metabolism and it is only in the later stages that this contribution declines to less than 50 per cent.

Another early view was that cardio myopathy contributed to the demise of stress-prone pigs. There is, however, no functional evidence for this for the hearts of pigs suffering MH appear to respond adequately to metabolic demand even when there is substantial acidosis and hyperkalaemia. Only in the later stages does cardio-vascular function become limiting. At this time cyanosis of the skin, resulting from peripheral vasoconstriction, is associated with reduced heat loss, reduction in substrate provision and lactate clearance from tissues and as the biochemical changes reach extreme values, blood pressure falls and cardiac arrest occurs.

All these changes are familiar findings in the terminal stages of the Porcine Stress Syndrome.

HOW IS MEAT QUALITY AFFECTED?

The consequences of PSS post mortem can be either the acceleration of the normal biochemical changes which occur during the rigor process or their curtailment, leading in the first instance to pale, soft and exudative (PSE) meat and, in the second, to dark, firm and dry (DFD) meat.

In the living animal, metabolism is sustained through adenosine triphosphate (ATP) production, predominantly via aerobic means, using glucose and fatty acids as substrates. At times, demand is such that aerobiosis is insufficient to meet metabolic needs for ATP. This occurs in the classical fight/flight reactions when anaerobic glycolysis provides for an increasing proportion of energy needs. Though this process is only approximately 5% as efficient in the production of ATP as aerobic processes, it is nevertheless an important source of energy for 'emergency' needs and the sole mechanism employed in the transition of living muscle to dead meat. The two systems in muscle cells which control these changes are the glycolytic system which converts glycogen to lactate and those processes which lead to the dephosphorylation of ATP. Acidification occurs as a consequence of glycolysis; rigor, as a result of the loss of ATP. The series of reactions which regulate all this is described by Bendall (1973).

It is important to recognise this interdependence of glycolysis and the production of lactate and the enzymic processes in living muscle whose function is to synthesis ATP from adenosine diphosphate (ADP) and inorganic phosphate in anaerobic conditions. In 'fast' muscles, the production of sufficient pyruvate for mitochondrial oxidation requires only a small fraction of the glycolytic enzymes present; the large excess of these enzymes exists for anaerobic circumstances. The freely reacting adenine mononucleotides exist almost entirely as ATP and glycolysis can occur only as long as ADP is being produced by an ATPase system. Ultimately, therefore, the rate of glycolysis is determined by ATPase activity and not directly by the activities or amounts of glycolytic enzymes or hormones.

In the initial stages of a surge in demand for ATP or in the post mortem period, phosphocreatine, which is present at high levels initially, provides for the re-conversion of ADP (produced by ATPase activity) to ATP. Glycolytic phosphorylation of ADP becomes increasingly important as the concentration of phosphocreatine falls. The first reactions of glycolysis produce enough substrate to remove ADP at the rate it is formed but there is no suggestion that the concentration of ADP is rate limiting. Control is achieved via feedback on both phosphorylase and phosphofructokinase activity by the adenosine monophosphate (AMP) level which in turn is regulated by AMP deaminase. This enzyme is also primarily responsible for the simultaneous accumulation of inosine monophosphate (IMP) and depletion of AMP especially at pH<6.5. The action of myokinase restores the lost AMP via ADP which in turn is restored by ATPases and a consequent reduction in ATP concentration.

Glycolysis stops and the ultimate pH is reached when there is no freely reacting adenine mononucleotide in the muscle. Glycolysis will also stop at an abnormally high pH (DFD meat) when all glycogen has been depleted from muscle. This implies inadequate stores of glycogen at death for normal rigor will continue to the usual ultimate pH (~5.5 at which there may be 'residual' glycogen in muscle. In this case the exact determinant of ultimate pH i.e. the point at which glycolysis stops is not clear. It seems likely however that AMP deaminase is primarily responsible since it represents the sole cause of loss of adenine nucleotides.

From this it can be concluded that variations in the rate and extent of pH fall are attributable to the controls exercised before and after death by ATPases and AMP deaminase. The problems arise in identifying the physiological basis for these activities.

The practical consequences of these biochemical changes are the PSE and DFD conditions of pigmeat. Rigor mortis takes place in pig muscle over a period of about 10 hours during which the pH has fallen from about 7.3 to 5.5. On occasion this change may occur within a few minutes of death and the meat develops the pale, soft and exudative characteristics. The classical explanation for this has been 'thermoprotonic' stress (Scopes, 1964) which occurs when the pH of muscle falls below 6.0 and its temperature is $>30^{\circ}C$. Under these conditions extensive denaturation of the soluble and structural proteins occurs which leads to a loss of their water binding abilities and the precipitated proteins interfere with the optical properties of the muscle to cause more incident light to be reflected and a paler appearance.

Penny (1975) suggested that during normal rigor, drip came from the extracellular water compartment which became enlarged as the muscle fibres shrank. This, in effect, has the same basis as Wilding et al.'s (1984) finding that rabbit l.dorsi fibres swelled to two to three times their usual diameter when immersed in hypertonic salt solutions and, like Offer & Trinick's (1983) observations, was explained by myofibrillar swelling. Wilding et al. (1984) also concluded that myofibrillar swelling was constrained by the endomysial sheaths around fibres.

It is now proposed (see Offer, 1984) that drip formation during rigor mortis represents the redistribution of extra and intra-cellular water and the expulsion of part of the extracellular component out of the cut ends of meat close to the perimysium, probably as a result of internal pressure exerted by the connective tissue network and disruption of cell membranes (Currie & Wolfe, 1983).

This explanation does not require there to be a change in the total water content of muscle prior to an animal's death. It is, however, quite possible, indeed likely, that quite sizeable changes occur in the fluid compartments in the body prior to death, especially if there are profound metabolic and physiological disturbances happening simultaneously (Guyton et al., 1975). Lister et al. (1985 - in preparation) recently examined this proposition in halothane sensitive pigs of the Lacombe breed.

During MH reactions precipitated by halothane and suxamethonium administration, the typical changes in muscle and rectal temperature occurred, the haematocrit rose and plasma volume fell. The pH of biopsy samples of muscle fell and amounts of expressible juice and the turbidity of extracts of muscle increased. Moreover, the water content of the muscle samples at death was higher in reacting pigs than in non-reacting control animals. Thus at least a proportion of the drip released by PSE muscle may well be attributable to the increase in intracellular fluid volume which accompanies the development of MH and, probably, PSS.

DFD meat is the condition which affects meat which remains at pH>6.0 - 6.3 when there is insufficient glycogen available in muscle at death to allow the full development of acidification to occur. The premature depletion of muscle glycogen in stressed animals is not easily explained. It is not induced simply by fasting, exercise or a combination of the two; mixing groups of animals even for a short period may, however, bring it about. Tentative explanations may be found by examining the mechanisms which allow muscle to function under a variety of physiological states.

The concentration of glucose in the blood of animals is maintained within fairly constant limits even though there might be fluctuations in the storage and mobilisation of nutrients which are associated with feeding and activity. Even under resting conditions, however, activity can only be maintained for relatively short periods if carbohydrates (derived from liver glycogen in the main) are the sole source of energy. For longer periods of activity, the long chain free fatty acids (FFA) mobilised from adipose tissue are employed (Newsholme, 1980) which progressively, as activity intensifies, cause a reduction in glucose utilisation by muscle. The control over this balance is thought to be exercised by fatty acid oxidation and due in part to the inhibitory effect of citrate on the enzyme phosphofructokinase. A further increase in the workload will reduce the concentration ratio ATP/ADP and glycolysis is stimulated at the expense of stores of muscle glycogen. A continuing ability to mobilise fatty acids and the maintenance of aerobiosis is thus of the utmost importance for sustained activity.

Hall et al. (1980) showed that in MH reactions, the continued use of free fatty acids for energy purposes can be prejudiced. Lipolysis is stimulated during MH under the influence of catecholamines, but plasma FFA concentrations may fall as a consequence of their re-esterification during the developing acidosis. Because the utilis-ation of FFA is determined by their concentration in plasma, the continued energy demand of the reaction must be met ultimately by muscle glycogen.

Thus one might conclude that the loss of muscle glycogen and predisposition to DFD by stressed animals is a natural consequence of continued muscular stimulation when the supply of other energy substrates, notably FFA, is reduced.

WHAT MAKES PIGS STRESS PRONE?

It can be concluded from much of the above that many of the symptoms of PSS are the consequences of profound muscular stimulation. It may well be also that such stimulation not only provides the symptoms of PSS but also the ease with which this occurs, conditions whether pigs become stress susceptible or resistant.

The rate of glycolysis in muscle, occurring either before death or during rigor mortis, is determined by ATPase activity and not directly by the activities or amounts of glycolytic enzymes or hormones. Stress proneness, therefore, amounts to the ease with which muscle is stimulated and how readily recovery can be established.

There seems to be general agreement, though a dearth of evidence, that the ultimate triggering mechanism resides in muscle itself. However, no major morphological differences have been identified in muscle except evidence of myofibrillar damage and regeneration (Muir, 1970; Venable, 1973). Swatland and Cassens (1972) described various abnormalities in the peripheral innervation of striated muscle though no functional abnormalities have been recognised.

Defects in the ability of muscle components to handle Ca^{++} have long been suspected but rarely demonstrated, though the role of Ca^{++} in the stimulation of myofibrillar ATPases and phosphorylase kinase must be of potential importance. The sarcoplasmic reticulum was thought to have impaired function though it is likely that this could be explained by experimental artefact (Greaser et al., 1969) (though see Heffron - this volume). Halothane sensitive pigs appear to have higher sarcoplasmic Ca^{++} levels than normal (Cheah & Cheah, 1984) and this has been linked with significantly higher activity of Ca^{++} activated phospholipase A_2 (EC 3.1.1.4) (Cheah & Cheah, 1981). This in turn leads to the enhanced release of mitochondrial Ca^{++}. More recently Cheah & Cheah (1984) suggested that the enhanced phospholipase A_2 activity was due to the higher concentrations of endogenous calmodulin in sensitive pigs.

Even via these postulated mechanisms, a means has yet to be identified of switching on the aberrant behaviour of muscle in MH or PSS. Mormède and Dantzer (this volume) propose that behavioural responses to stressful environmental factors in intensive pig husbandry are prime factors in the aetiology of PSS. It is certainly true that environmental stressors affect the incidence of PSS and PSE meat but anaesthesia or tranquillisation substantially modifies the incidence of PSE only amongst those animals which are not genetically prone to the PSS condition (Briskey & Lister, 1969). What matters then, is the inherent mechanism which sets the threshold of sensitivity in pigs to make them more or less liable to develop PSS.

The role which the sympathetic nervous system and the adrenal gland play in stress responses generally (see Usdin, Kvetnansky and Kopin, 1976) and MH in particular (Hall, 1976) has been very well

documented. In MH, catecholamines appear primarily to promote the
developing reaction via $\alpha + \beta$ adrenergic responses, but pre-treatment
with α adrenergic blockers or agonists can prevent or initiate a fatal
MH response (Lister et al., 1976; Hall et al., 1977). The combination
of effects mediated by the main adrenoreceptors can account for the
majority of the metabolic events seen in MH and PSS. β receptors, for
example, are associated, inter alia, with increased liver glycogenolysis and vasoconstriction, β_1 receptors with heart rate and fat
mobilisation, and β_2 receptors with muscle glycogenolysis and vaso-
dilation.

The use of receptor specific drugs has allowed us to clarify the
particular mechanisms of relevance within PSS and MH. α blockade
clearly prevents the development of MH in sensitive pigs whereas β
blockade with propranolol (Lister et al., 1976) or carazolol (S.Lens -
personal communication) does not. The development of PSE meat can be
impeded by α blockade, though not substantially, and treatment with
propranolol may also have a slight beneficial effect (Lister, 1974).
Warriss & Lister (1982) have provided clear evidence, however, of the
effectiveness of carazolol in reducing the incidence of PSE in stress-
susceptible Pietrains. Carazolol, like propranolol, is described as a
non-specific β blocker but Warriss & Lister's findings allow a clear
distinction to be drawn between MH and aspects of PSS such as PSE
meat. Whatever the specific mechanism is which triggers MH in the
living pig,it is not the same as the trigger for the production of PSE
meat although that might represent a further part of the continuum of
adrenergic reactions which constitute PSS.

The contribution identified for the sympathetic nervous system in
MH and PSS suggests mechanisms which link up with the other notable
feature of the syndromes i.e. the leanness of susceptible individuals.
Pietrain pigs are thought to develop their leanness through enhanced
responsiveness to the fat mobilising effect of noradrenaline (Wood
et al., 1977). For such a mechanism to contribute to leanness in the
growing animal, the adipose tissue of lean pigs would have to be
exposed to levels of noradrenaline which were higher than those found
in fatter types, or the appropriate adrenergic receptors present in
greater concentrations. Gregory & Lister (1981), using a Valsalva-

like manoeuvre and measurements of responses to intravenous tyramine, noradrenaline and phenylephrine in lean, stress sensitive (Pietrain) and fatter, stress resistant (Gloucester) pigs, were able to show that the sympathetic nervous system of Pietrains was more responsive than that in Gloucesters and that this was attributable to a higher pre-adrenoreceptor responsiveness in Pietrains. They further proposed that this led to the greater leanness, susceptibility to myocardial failure and propensity to stress-induced metabolic acidosis of Pietrains. Further support for this proposal has recently been provided via the studies of Böcklen, Flad, Müller and von Faber (1985 - in press) which showed that there were >30% more β adrenergic receptors in the muscle, hearts and fat tissues of Pietrains compared with Large White pigs which, they concluded, explained the differences of meat quality and carcase fatness between the two breeds.

CONCLUSION

Although there are several elements missing from the chain which links growth, body type and stress susceptibility, we now have enough evidence to conclude that propensity for lean growth (more particulary reduction in fat) and predisposition to developing PSS are directly associated via the sympathetic nervous system and the adrenal medulla. The determinant of the setting of the threshold of sensitivity to stress amongst pigs has not been identified, neither has that which in muscle is ultimately responsible for the final stimulation of ATPase activity. We still also need to know why it is that some animals are able to recover from perturbations of their metabolism whereas in others, a cascade of metabolic insults brings about their demise.

14

REFERENCES

Bendall, J.R. 1973. In: Structure and Function of Muscle
(Ed: Bourne, G.H.) (Academic Press, New York) 2, 243-309.

Berman, M.C., Harrison,C.G., Bull, A.B. and Kench, J.E. 1970.
Changes underlying halothane-induced malignant hyperpyrexia in
Landrace pigs. Nature (Lond.) 225, 653-655.

Briskey,E.J. 1964. Etiological status and associated studies of pale,
exudative porcine musculature. Adv. Fd. Res. 13, 89-178.

Briskey, E.J. and Lister, D. 1969. Influences of stress syndrome on
chemical and physical characteristics of muscle post mortem. In:
The Pork Industry: Problems and Progress. (Ed. Topel, D.G.) (Iowa
State University Press, Ames, Iowa) pp.177-186.

Cassens, R.G., Marple, D.N. and Eikelenboom, G. 1975. Animal
Physiology and Meat Quality. Adv. Fd. Res. 21, 71-155.

Cheah, K.S. and Cheah, A.M. 1981. Skeletal muscle mitochondrial
phospholipase A_2 and the interaction of mitochondria and
sarcoplasmic reticulum in porcine M.H. Biochem. Biophys. Acta.
638, 40-49.

Cheah, K.S. and Cheah, A.M. 1984. Endogenous calmodulin, Ca^{2+} and
phospholipase A_2 activity and their relationships to halothane
sensitivity in young and adult pigs. In: Proc. 30th European
Meeting of Meat Research Workers, Bristol) pp.106 -107.

Currie, R.W. and Wolfe, F.H. 1983. An assessment of Extracellular Space
measurements in post-mortem muscle. Meat Sci. 8, 147-161.

Forrest, J.C., Will, J.A., Schmidt, G.R., Judge, M.D. and Briskey, E.J.
1968. Homeostasis in animals (Sus domesticus) during exposure to a
warm environment. J. Appl. Physiol. 24, 33-39.

Greaser, M.L., Cassens, R.G., Hoekstra, W.G. and Briskey, E.J. 1969.
The effects of pH-temperature treatments on the calcium accumulating ability of purified sarcoplasmic reticulum. J. Food Sci. 34,
633-637.

Gregory,N.G. and Lister, D. 1981. Autonomic responsiveness in stress-
sensitive and stress-resistant pigs. J. Vet. Pharmacol. Therap.
4, 67-75.

Gronert, G.A. 1980. Malignant Hyperthermia Anaesthesiology 53,
395-423.

Guyton, A.C., Taylor, A.E. and Granger, H.J. 1975. Dynamics and
Control of Body Fluids. Saunders, Philadelphia.

Hall, G.M. 1976. A study of the metabolic changes and endocrine
factors in malignant hyperthermia in the pig. PhD Thesis,
University of London.

Hall, G.M., Bendall J.R., Lucke, J.N. and Lister, D. 1976. Porcine malignant hyperthermia. II. Heat production. Brit. J. Anaesth. 48, 305-308.

Hall, G.M., Lucke, J.N. and Lister, D. 1977. Porcine malignant hyperthermia. V. Fatal hyperthermia in the Pietrain pig associated with infusion of alpha adrenergic agonists. Brit. J. Anaesth. 49, 855-863.

Hall, G.M., Lucke, J.N. and Lister, D. 1980. Malignant Hyperthermia - Pearls out of swine. Br. J. Anaesth. 52, 165-171.

Harthoorn, A.M., van der Walk, K. and Young, E. 1974. Possible therapy for capture myopathy in captured wild animals. Nature (Lond.) 247, 577.

Judge, M.D. 1969. Environmental Stress and Meat Quality. J. Anim. Sci. 28, 755-760.

Kallweit, E. 1969. Effects of environmental temperature and exercise ante-mortem on blood and meat quality characteristics of pigs. In: Recent Points of View on the Condition and Meat Quality of Pigs for Slaughter. (Eds. Sybesma, W., van der Wal, P.G., and Walstra, P.) (IVO Zeist) p.143.

Kastenschmidt, L.L., Beecher, G.R., Forrest, J.C., Hoekstra, W.G. and Briskey, E.J. 1965. Porcine muscle properties. A. Alteration of glycolysis by artificially induced changes in ambient temperature. J. Food Sci. 30, 565-572.

Lister, D. 1974. The Stress Syndrome and Meat Quality. In: Proc. of 20th European Meeting of Meat Research Workers. Dublin, Rapporteurs Papers, pp.17-27.

Lister, D., Christopherson, R.J., Murray, A.C., Wolfe, F.H. and Thomson, J.R. 1985. (In preparation).

Lister, D., Gregory, N.G. and Warris, P.D. 1981. Stress in meat animals. In: Developments in Meat Science - 2. (Ed. R.A. Lawrie) (Applied Science Publishers, London) p.61.

Lister, D., Hall, G.M. and Lucke, J.N. 1976. Porcine Malignant Hyperthermia III. Adrenergic Blockade. Brit. J. Anaesth. 48, 831-838.

Lucke, J.N., Hall, G.M. and Lister, D. 1979. Malignant Hyperthermia in the pig and the role of stress. Ann. N.Y. Acad. Sci. 317, 326-335.

Muir, A.R. 1970. Normal and regenerating skeletal muscle fibres in Pietrain pigs. J. Comp. Pathol. 80, 137-143.

Nelson, T.E., Jones, E.W., Henrickson, R.L., Falk, S.N. and Kerr, D.D. 1974. Porcine malignant hyperthermia: Observations on the occurrence of pale, soft, exudative musculature among susceptible pigs. Amer. J. Vet. Res. 35, 347-350.

16

Newsholme, E.A. 1979. The control of fuel utilisation by muscle during exercise and starvation. Diabetes 28 (Suppl.) 1-7.

Offer, G.M. 1984. Progress in the biochemistry, physiology and structure of meat. In: Proc. 30th European Meeting of Meat Research Workers, Bristol.pp.87-94.

Offer, G.M. and Trinick, J. 1983. On the mechanism of water holding in meat: the swelling and shrinking of myofibrils. Meat Sci. 8, 245-281.

Penny, I.F. 1975. Use of a centrifuging method to measure the drip of pork longissimus dorsi slices before and after freezing and thawing. J. Sci. Fd. Agric. 26, 1593-1602.

Scopes, R.K. 1964. The influence of post mortem conditions on the solubilities of muscle proteins. Biochem. J. 91, 201-207.

Swatland, H.J. and Cassens, R.G. 1972. Peripheral innervation of muscle from stress susceptible pigs. J. Comp. Pathol. 82, 229-236.

Topel, D.G., Bicknell, E.J., Preston, K.S., Christian, L.L. and Matsushima, C.Y. 1968. Porcine Stress Syndrome. Mod. Vet. Pract. 49, 40-41, 59-60.

Usdin, E., Kvetnanský, R. and Kopin, I.J. (eds.). 1976. Catecholamines and Stress. Pergamon Press, Oxford & New York.

Venable, J.H. 1973. Skeletal muscle structure in Poland China pigs suffering from malignant hyperthermia. In: International Symposium on MH. (Eds. Gordon, R.A., Britt, B.A. and Kalow, W.) (Charles C. Thomas, Springfield, Ill). p.208.

Warriss, P.D. and Lister, D. 1982. Improvement of meat quality by beta-adrenergic blockade. Meat Sci. 7, 183-187.

Wilding, P., Hedges, N. and Lillford, P.J. 1984. Abstracts of 8th International Biophysics Congress, Bristol.

Wood, J.D., Gregory, N.G., Hall, G.M. and Lister, D. 1977. Fat mobilization in Pietrain and Large White Pigs. Brit. J. Nutr. 37, 167-186.

CALCIUM RELEASING SYSTEMS IN MITOCHONDRIA AND SARCOPLASMIC RETICULUM

WITH RESPECT TO THE AETIOLOGY OF MALIGNANT HYPERTHERMIA: A REVIEW

J. J. A. Heffron

Department of Biochemistry,
University College Cork,
Cork, Ireland.

ABSTRACT

The recent literature dealing with biochemical measurements of calcium release from both porcine and human skeletal muscle mitochondria and
sarcoplasmic reticulum is reviewed. While calcium release under anaerobic conditions is increased in porcine MH mitochondria, the phenomenon
cannot be clearly related to the primary pathogenesis of the MH syndrome.
It is shown that other factors might be responsible for the observed
calcium release from mitochondria obtained from postmortem MH muscle. No
major differences were detected in aerobic calcium release pathways or in
calcium retention in human MH mitochondria. The calcium-induced calcium
release mechanism of porcine and human MH sarcoplasmic reticulum appears
to be abnormally sensitive to calcium but at concentrations which are
about one hundred times greater than those considered to obtain physiologically. Rather than indicating the site of the fundamental defect in MH,
this finding may be indicative of a more widespread membrane lesion. It
is emphasised that future studies should have greater regard for better
experimental design and for more rigidly defined assay conditions.

INTRODUCTION

The true incidence of malignant hyperthermia (MH) in man is about 1:

40000 of the unselected anaesthetised population (Ellis & Heffron, 1985).

Whilst it is a rare syndrome, MH is of considerable social significance

because it is autosomally dominantly inherited and may result in the

death of an otherwise, apparently, healthy individual. Although halothane

appears to be the most potent causative agent, it seems that all inhala-

tional anaesthetics can trigger MH (Britt et al., 1980) with the probable

exception of nitrous oxide (Gronert and Milde, 1981). Of the muscle

relaxants only succinylcholine is definitely known to cause MH. A

similar MH syndrome occurrs in many breeds of pig, most notably, the

Poland China, Pietrain and Landrace breeds (Gronert, 1980). However,

porcine MH is triggered by physical and psychological stressors in addi-

tion to inhalational anaesthetics and succinylcholine. This is not so in

man (Fletcher et al., 1981) though some reports have indicated that stress

can induce awake episodes of MH in humans (Wingard, 1980; Gronert et al.,

1980). Numerous differences between the human and porcine MH syndromes

have been reported and are considered in some detail in the reviews by
Gronert (1980) and Mitchell and Heffron (1982). In general, the differences appear not to be of a fundamental nature, a view which has led to the
acceptance of porcine MH as the model for the human one and which has
justified the use of porcine tissue in fundamental investigations of the
primary lesion in MH. This is a reasonable development but the differences referred to above should be borne in mind when making interpretations
of new findings particularly those of a biochemical or ultrastructural
nature.

So far, most experimental findings in human and porcine MH suggest
that the regulation of the intracellular free Ca^{2+} concentration in
skeletal muscle is defective. Although this still remains a theory
(Gronert, 1980; Heffron, 1984), very recent evidence reported by Lopez
et al. (1985) shows that the "resting" free Ca^{2+} in human MH muscle
fibres is three times greater than in normal muscle fibres. This short
review will therefore focus on recent studies in which dysfunction of two
of the major organelles, the mitochondrion and sarcoplasmic reticulum (SR),
involved in intracellular calcium homeostasis have been reported. While
the plasma membrane is undoubtedly a regulator of intracellular calcium
concentration, no calcium transport studies on this membrane from MH
muscle have been published. Since 1980, several useful general reviews of
the literature published on both human and porcine MH since its first
clear description by Denborough and Lovell in 1960 have appeared (Gronert,
1980; Hall et al., 1980; Mitchell and Heffron, 1980, 1982; Gallant and
Aherne, 1983; Ellis and Heffron, 1985). Here I have tried to concentrate
on the Ca^{2+} releasing systems of skeletal muscle in so far as they may be
related to the fundamental cause of MH, an area which has not been
specifically or critically reviewed before.

INTRACELLULAR REGULATION OR IONIZED CALCIUM

The ionized calcium concentration in skeletal muscle cells is
regulated by one or more of the following systems: (1) the uptake and
release systems of the SR: (2) the Ca^{2+} pump of the sarcolemma or plasma
membrane; and (3) the Ca^{2+} uptake and release systems of the mitochondria
(see Fig. 1). The relative contribution of the various Ca^{2+} transporting
systems to cellular calcium homeostasis is probably in the order set out
(Martonosi, 1983) though it may vary with muscle fibre type.

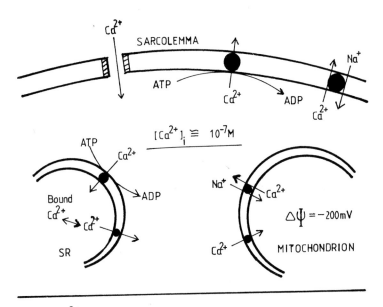

Fig. 1 Ca^{2+} - transporting systems in the skeletal muscle fibre. They are located in the sarcolemma, sarcoplasmic reticulum (SR) and mitochondrion. The physiological mechanism of Ca^{2+} release from SR is not known yet; mitochondrial Ca^{2+} release is stimulated by physiological concentrations of Na^+.

In skeletal muscle cells, the SR is undoubtedly the primary system which releases and sequesters Ca^{2+} during excitation-contraction coupling and relaxation (see Martonosi, 1983, 1984 for exhaustive reviews). On the other hand, the relative importance of the sarcolemmal Ca^{2+} pump in controlling the intracellular ionized Ca has only been studied in very recent times principally because of earlier methodological difficulties in obtaining pure sarcolemmal membranes or vesicular preparations. From the time that mitochondria were first shown to accumulate Ca^{2+} in an energy-dependent manner, their role in controlling the intracellular ionized calcium has been intensely debated (Denton and McCormack, 1980; Martonosi, 1983). Because mitochondria have a relatively large Ca^{2+} storage capacity and a relatively low Ca^{2+} affinity, it has been convenient to suggest that mitochondria may provide a last line of defence against the acute toxic effects of increased sarcoplasmic Ca^{2+} concentration (for example, Martonosi, 1983, 1984). This view completely ignores the very strong evidence which indicates that the primary role of the mitochondrial calcium transporting systems is to control the concentration of Ca^{2+} in

the matrix (Denton and McCormack, 1985). This is essential for rate
control of the citric acid cycle oxidation of cellular fuels since the
two principal regulatory enzymes of the cycle, isocitrate dehydrogenase
and 2-oxoglutarate dehydrogenase, are Ca^{2+}-dependent enzymes (Denton and
McCormack, 1980). Furthermore, the sensitivity of the enzymes to Ca^{2+} is
well within the accepted physiological range of 10nM to 5μM. It is unnecessary to invoke any other physiological role for mitochondrial Ca^{2+}
transport. Mitochondria possess an electrophoretic, carrier mediated Ca^{2+}
uptake system and an independent Na^{+}-stimulated Ca^{2+} efflux system
(Crompton et al., 1978) both of which are extremely active in skeletal
muscle mitochondria under simulated physiological assay conditions
(Allshire and Heffron, 1984). Decreased mitochondrial Ca^{2+} uptake or
enhanced Na^{+}-stimulated Ca^{2+} efflux would result in a diminished ability
of mitochondria to oxidise pyruvate and would explain the generation of
non-hypoxic lactate and the unexpectedly small increase in whole-body
oxygen consumption observed during active MH in swine (Gronert et al.,
1977; Ahern et al., 1985).

MH AND MITOCHONDRIAL CALCIUM EFFLUX

Ca^{2+} efflux from mitochondria of MH porcine muscle has only been
studied by K. S. Cheah and his group (Cheah, 1984). Basically, anoxia-
induced Ca^{2+} efflux was shown to be greater in muscle mitochondria from
halothane-sensitive pigs than halothane-insensitive pigs (Cheah and Cheah,
1978, 1979); 2% halothane increased this efflux in the mitochondria from
the halothane-sensitive animals only, on the basis of which the authors
suggested that their observation offered an ultimate explanation for the
aetiology of MH. This has not proved to be the case and in any event it
was unlikely to be so since there is adequate evidence showing that
tissue perfusion is maintained during the onset and early phase of
porcine MH at least (Gronert et al., 1977; Ahern et al., 1985). The
significance of this increased anaerobically-induced Ca^{2+} efflux in the
aetiology of MH or the porcine syndrome remains unclear; perhaps the
time of onset of the efflux may be important in determining the reversibil-
ity of the established syndrome and may indicate why it is so essential to
commence active treatment with dantrolene in the early phase if therapy is
to be successful. More recently, these authors have shown that this
enhanced Ca^{2+} efflux in MH mitochondria is probably caused by increased
phospholipase A_2 activity which liberates long-chain unsaturated fatty

acids from the mitochondria (Cheah and Cheah, 1981a, b). The products of phospholipase A_2 activity, free fatty acids and lysophospholipids, are well known promoters of mitochondrial membrane Ca^{2+} permeability in addition to the established role of fatty acids as protonionophores. Very recently, Cheah (1984) has reported that the increased phospholipase A_2 activity of the MH mitochondria may be caused by the greater than normal amount of calmodulin, a Ca^{2+} regulator protein, in MH muscle mitochondria. It is still not clear if calmodulin is present at all in mitochondria (Ruben and Rasmussen, 1981) because of the lack of specificy of the calmodulin assays used by different workers. Cheah (1984) does not indicate the method of calmodulin assay used in his laboratory. At this time it is reasonable to state that the use of the putative calmodulin antagonists such as trifluoperazine or chlorpromazine to identify calmodulin in an enzyme system, be it membrane-bound or otherwise, is not an adequate criterion for calmodulin's presence; the protein should be detected and quantitated by either affinity chromatography and phosphodiesterase activation or by radioimmunoassay. Indeed, there is conflicting data in the literature as to whether or not the Ca^{2+} activation of phospholipase A_2 is mediated by calmodulin and a very recent study shows fairly conclusively that it is not (Withnall et al., 1984). Another explanation of the increased phospholipase A_2 activity in mitochondria may be deduced from the work of Parce et al. (1978) who showed that the previously inactive enzyme becomes active as the ATP level falls to zero. Although this was observed with liver mitochondria, the result can be legitimately extrapolated to muscle mitochondria, particularly those derived from muscle which undergoes such rapid postmortem glycolysis and ATP depletion (cf. Somers et al., 1977). Other limitations in using postmortem muscle for preparation of physiologically intact mitochondria from MH pigs have been discussed by Brooks and Cassens (1973) and Ellis and Heffron (1985). Only one brief report of Ca^{2+} retention and release by muscle mitochondria from humans has appeared (Heffron, 1984). MH susceptibility was determined by the caffeine contracture test; no significant differences in Ca^{2+} retention times, basal or Na^+-stimulated Ca^{2+} efflux or fatty acid contents of MH or normal mitochondria was noted. On account of the small amounts of mitochondria which can be isolated from human muscle biopsies it was not possible to examine anaerobic Ca^{2+} efflux or the effect of halothane

as had been done by Cheah and Cheah in their experiments already described. Nevertheless, the results do indicate that there are several
differences between the Ca^{2+} transporting properties of human and
porcine MH mitochondria in addition to earlier published differences in
respiratory characteristics as discussed in the recent major reviews.

MH AND CALCIUM EFFLUX FROM SARCOPLASMIC RETICULUM

There is still disagreement as to whether SR Ca^{2+} uptake is unalter-
ed or somewhat decreased in MH (Ellis and Heffron, 1985). Differences
are undoubtedly due to the quality of the isolated SR and to gross
differences in assay conditions such as temperature, ionised calcium and
ATP concentrations. The most important factor influencing isolated SR
quality is the pH of the muscle sample at the time of and during homogenization and ultracentrifugation. In a recent Abstract, O'Brien et al.
(1985) have shown that it is essential to maintain the pH of the homogenate during the above isolation procedures for porcine SR, otherwise up to
90% of the Ca^{2+} sequestering ability of the SR is lost. This finding
indicates that previous studies, in which this pH effect was not taken
account of, may have produced data which are artifacts of the isolation
procedure. Obviously this applies to MH porcine muscle and it remains to
be established if human MH biopsy muscle exhibits such rapid postmortem
glycolysis. O'Brien et al.'s finding provides further substance to
support this author's view of the results obtained with mitochondria from
MH muscle obtained postmortem. Since 1983, four studies of the so-called
calcium-induced calcium release system of SR of MH and normal porcine
muscle have appeared (Ohnishi et al., 1983; Nelson, 1983, 1984; Kim et
al., 1984). One report of calcium-induced calcium release from SR of one
MH patient has appeared (Endo et al., 1983). All of the reports on MH
porcine muscle agree that the calcium threshold for calcium-induced
calcium release is significantly reduced compared with normal SR.
Dantrolene had no effect on this calcium release system (Nelson, 1984)
but it did partially block the rather transient halothane-induced calcium
release reported by Ohnishi et al. (1983). As already pointed out by
Nelson (1984), there is a number of experimental inconsistencies in
Ohnishi et al.'s paper which make it difficult to draw definite conclusions on the effect of dantrolene on SR calcium release. The data of Kim
et al. (1984) show that both the rate and extent of Ca^{2+} release by

halothane, by external Ca^{2+}, by halothane and external Ca^{2+} combined and by membrane depolarization are significantly increased in MH SR compared with normal SR. Nelson (1984) and Ohnishi et al. (1983) did not find any difference in the amount of Ca^{2+} released by added Ca^{2+}. It is also important to note that only in the case of the animals used by Ohnishi et al. were the MH- susceptible and MH-resistant pigs littermates. The calcium-induced calcium release mechanism of human MH muscle was also found to be more sensitive to calcium (Endo et., 1983) while halothane accelerated this Ca^{2+} release to a similar extent in both the MH and normal muscle fibres. These workers employed the so-called skinned fibre technique to demonstrate calcium-induced calcium release whilst all studies on MH porcine muscle used isolated 'heavy' SR (fraction enriched in terminal cisternae). In the report of Endo et al. it is not clear how MH was diagnosed or confirmed although there are now clear guidelines for same in both the United States and Europe (Rosenberg and Reed, 1983; Ellis et al., 1984). In most of these studies no mention is made of the actual halothane concentrations used in the various assay and test media. This is an important parameter for all of these studies and should be rigidly monitored by appropriate analysis of the anaesthetic concentration in the various aqueous media used.

FUTURE STUDIES

There is good agreement that the calcium-induced calcium release mechanism of SR of MH muscle is abnormally sensitive to Ca^{2+} but the many discrepancies noted above must be resolved in future studies. Because this calcium release mechanism is most probably not the primary one by which the normal excitation-contraction cycle is initiated (Ohnishi et al., 1983), it remains to be established what the significance of the present findings are to the pathogenesis of MH. As with the use of caffeine in the MH diagnostic contracture test, the increased sensitivity of the calcium-induced calcium release system of MH SR may merely be another manifestation of some subtler alteration in the SR and perhaps other membrane systems. This interpretation is enhanced by the recent demonstrations of the normality of the calcium-binding proteins of the contractile apparatus in both porcine and human muscle (Lorkin and Lehmann, 1983; Endo et al., 1983). In view of the findings of Denton and McCormack (op.cit.), further studies of the mitochondrial calcium transport cycle should be worth undertaking in MH porcine muscle. Future studies should pay proper attention to ensuring that control and MH animals are of the same breed at least and, where halothane is used in in vitro studies the concentrations used should be in the clinical range and should be appropriately analysed. Finally, the calcium transporting characteristics of the sarcolemma in MH must be examined in similar detail to those of SR and mitochondria.

I wish to thank P.J. O'Brien, University of Minnesota, for a pre-print of his paper presented at the Americal Society of Anesthesiologists' Meeting, October 1985.

REFERENCES

Aherne, C.P., Milde, J.H. and Gronert, G.A. 1985. Electrical stimulation triggers porcine malignant hyperthermia. Res. Vet. Sci. 39, 257-258.

Allshire, A.P. and Heffron, J.J.A. 1984. Uptake, retention and efflux of calcium by mitochondrial preparations from skeletal muscle. Arch. Biochem. Biophys. 228, 353-363.

Britt, B.A., Endrenyi, L., Frodis, W., Scott, E. and Kalow, W. 1980. Comparison of effects of several inhalational anaesthetics on caffeine-induced contractures of normal and malignant hyperthermia skeletal muscle. Can. Anaesth. Soc. J. 27, 12-15.

Brooks, G.A. and Cassens, R.G. 1973. Respiratory functions of mitochondria isolated from stress-susceptible and stress-resistant pigs. J. Anim. Sci. 37, 688-691.

Cheah, K.S. 1984. Skeletal muscle mitochondria and phospholipase A in malignant hyperthermia. Biochem. Soc. Trans. 12, 358-360.

Cheah, K.S. and Cheah, A.M. 1978. Calcium movements in skeletal muscle mitochondria of malignant hyperthermic pigs. FEBS Lett. 95, 307-310.

Cheah, K.S. and Cheah, A.M. 1979. Mitochondrial calcium, erythrocyte fragility and porcine malignant hyperthermia. FEBS Lett. 107, 265-268.

Cheah, K.S. and Cheah, A.M. 1981a. Mitochondrial calcium transport and calcium activated phospholipase in porcine malignant hyperthermia. Biochim. Biophys. Acta 634, 70-84.

Cheah, K.S. and Cheah, A.M. 1981b. Skeletal muscle phospholipase A and the interaction of mitochondria and sarcoplasmic reticulum in porcine malignant hyperthermia. Biochim. Biophys. Acta 638, 40-49.

Crompton, M., Moser, R., Ludi, H. and Carafoli, E. 1978. The interrelations between the transport of sodium and calcium in mitochondria of various mammalian tissues. Eur. J. Biochem. 82, 25-31.

Denborough, M.A. and Lovell, R.R.H. 1960. Anaesthetic deaths in a family. Lancet 2, 45.

Denton, R.M. and McCormack, J.G. 1980. On the role of the calcium transport cycle in heart and other mammalian mitochondria. FEBS Lett. 119, 1-8.

Denton, R.M. and McCormack, J.G. 1985. Physiological role of calcium transport by mitochondria. Nature 315, 635.

Ellis, F.R., Halsall, P.J., Ording, H., Fletcher, R., Ranklev, E., Heffron, J.J.A., Lehane, M., Mortier, W., Steinbereitner, K., Sporn, P., Theunynck, D. and Verburg, R. 1984. A protocol for the investigation of malignant hyperpyrexia susceptibility. Br. J. Anaesth. 56, 1267-1269.

Ellis, F.R. and Heffron, J.J.A. 1985. Clinical and biochemical aspects of malignant hyperpyrexia. Rec. Adv. Anaesthesia Analgesia No. 15, 173-207.

Endo, M., Yagi, S., Ishizuka, T. et al. 1983. Changes in the calcium-induced calcium release mechanism in the sarcoplasmic reticulum of the muscle from a patient with malignant hyperthermia. Biomed. Res. 4, 83-92.

Fletcher, R., Ranklev, E., Olsson, A.K. and Leander, S. 1981. Malignant hyperthermia syndrome in an anxious patient. Br. J. Anaesth. 53, 993-995.

Gallant, E.M. and Ahern, C.P. 1983. Malignant hyperthermia: responses of skeletal muscles to general anaesthetics. Mayo Clin. Proc. 58, 758-763.

Gronert, G.A. 1980. Malignant hyperthermia. Anesthesiol. 53, 395-423.

Gronert, G.A., Heffron, J.J.A., Milde, J.H. and Theye, R.A. 1977. Porcine malignant hyperthermia: role of skeletal muscle in increased oxygen consumption. Can. Anaesth. Soc. J. 24, 103-109.

Gronert, G.A., Thompson, R.L. and Onofrio, B.M. 1980. Human malignant hyperthermia: awake episodes and correction by dantrolene. Anesth. Analg. 59, 377-378.

Gronert, G.A. and Milde, J.H. 1981. Hyperbaric nitrous oxide and malignant hyperthermia. Br. J. Anaesth. 53, 1238.

Hall, G.M., Lucke, J.N. and Lister, D. 1980. Malignant hyperthermia-pearls out of swine. Br. J. Anaesth. 52, 165-171.

Heffron, J.J.A. 1984. Mitochondrial and plasma membrane changes in skeletal muscle in the malignant hyperthermia syndrome. Biochem. Soc. Trans. 12, 360-362.

Kim, D.H., Sreter, F.A., Ohnishi, S.T. et al. 1984. Kinetic studies of calcium release from sarcoplasmic reticulum of normal and malignant hyperthermia susceptible pig muscles. Biochem. Biophys. Acta 775, 320-327.

Lopez, J.R., Alamo, L., Caputo, C., Wininski, J. and Ledezma, D. 1985 Intracellular ionised calcium concentration in muscles from humans with malignant hyperthermia. Muscle and Nerve 8, 355-358.

Lorkin, P.A. and Lehmann, H. 1983. Malignant hyperthermia in pigs: a search for abnormalities in calcium binding proteins. FEBS Lett. 153, 81-87.

Martonosi, A.M. 1983. The regulation of cytoplasmic calcium concentration in muscle and non-muscle cells. In "Muscle and Non-muscle Motility" (Ed. A. Stracher). (Academic Press, New York). pp. 233-357.

Martonosi, A.M. 1984. Mechanisms of calcium release from sarcoplasmic reticulum of skeletal muscle. Physiol. Revs. 64, 1240-1320.

Mitchell, G. and Heffron, J.J.A. 1980. Porcine stress syndromes: a mitochondrial defect? S. Afr. J. Sci. 76, 546-551.

Mitchell, G. and Heffron, J.J.A. 1982. Porcine stress syndromes. Adv. Food Res. 28, 167-230.

Nelson, T.E. 1983, Abnormality in calcium release from skeletal sarcoplasmic reticulum of pigs susceptible to malignant hyperthermia.

J. Clin. Invest. 72, 862-870.

Nelson, T.E. 1984. Dantrolene does not block calcium-induced calcium release from a putative calcium channel in sarcoplasmic reticulum malignant hyperthermia and normal pig muscle. FEBS Lett. 167, 123-126.

O'Brien, P.J., Michelson, J.M., Gronert, G.A. and Louis, C.F. 1985. Malignant hyperthermia susceptibility: increased calcium sequestering activity of muscle sarcoplasmic reticulum. Abstracts American Soc. Anesthesiologists' Meeting, San Francisco, October 1985.

Ohnishi, S.T., Taylor, S.R. and Gronert, G.A. 1983. Calcium-induced calcium release from sarcoplasmic reticulum of pigs susceptible to malignant hyperthermia. The effects of halothane and dantrolene. FEBS Lett. 161, 103-107.

Parce, J.W., Cunningham, C.C. and Waite, M. 1978. Mitochondrial phospholipase A activity and mitochondrial aging. Biochemistry 17, 1634-1639.

Rosenberg, H. and Reed, S. 1983. In vitro contracture tests for susceptibility to malignant hyperthermia. Anesth. Analg. 62, 415-420.

Ruben, L. and Rasmussen, H. 1981 Phenothiazine and related compounds disrupt mitochondrial energy production by a calmodulin-independent reaction. Biochim. Biophys. Acta 637, 415-422.

Somers, C.J., Wilson, P., Ahern, C.P. and McLoughlin, J.V. 1977. Energy phosphate turnover and glycolysis in skeletal muscle of the Pietrain pig: the effects of premedication with azaperone and pentobarbitone anaesthesia. J. Comp. Path. 87, 177-183.

Withnall, M.T., Brown, T.G. and Diocee, B.K. 1984. Calcium regulation of phospholipase A is independent of calmodulin. Biochem. Biophys. Res. Comm. 121, 507-513.

Wingard, D.W. 1980. A stressful situation. Anesth. Analg. 59, 321-322.

CONTRACTION AND METABOLISM TRAITS IN SKELETAL MUSCLE BIOPSIES FROM HALOTHANE POSITIVE PIGS AS STUDIED BY MECHANICAL MEASUREMENTS AND ^{31}P NMR

G. Kozak-Reiss*, F. Desmoulin**, P. Canioni**, P. Cozzone**,
J.P. Gascard*, G. Monin***, J.M. Pusel**, J.P. Renou***, A. Talmant***

*Département de Physiologie Humaine, CCML, Faculté de Médecine Paris-Sud
UA CNRS 1159 - 133, avenue de la Résistance
92350 Le Plessis-Robinson, France
** Institut de Chimie Biologique - Place Victor-Hugo
13003 Marseille, France
*** Station de Recherche sur la Viande - INRA
63122 Theix par Ceyrat, France

ABSTRACT

The effects of caffeine and A23187 ionophore (calcimycine) on some traits of contraction and metabolism in skeletal muscle biopsies from halothane positive (HP) and halothane negative (HN) piglets were studied using simultaneous mechanical and NMR measurements on the same sample. Before applying caffeine or A23187, pH and CP level were much lower in HP pig muscle than in HN pig muscle. Two Pi peaks were observed, although not constantly, in HP pig muscle, indicating two pH compartments. Both caffeine and A23187 provoked in HP pig muscle, as compared to HN pig muscle, a stronger contracture, a decrease in twitch tension (increased in HN muscle), a faster pH decrease and a faster depletion of creatine phosphate. It is proposed that, in HP pig muscle, the mechanical response could be affected not only by the changes in free intracellular Ca^{2+} but also by acidosis and CP depletion induced by caffeine and A23187.

INTRODUCTION

In humans, malignant hyperthermia (MH) constitutes a severe complication of general anaesthesia. The syndrome is characterized by a sudden rise in central temperature, metabolic acidosis and notably an intense stiffness of skeletal musculature.

Some strains of pigs exhibit similar and possibly identical incidents. They may serve as animal models for this disease. The aetiology of MH is yet unknown, but the acute syndrome is related to an abnormally sustained increase of calcium in the skeletal muscle fibre cytoplasm. In humans, pharmacological tests using caffeine and the triggering agent halothane are used to determine MH susceptibility. In pigs as in humans, caffeine and the calcium ionophore A23187 produce stronger contractures in MH muscles (Reiss et al., 1985). Caffeine and A23187 enhance the concentration of free cytoplasmic calcium by different ways. The first acts directly on the sarcoplasmic reticulum Ca^{2+} release (Endo, 1975) ; the second seems to act by permitting a calcium influx from the extracellular space (Mac Laughlin et

al., 1975). In MH muscle biopsies, A23187 produces a large contracture
resembling a "rigor state" and biochemical data indicate a total depletion
of creatine phosphate and ATP(Reiss et al., 1985). Phosphorus-31 nuclear magnetic resonance (NMR) is a powerful non-destructive method for determining
variations in intracellular pH and concentration of phosphorylated metabolites in muscle samples or biopsies. In the present study, NMR was applied
to stimulated muscle biopsies from normal and MH pigs in the presence of
caffeine or A23187.

MATERIAL AND METHODS

Animals and sampling

Three halothane-negative (HN) and three halothane-positive (HP) Pie-
train piglets were identified according to the technique described by
Ollivier et al. (1978). The HP piglets were considered as MH sensitive.
They were used in the experiments at a liveweight of 45-50 kg.

The animals were anaesthetized by an intravenous injection of Pentho-
tal ; the anaesthesia was maintained for 4 to 6 h using pentobarbital.
Muscle strips of about 1 g were obtained from the left M. biceps femoris
at intervals of approximately 1 h (time needed to perform one test).

NMR and mechanical measurements

As quickly as possible after excision, the sample was put in a Krebs
solution (118 mM NaCl, 24.6 mM $NaHCO_3$, 5.6 mM KCl, 0.77 mM KH_2PO_4, 0.77 mM
$MgSO_4$, 2.3 mM $CaCl_2$, 11 mM glucose, 2 mM mannitol, pH 7.4) maintained at a
temperature of 37°C and gassed with carbogen (95 % O_2 , 5 % CO_2).

The muscle strip was attached to the device shown in Fig. 1 and
transfered into a NMR tube (20 mm in diameter). All these operations were
made in oxygenated Krebs. The muscle was then superfused with oxygenated Krebs medium (37°C, 75 ml min^{-1}) and placed into the NMR spectrometer.
The upper part of the strip was connected to the lever arm of a strain
gauge force transducer and stimulated with silver electrodes (see Fig. 1).
The strip was pretensioned to the length at which the maximum twich tension
was obtained. Then the NMR and mechanical measurements were initiated.
Throughout the whole experiment, the muscle was submitted to supramaximal
stimulation at a frequency of 0.1 Hz.

After a control period of about 20 min, normal Krebs medium was replaced by a solution to be tested, i.e. either 8 mM caffeine or 16 mM caffeine

4 µg.l^{-1} A23187 (calcimycine) in oxygenated Krebs medium for 16 min (testing period). Then normal Krebs medium was used again for a rinsing period of 20 min. The following mechanical parameters were recorded : resting tension, contracture tension and active tension, i.e. maximal isometric twitch and tetanic tensions.

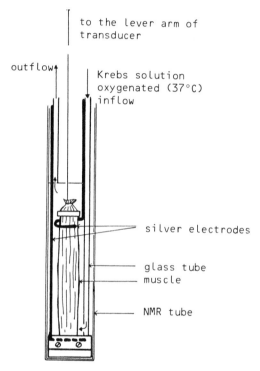

Fig. 1   NMR device for isolated muscle studies.
At the bottom, the muscle strip is mounted on a removable support consisting of a teflon clamp embedded within two glass tubes holding silver electrodes. The upper part of the muscle is flattened by a small piece of teflon and tied up with a string hooked to the transducer arm.

^{31}P NMR measurements

NMR experiments were performed on a conventional Nicolet NT200-WB spectrometer equipped with a superconducting magnet operating at 4.7 Tesla. ^{31}P NMR spectra were recorded at 80.9 MHz without proton decoupling. Typical spectra of superfused muscle biopsies were obtained in 4 min (20 mm diameter probe, muscle weight 1 ± 0.2 g) by accumulating 40 free induction decays resulting from 90° radio frequency pulses applied at 6 sec intervals.

ATP, CP and Pi values were derived from the comparison of the 1st spectrum with the results of biochemical ATP determination.

Biochemical measurements

A 1 g muscle sample was taken at the beginning of the first test and immediately frozen using tongs precooled in liquid nitrogen. The frozen tissue was kept in liquid nitrogen until extraction with frozen 0.6 M perchloric acid. Adenosine triphosphate (ATP) and creatine phosphate (CP) were enzymatically determined according to the techniques described by BERGMEYER (1974).

RESULTS

1 – ^{31}P NMR spectra and mechanical data of superfused pig muscle biopsies in normal Krebs medium

Typical spectra from HN and HP pig muscle recorded during the control period are shown in Fig. 2.

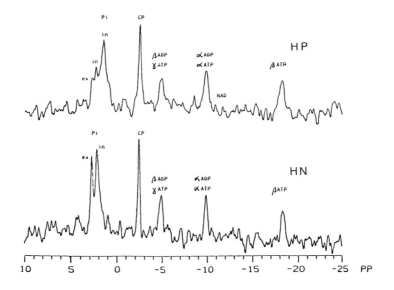

Fig. 2 ^{31}P NMR spectra of strips of M. biceps femoris from HN and HP Pietrain pigs.
Spectra were recorded at 80.9 MHz for 6 min (60 scans). Stimulated muscle strips were bathed in oxygenated (95 % O_2 , 5 % CO_2) Krebs solution at pH 7.45, temperature 37°C.
Resonances of inorganic phosphate (Pi), creatine phosphate (CP) at – 2.45 ppm, γ, α and β ATP (adenosine triphosphate). In HN pig muscle, chemical shifts of Pi indicate two pH corresponding to extracellular

Pi at + 2.88 ppm (pH 7.47) and intracellular Pi at + 2.43 ppm (pH 7).In HP pig muscle, the Pi resonance displays three components ; a low field peak corresponding to the extracellular Pi at + 2.88 ppm (pH 7.47), and a high field peak with a shoulder at + 2.43 ppm and + 1.72 ppm respectively corresponding to two pH values of 7 and 6.49.

Although the overall spectra were comparable, distinct differences were observed. The CP peak wqs substantially lower in HP animals, indicating a lower content which was confirmed by biochemical analysis (see Table I).

TABLE I   Concentrations of phosphorylated metabolites and intra-cellular pH in HN and HP pig muscle.
Number of samples is given in brackets. Values are derived from NMR measurements.

(Student test: P < 0.001*** ;   P < 0.01** ;   NS, non significant)

	ATP μmole.g^{-1} wet weight	CP μmole.g^{-1} wet weight	Pi μmole.g^{-1} wet weight	pH U pH
HN	4.53 ± 0.57 (9)	16.93 ± 3.24 (9)	7.69 ± 2.25 (9)	6.99 ± 0.07 (9)
HP	4.35 ± 0.47 (7)	10.68 ± 3.77 (8)	12.30 ± 4.10 (9)	6.82 ± 0.08 (4) **
	NS	***	***	6.49 ± 0.01 (3) ***
				6.26 ± 0.08 (3) ***

In HP animals, the intracellular Pi peak was increased and frequently displayed an heterogeneity indicative of the presence of two pH compartments. Consequently, two values of intracellular pH were calculated whenever hete-rogeneity was unequivocally apparent. However, it was not possible to mea-sure the individual Pi contents of each compartment. Measured   pH was gene-rally much lower in HP animals. When two values could be computed, one of them was close to the value measured in HN pigs (see Fig. 6). The decrease in pH and in CP showed a similar rate in both types of pigs (Fig. 5 and 6).

The contraction strength was satisfactorily maintained in both types of animals throughout the whole control period (Fig. 3, 5 and 6).

2 - Effects of caffeine and of ionophore A23187

a) Caffeine

Upon perfusion with caffeine, a contracture developed in both types of muscles though stronger in HP animals (Fig. 3).

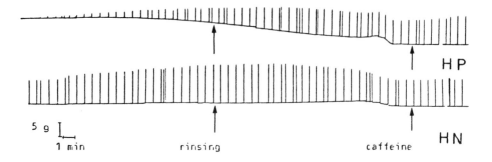

Fig. 3   Recording of contractions of isolated M. biceps femoris
from HP and HN pigs.
Muscle wet weight : HN : 1.8 g ; HP : 2.06 g
The record extends from right to left. Beginning of caffeine and
rinsing periods are indicated by vertical solid arrows.

Moreover, the contracture tension was increased by increasing the caffeine concentration (Table 2).

TABLE 2: Effect of caffeine and A23187 on mechanical parameters
of HN and HP muscle fibres.
Values are presented as means ± standard deviation. Number of samples
measured is given in brackets.

Normal Krebs	Caffeine 8 mM g	Caffeine 16 mM g	A23187 g
7.7 ± 0.71 (9)	8.14 ± 1.12 (3)	10.1 ± 3.44 (3)	5.27 ± 0.5 (3)
0	0.55 ± 0.05	1.5 ± 0.74	0.24 ± 0.12
6.04 ± 1.66 (8)	3.79 ± 1.93 (3)	3.42 ± 2.51 (2)	2.26 ± 1.22 (3)
0	1.19 ± 0.05	3.51 ± 0.22	0.62 ± 0.12

Simultaneously the twitch tension was increased in HN animals (caffeine
potentiation) and by contrast, was sharply reduced in HP pigs (Fig. 4).

Creatine phosphate showed a very sharp reduction, particularly in HP
pigs where the zero level was practically reached at the beginning of the
rinsing period. ATP content was kept approximately constant over the course
of the experiment. During the testing period, two Pi peaks began to be distinguishable ; the two pH values dropped faster in HP than in HN pigs.

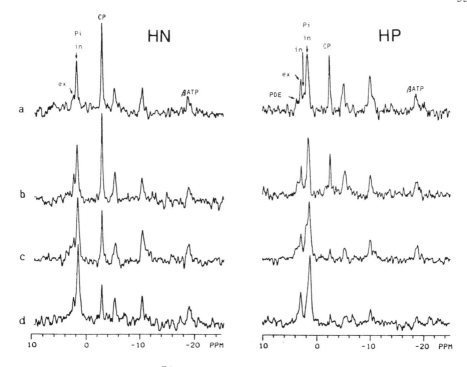

Fig. 4   Sequence of ^{31}P NMR spectra for strips of M. biceps femoris
from HP and HN pigs showing changes in phosphorylated metabolites
during caffeine contracture.
Each spectrum is the block average of 4 x 20 scans (8 min). CP peak
is set at − 2.45 ppm. PDE : phosphodiesters. Note the respective
changes in the CP and Pi peaks.
a − Muscle at rest, in normal Krebs medium. Two intracellular pH
compartments are distinguishable in HP muscle.
HP muscle : pH ex : 7.5 ; pH in : 6.51 and 6.39 ; ATP : 5.4 µmol.g^{-1} ;
Pi in : 16.6 µmol.g^{-1} (ex, extracellular ; in, intracellular).
HN muscle : pH ex : 7.5 ; pH in : 6.96 ; CP : 20.8 µmol.g^{-1} ; ATP :
4 µmol.g^{-1} ; Pi in : 10.8 µmol.g^{-1}.

b − Stimulated muscle 20 min stay in normal Krebs medium.
c − Stimulated muscle in 16 mM caffeine Krebs medium at beginning of
contracture.
d − Stimulated muscle in 16 mM caffeine Krebs medium at the end of
the listing period. Note the PC drop close to zero level.

b) A23187

In the experiments described here, A23187 produced more restricted
effects than in a previous work (Reiss et al., 1985). The contracture was
weak in HN muscle, and no twitch potentiation was observed (Fig. 6). HP
muscles developed a noticeable contracture accompanied by a considerable
reduction of the twitch tension in a similar manner as previously reported
(Reiss et al., 1985). There was a drastic decrease of CP level in HP muscle

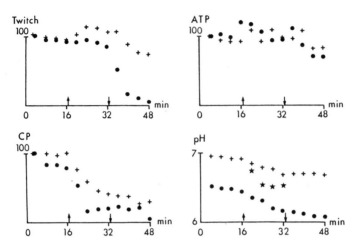

Fig. 5   Effect of 16 mM caffeine upon the dynamics of twitch tension,
ATP and creatine phosphate (CP) contents and intracellular pH of
strips of M. biceps femoris of HN and HP pigs. Values are expressed
as percentage of the value obtained at the beginning of the experiment in normal Krebs medium (+ HN ; ● HP). pH is expressed in pH units
(+ HN ; ● HP). Where two pH compartments were distinguishable in HP
muscle, the second peak is indicated by a star (*). Each spectrum
point is derived from the average of 40 scans accumulated during
4 min : 1st to 4th, control period ; 5th to 8th, testing period ;
9th to 12th, rinsing period. The testing period is included between
the two solid arrows.

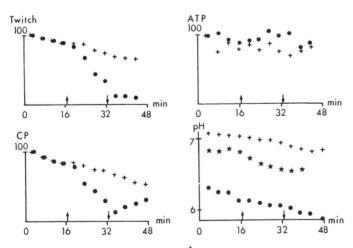

Fig. 6   Effect of A23187 (4 μg.mL^{-1}) upon dynamics of twitch tension,
ATP and creatine phosphate (CP) contents and intracellular pH of HN
and HP biceps femoris muscle strips (see legend Figure 5).

as compared to HN muscle. ATP level was maintained during the whole experiment starting from the control period. Two pH values were distinguishable in samples from two HP animals. These values seemed to decrease faster than the one observed in HN pig muscle.

c) Rinsing period

During this period, contracture was suppressed in HN muscles. Simultaneously, the twitch tension recovered to control value in HN whereas it continued to decrease in HP pigs and was nearly abolished at the end of the period (Fig. 3). The pH seemed to stabilize in HN muscle, but continued to decrease in HP muscle to reach close to or even below pH 6.0.

DISCUSSION

During the control period, pH as well as ATP and CP contents slowly decreased in all samples. However, the observation that pH and CP were noticeably lower in HP muscle at the beginning of the experiment indicates that probably the physiological conditions were altered in this muscle. By contrast, HN muscle seemed to be preserved in a more satisfactory condition in our procedure. In fact, according to the results obtained on "in situ" frozen samples from HN Large White and HP Pietrain pigs by Hall and Lucke (1983) there should be no difference in CP and ATP levels between both types of muscles. In the same way, using "in vivo" ^{31}P NMR spectroscopy, Roberts et al. (1983) did not observe any difference in CP levels between HP and HN piglets. The differences found here may be due to differential sensitivity of HP and HN fibres to the cutting procedure. Many authors reported similar differences on samples frozen after cutting (Nelson, 1973; Veerburg et al., 1984 ; Kozak-Reiss et al., 1985). If that were the case, it would seem difficult to correct for this artifact (Meyer et al., 1985).

When performed in the same way as in humans, the caffeine contracture studies show clear differences in HP and HN pigs. In HN muscles, 8 mM as well 16 mM caffeine produced twitch potentiation accompanied by a contracture. By contrast, in HP muscles, the twitch strength was reduced from the onset of the caffeine perfusion, and continued to decline after the muscle was placed in the rinsing solution. The NMR spectra showed a considerable decrease in CP content with a concomittant increase in Pi content, even though the ATP level was maintained. A similar effect was observed with A23187. Results are consistent with our previous observations.

In the presence of caffeine and A23187, an impairment of excitation-contraction coupling was observed in HP muscle. It can be partly related to the critically low level of CP (Mainwood et al., 1982) rather than to ATP level which was maintained. This latter point does not agree with Nelson (1973) who reported on the basis of biochemical assays a considerable depletion of CP and ATP, with production of inorganic phosphate and lactate in halothane treated HP muscle strips. With the A23187, a similar reduction of CP and ATP was observed in HP pig muscle (Kozak-Reiss et al., 1985). The discrepancy between previous reports and the results of the present NMR study pertains mostly to the ATP level which was considerably lower in Nelson's study and seemed only slightly affected in the present experiment.

Intracellular pH is the second and perhaps the most important factor affecting the mechanical response. It has been often reported that intracellular pH affects the binding and release of Ca^{2+} from the sarcoplasmic reticulum (Lea and Ashley, 1982) and the sensitivity of the contractile elements to Ca^{2+}. In skinned frog skeletal muscle fibres, Ca^{2+} activated force is pH dependent i.e. force is near zero at pH 5.5 and increases to maximal at pH 7.5 (Robertson and Kerrick, 1979). Comparing the effects of fatigue and low intracellular pH in single frog skeletal muscle fibres, Edman and Mattiazzi (1981) showed that in a non-fatigued fibre , the drop in mechanical performance during fatigue can be reproduced by reducing the pH. The same results were obtained in whole muscle by Curtin and Rawlinson (1984) using frog Sartorii. All these data support the idea that both a reduced state of activation of the contractile system and a specific reduction of cross bridge number and/or turnover rate are linked to an increased intracellular H^+ concentration. The fact that in HP pig muscle, the mechanical output was reduced, not only in caffeine and A23187 media but also in Krebs rinsing solution, can be related both to the reduction of CP level and to the substantial drop in intracellular pH.

A question rises regarding the origin of the two intracellular Pi compartments observed in the spectra of HP muscle and corresponding to two pH values. This observation could result from an heterogeneity in the physiological state of fibres, with central fibres being more anoxic and then acidotic. Alternatively, the two compartments could correspond to different types of fibres in biceps femoris (80 % fast twitch glycolytic, 20 % slow twitch oxidative) as was noted in cat biceps (Meyer et al., 1985). This may be related to the observation that red fibres are more sensitive to halothane that white fibres (Van den Hende, 1979 ; Kozak-Reiss, unpublished).

In the presence of caffeine and A23187, an impairment of excitation affect the cell metabolism of HP pig muscle to such an extent that the mechanical response to stimulation could be affected not only by the changes in free Ca^{2+} levels, but also by the induced acidosis.

Acknowledgement

The authors are grateful to Dr Sellier (I.N.R.A.), who provided the animals used in the experiment.

REFERENCES

Bergmeyer, H.V. 1974. Methods of enzymatic analysis. Academic Press, New York.

Bernard, M., Canioni P. and Cozzone P.J. 1983. Etude du métabolisme cellulaire in vivo par résonance magnétique nucléaire du phosphore-31. Biochimie, 65 : 449-480.

Curtin, N.A. and Rawlinson S.R. 1984. Effect of carbon dioxide on force during shortening of isolated skeletal muscle from frog. J. Physiol., (London), 354.

Edman, K.A.P. and Mattiazzi A.R. 1981. Effects of fatigue and altered pH on isometric force and velocity of shortening at zero load in frog muscle fibres. J. of Muscle Res. and Cell Motility, 2 : 321-334.

Endo, M. 1975. Mechanism of action of caffeine on the sarcoplasmic reticulum of skeletal muscle. Proc. Japan Acad. : 479-484.

Hall, G.M. and Lucke J.N. 1983. Porcine malignant hyperthermia. IX. Changes in the concentrations of intramuscular high-energy phosphates, glycogen and glycolytic intermediates. Br. J. Anaesth., 55 : 635-640.

Kozak-Reiss, G., Monin G. and Lacourt A. 1985. Métabolisme énergétique et contraction du muscle squelettique de porc susceptible à l'hyperthermie maligne. J. Physiol. (Paris) 79, 5 : 78A.

Kozak-Reiss G., Monin G., Desmoulin F., Canioni P., Pusel J.M. and Cozzone P. 1985. Etude du métabolisme énergétique par R.M.N. du P31 du muscle squelettique de porc pendant la contraction "in vitro". Effet de la caféine et de l'ionophore A23187 sur l'animal sensible à l'Hyperthermie Maligne. Regards sur la Biochimie, 1, 36.

Kozak-Reiss, G., Gascard J.P., Monin G., Lacourt A., Lacourt P., Talmant A. and Mejenes-Quijano A. 1985. Evolutions morphologique, biochimique et contractile du muscle intercostales externi de porcs normaux et sensibles à l'halothane au cours de la croissance. XXXI Congrès Européen des Chercheurs en Viande. Albéna (Bulgarie), 4-33, p. 243-347. 25-31 août 1985.

Kushmerick, M.J. 1985. Patterns in mammalian muscle energetics. J. Exp. Biol., 115 : 165-177.

Lea, T.J. and Ashley C.C. 1982. The effect of pH on the rate of relaxation of isolated barnacle myofibrillar bundles. Biochimica et Biophysica Acta, 681 : 130-137.

Mac Laughlin, S. and Eisenberg M. 1975. Antibiotics and membrane biology. Ann. Rev. Biophys. Biochem., 4 : 335-366.

Mainwood, G.W., Alward M. and Eiselt B. 1982. Contractile characteristics of creatine-depleted rat diaphragm. Canadian Journal of Physiology and Pharmacology, 60, n°2 : 120-127.

38

Meyer, R.A., Brown T.R. and Kushmerick M.J. 1985. Phosphorus nuclear magnetic resonance of fast- and slow-twitch muscle. Am. J. Physiol., 248 : C279-C287.

Nelson, T.E. 1973. Porcine stress syndrome. In : International Symposium on Malignant Hyperthermia (Edited by R.A. Gordon, B.A. Britt, W. Kalow, C.C. Thomas)(Publisher, Springfield, Illinois).

Ollivier, L., Sellier P. and Monin G. 1978. Fréquence du syndrome d'hyperthermie maligne dans des populations porcines françaises ; relation avec le développement musculaire. Ann. Génét. Sél. Anim., 10, n°2 : 191-208.

Reiss, G., Monin G. and Lauer C. 1985. Comparative effects of the ionophore A23187 on the mechanical responses of muscle in normal Pietrain pigs and pigs with malignant hyperthermia. Can. J. Physiol. Pharmacol. (in press).

Roberts, T.J., Burt T., Gouylai L., Chance B., Sreter F. and Ryan J. 1983. Immediate uncoupling of high energy oxidative phosphorylation in muscle of malignant hyperthermic swine determined non-invasively by whole body ^{31}P nuclear magnetic resonance. Anesthesiology, 59, n°3 : A230.

Robertson, S.P. and Kerrick W.G. 1979. The effect of pH on Ca^{2+}-activated force in frog skeletal muscle fibers. Pflügers Archiv, 380 : 41-45.

Van den Hende, C. 1979. Isometric contraction of skeletal muscles of MH susceptible and resistant Belgian Landrace pigs. Acta Agric. Scandin., suppl. 21 : 339-342.

Veerburg, M.P., Oberlemans F.T.J.J., Vanbennekom C.A., Gielen M.J.M., De Bruyn C.H.M.M. and Crul J.F. 1984. In vivo induced malignant hyperthermia in pigs. I. Physiological and biochemical changes and the influence of Dantrolene sodium. Acta Anaesth. Scand., 28 : 1-8.

MALIGNANT HYPERTHERMIA: NEUROCHEMICAL ASPECTS

J. V. McLoughlin

Department of Physiology
Trinity College
Dublin 2
Ireland

ABSTRACT

There is conflicting evidence about the role which the peripheral sympathetic nervous system plays in drug-induced malignant hyperthermia and very little information on the possible involvement of the central nervous system. In this study, the concentrations of noradrenaline, dopamine and their non-O-methylated metabolites were examined in two regions of the brain, the hypothalamus and the corpus striatum. Differences between MH susceptible and non-susceptible animals were found only in the concentrations of two metabolites. The activity of monoamine oxidase in the hypothalamus, corpus striatum, liver, kidney, heart, skeletal muscle and intestinal mucosa, and of catechol-O-methyl transferase in kidney were not significantly different between the two types of animal.

INTRODUCTION

The malignant hyperthermia (MH) occurs in man and the

pig when susceptible individuals are exposed to chemical

trigger agents, particularly the fluorinated inhalent anaes-

thetic halothane and the myorelaxant succinylcholine, although

other fluorinated anaesthetics as well as cyclo-propane and

diethyl ether have been associated with the development of

the condition. The syndrome is characterised by muscular

rigidity, hyperthermia, metabolic and respiratory acidosis,

tachycardia, arrythmias, alterations in electrolyte balance

in blood and haemodynamic changes which may lead to death.

Predisposition to MH is determined genetically in both species.

In man, it is associated with the presence of dystrophic

changes in skeletal muscle and familial histories of neuro-

muscular disorders; with hypermobility of joints and a

tendency for joint dislocation; and with unusual cardiac murmurs, symmetrical hypertrophy of the myocardium and a tendency to develop arrythmia in response to excitement and physical exercise. In the pig, susceptibility to MH is associated with very high rates of high energy phosphate turnover and glycolysis in skeletal muscle post-mortem. In both species, drug-induced MH appears to be associated with susceptibility to a variety of stressors. This relationship is well established in the case of the pig where susceptibility to experimentally-induced physiological and environmental stressors has been demonstrated. The evidence for a similar stress-susceptibility in man is mainly clinical and less conclusive.

In the pig three conditions occur which are closely related to each other, i.e. pale soft exudative (PSE) muscle, drug-induced malignant hyperthermia and the stress syndrome. The processes which cause skeletal muscle to become pale soft and exudative are well established and consist essentially of a critical combination of low pH and high temperature (above 30 C) as the muscle goes into rigor mortis. This causes the denaturation of the sarcoplasmic and myofibrillar proteins, dislocation of the myofibrillar structure and rupture of the sarcolemma which makes the muscle become pale and exude fluid. Drug-induced MH is a pharmacogenetical disorder, that is, an individual may have an inherent susceptibility to the condition but it requires the presence of certain pharmacological agents to initiate the syndrome. Drugs which initiate MH have been identified and drugs which alleviate it are known but the molecular and cellular mechanisms which lead to the MH are still uncertain. However, much evidence suggests that the trigger site may be in skeletal muscle and that it may involve a defect in the transport, storage, release or uptake of the calcium ion by intracellular organelles. Whether such are, in fact, the primary cause of MH has not been clarified.

The stress syndrome is a much more complex entity than either PSE muscle or drug-induced MH and it is hardly surprising that despite the considerable amount of research attention devoted to it little is known about the physiological mechanisms involved. The stress syndrome seems to be a failure of the homeostatic mechanisms of an animal to cope adequately with a wide variety of environmental and physiological stressors. It is difficult to explain or understand such a physiological deficiency except in terms of the entire neuroendocrinological axis and its integration with physiological drives, behavioural responses and autonomic and somatic motor activity. Hence, it is unlikely that experiments which involve the use of markers such as PSE muscle or an MH response to a drug will tell us much about the fundamental nature of stress susceptibility. The contribution of biochemistry and physiology to the understanding of disease processes is probably least in the case of the central nervous system. Nevertheless, certain disorders are now associated with defects in neurohumeral transmission, e.g., Parkinson's disease in man which involves a functional deficiency in dopaminergic mechanisms in the basal ganglia. The affective disorders and schizophrenia also appear to involve defects of

monoamine function although studies on the action of drugs,
metabolites in the urine and cerebrospinal fluid and the
examination of brains post mortem have not provided un-
equivocal evidence for the monoamine hypothesis for these
disorders.

THE SYMPATHETIC NERVOUS SYSTEM AND MH

There is evidence which indicates that the peripheral
sympathetic nervous system is involved in MH but a conflict
of views as to whether this involvement is primary or secondary
to some other initiating factors. The concentration of
catecholamines in plasma increases markedly during drug-
induced MH (Gronert and Theye, 1976; Lucke, Hall and Lister,
1976; Lister, Hall and Lucke, 1976) and also when an MH-like
syndrome is induced by electrical stimulation of the lumbro-
sacral and brachial plexus (Ahern, Milde and Gronert, 1985).
Whether there are differences in the concentration of catecholamines in plasma between normal pigs and MH-susceptible pigs
in the absence of any triggering agents is not certain. Ahern
et al. (1985) reported that plasma of normal pigs contained
0.55 ± 0.35 $\mu g/l^{-1}$ NA, MH-susceptible 1.05 ± 0.37, although
the differences were not statistically significant. Wheatley
and McLoughlin (unpublished) found 0.04 ± 0.02 $\mu g/l^{-1}$ NA in
plasma from normal pigs and 0.51 ± 21 $\mu g/l^{-1}$ for MH susceptible pigs. Wheatley and McLoughlin also found significant
differences in the concentration of DA between normal
(0.07 ± 0.02 $\mu g/l^{-1}$) and MH susceptible (0.54 ± 0.11 $\mu g/l^{-1}$)
pigs.

The use of adrenergic antagonists during episodes of MH

influences the development of the syndrome but the conclusions which can be drawn from such experiments are limited because of the variety of pharmacological agents used. Studies on preparations which were effectively "sympathetically denervated" either by means of adrenalectomy coupled with blockade by bretylium (Lucke, Denny, Hall,Lovell and Lister, 1978) or by total spinal anaesthesia (Gronert, Milde and Theye, 1977) gave conflicting results. The former experiments supported the view that the integrity of the sympathetic nervous system was necessary for the development of MH, the latter that it was not. There are also conflic-ting reports on the effect of epidural block on the develop-ment of MH (Kerr, Wingard and Gatz, 1975; McLoughlin, Somers, Wilson and Ahern, 1976). A greater autonomic responsiveness has been reported in stress susceptible compared to stress-resistant breeds of pig (Gregory and Lister, 1981). Studies using an isolated perfused caudal muscle preparation provided clear evidence that metabolic responses (increase in oxygen consumption and lactate formation) occur in MH susceptible muscle in the absence of neural and hormonal influences (Gronert, Milde and Taylor, 1980). These metabolic changes were induced by elevated temperature and the presence of the cholinergic agonist carbachol but not by α- and β-adrenergic agonists. However, these experiments also raise a question about the role which acetylcholine, released at the somatic neuromuscular junction of the intact animal, may play in the initiation of MH.

CENTRAL NERVOUS SYSTEM AND MH

A wide range of drugs which have a central action modify MH initiated by halothane. The induction of anaesthesia with barbiturates, ketamine or althesin prior to maintainance with halothane delays the clinical manifestation of MH although metabolic changes in skeletal muscle may have been initiated already (Somers, Wilson, Ahern and McLoughlin, 1977). The delaying action of these anaesthetics may simply be a consequence of their effects on the dose of halothane required to maintain a satisfactory depth of anaesthesia. However, other drugs have a similar protective action, in particular phenothiazines and butyrophenones (McLoughlin, Somers, Ahern and Wilson, 1978; McGrath, Rempel, Addis and Crimi, 1981; Wingard and Gatz, 1978). The butyrophenone azaperone was developed specifically for use in the pig and premedication with this drug reduces the incidence of death during transportation and of PSE muscle post mortem. Azaperone also delays the onset of halothane-initiated MH (McLoughlin et al., 1978). Spiperone administered to MH-susceptible animals daily over a period of seven days and then discontinued blocked a positive response (muscular rigidity with 5 minute inhalation of halothane (5% in O_2)) in these animals for up to four weeks. While such observations may contribute to the pretreatment of patients who may be susceptible to MH they cannot provide any very useful information about central mechanisms or pathways which may be involved in MH or stress reactions. This is because such drugs have widespread psychophysiological and behavioural effects. The variety of central and peripheral actions of the drugs make it virtually impossible to even speculate on the relevance of their actions to central neural functions and MH. Nevertheless, factual information about neurotransmitter function in the central nervous system of stress susceptible and normal animals might contribute something to the understanding of the role of the CNS in the stress syndrome and MH.

Monoamines in the brain

The use of histochemical fluorescence techniques has shown that the monoamines noradrenaline (NA), dopamine (DA) and 5-hydroxy-tryptamine (5HT) are located in neurons in specific regions of the brain. These neurons have high concentrations of monoamine in their widely ramifying terminal axons and relatively low concentrations in their cell bodies. The neurons are mainly located in discrete clusters in the brain stem, the cells (NA and DA) in the caudal brain stem giving rise to descending fibres which terminate in the grey matter of the spinal cord, those in the rostral brain stem to ascending fibres which pass through the hypothalamus and have a widespread terminal distribution which includes nuclei of the hypothalamus and the neocortex. The DA-containing neurons are mainly located in large clusters in regions of the substantia nigra and in the ventral mesencephalon. Their axons terminate in the lateral hypothalamus, the corpus striatum, nuclei in the limbic forebrain and in regions of the frontal cortex. Some cell bodies in the caudal brain stem send axons to the hypothalamus which appear to release adrenaline from their terminals.

Corpus striatum and hypothalamus

The corpus striatum and the hypothalamus were used to study the concentrations of NA, DA and their metabolites in stress (MH)-susceptible and normal pigs (Bardsley, Wheatley, Fowler, McCrodden, McLoughlin and Tipton, 1982). These regions were chosen for several reasons. Firstly, both areas are rich in catecholamines. Secondly, the corpus striatum is involved in muscular control and the hypothalamus integrates neural and endocrine functions and maintains homeostasis. Thirdly, all antipsychotic drugs, including the phenothiazines, block DA receptors and increase the turnover rate of DA in the corpus striatum. The phenothiazines and the butyrophenones also influence neuronal function in the hypothalamus. Both chlorpromazine and haloperidol enhance prolactin secretion by an antagonistic action on DA inhibitory receptors and chlorpromazine also interferes with release of growth hormone and corticotropin releasing factor and with thermoregulation. The experimental animals were anaesthetised with sodium pentobarbitone and maintained for one hour under this anaesthetic before craniotomy. The anaesthetic and surgical procedures involved have been described by Bardsley et al. (1982). The results of the study did not show that there were significant differences in the concentrations of NA and DA in either the corpus striatum or hypothalamus between stress (MH) susceptible and normal pigs. There were differences in the concentrations of certain non-O-methylated metabolites. The MH susceptible pigs had higher concentrations of 3, 4-dihydroxymandelic acid (DHMA) in the hypothalamus

(0.33 $\pm$ 0.14 ng/mg protein) and higher concentrations of
dihydroxyphenylglycol (DHPG) in the corpus striatum
(0.24 $\pm$ 0.05 ng/mg protein) than had normal pigs(hypothalamus
0.19 $\pm$ 0.07 ng/mg protein,corpus striatum 0.1 $\pm$ 0.04 ng/mg
protein of DHMA and DHPG respectively). These differences may
not be related to susceptibility to either stress or MH and
such steady state concentrations of metabolites may not be
related in any direct way to neuronal function.

Monoamine oxidase

The action of monoamines is terminated by active reuptake
into axon terminals and by the action of two enzymes,
monoamine oxidase (MAO) and catechol-0-methyltransferase.
Williams (1976) proposed that malignant hyperthermia might
be due to a functional deficiency of MAO which led to a
prolongation of monoamine activity. MAO exists in two forms,
A and B,which have different substrate specificities and
different sensitivities to inhibitors. It is considered that
both forms of the enzyme metabolise tryamine and DA, while
MAO-A is responsible for the deamination of NA and 5HT and
MAO-B for deamination of β-phenethylamine. There is very
little of the MAO-A in the pig (Hall, Logan and Parsons,
1969; Tipton, 1971) so it was a possibility that this species,
and perhaps especially the MH susceptible breeds, might have
inadequacies in the deamination of monoamines. However, the
results of the study showed that there was no significant
difference in MAO activity in the hypothalamus and corpus
striatum between MH susceptible and non-susceptible using as
substrates phenethylamine and 5HT. MAO activity was also

measured in the liver, kidney, heart, intestinal mucosa and cerebral cortex between the two types of pig but no significant differences in the activity of the enzyme were found. The activity of catechol-O-methyl transferase was measured in kidney but no differences were found between the susceptible and non-susceptible animals.

CONCLUSIONS

The virtual absence of monoamine oxidase type A in pig tissues does not adversely affect the ability of this species to deaminate monoamines. A functional deficiency of monoamine oxidase does not appear to be associated with susceptibility to MH. Steady state concentrations of noradrenaline, dopamine and their non-O-methylated metabolites are not notably different in the hypothalamus and corpus striatum.

REFERENCES

Ahern, C.P., Milde, J.H. and Gronert, G.A. 1985. Electrical stimulation triggers porcine malignant hyperthermia Research in Veterinary Science.39, 257-258.

Bardsley, M.E., Wheatley, A.M., Fowler, C.J., McCrodden,J.M., McLoughlin, J.V. and Tipton, K.F. 1982. Metabolism of monoamines in malignant hyperthermia-susceptible pigs. British Journal of Anaesthesia, 54, 1313-1317.

Gregory, N.G. and Lister, D. 1981. Autonomic responsiveness in stress-sensitive and stress-resistant pigs. Journal of Veterinary Pharmacology and Therapeutics, 4, 67-75.

Gronert, G.A. and Theye, R.A. 1976. Halothane-induced porcine malignant hyperthermia : metabolic and haemodynamic changes. Anaesthesiology, 4, 36-43.

Gronert, G.A., Milde, J.H. and Theye, R.A. 1977. Role of sympathetic activity in porcine malignant hyperthermia. Anaesthesiology, 47, 411-415.

Gronert, G.A., Milde, J.H. and Taylor, S.R. 1980. Porcine muscle responses to carbachal, α- and β-adrenoreceptor agonists, halothane or hyperthermia. Journal of Physiology, Vol. 307, 319-333.

Hall, D.W.R., Logan, B.W. and Parsons, G.H. 1969. Further studies on the inhibition of monoamine oxidase by M & B 9302 (clorgyline)-1, Biochemical Pharmacology, 18, 1447-1452.

Kerr, D.D., Wingard, D.W. and Gatz, E.E. 1975. Prevention of porcine malignant hyperthermia by epidural block. Anaesthesiology, 42, 307-311.

Lister, D., Hall, G.M. and Lucke, J.N. 1976. Porcine malignant hyperthermia 111: adrenergic blockade. British Journal of Anaesthesia, 48, 831-838.

Lucke, J.N., Hall, G.M. and Lister, D. 1976. Porcine malignant hyperthermia 1: metabolic and physiological changes. British Journal of Anaesthesia, 48, 297-304.

Lucke, J.N., Denny, H., Hall, G.M., Lovell, R. and Lister,D. 1978. Porcine malignant hyperthermia Vl: the effects of bilateral adrenalectomy and pretreatment with bretylium on the halothane-induced response. British Journal of Anaesthesia, 50, 241-246.

McGrath, D.V.M., Rempel, W.E., Addis, P.B. and Crimi, A.J. 1981. Acepromazine and droperidol inhibition of halothane-induced malignant hyperthermia(porcine stress syndrome) in swine. American Journal of Veterinary Research, 42, 195-198.

McLoughlin, J.V., Somers, C.J., Wilson, P. and Ahern, C.P., 1976. Halothane-initiated malignant hyperthermia in experimental animals. Irish Journal of Medical Science, 145, 99.

McLoughlin, J.V., Somers, C.J., Ahern, C.P. and Wilson, P., 1978. The influence of neuroleptic drugs on the development of muscular rigidity in halothane sensitive pigs. In Biochemistry of Myasthenia Gravis and Muscular Dystrophy, (Eds. Lunt, G.G. and Marchbanks, R.M.) (Academic Press, N.Y. and London) pp 361-365.

Somers, C.J., Wilson, P., Ahern, C.P. and McLoughlin, J.V., 1977. Energy phosphate turnover and glycolysis in skeletal muscle of the Pietrain pig: the effects of premedication with azaperone and pentobarbitone anaesthesia. Journal of Comparative Pathology, 90, 177-186.

Tipton, K.F. 1971. Monoamine oxidases and their inhibitors. In Mechanism of Toxicity, (Ed. Aldrich, W.N.) (Macmillan Press, London) p 13.

Williams, C.H. 1976. Some observations on the etiology of the fulminant hyperthermia-stress syndrome. Perspectives in Biological Medicine, 20, 120-125.

Wingard, D.W. and Gatz, E.E. 1978. Some observations on stress susceptible patients. In Malignant Hyperthermia, (Eds. Aldrete, J.A. and Britt, B.A.) (Green and Stratton, N.Y.), pp 363-372.

THYROID FUNCTION IN THE AETIOLOGY OF THE PSE CONDITION

Bruce W Moss

Agricultural and Food Chemistry Research Division
Department of Agriculture for Northern Ireland
and
Department of Agricultural and Food Chemistry
The Queen's University of Belfast

ABSTRACT

Thyroid function has been implicated in the aetiology of stress sensitivity and pale watery muscle (PSE) condition in pigs. The various approaches used to study thyroid function are discussed particularly in relation to interpretation of the findings. It is suggested that clinical terms such as hypo- or hyper-thyroid should not be applied in relation to stress sensitivity in pigs. Differences in some thyroid parameters between breeds appear to be paralleled by differences between sexes. Thus the leaner breed and leaner sex have lower thyroxine (T_4) and higher TSH/T_4 ratios. A compartmental model for studying the kinetics of thyroidal iodine metabolism is proposed. This model can be used to elucidate the problems involved in interpreting clinical approaches to studying thyroid function. There is evidence from studies on thyroidal iodine kinetics that the PSE condition is related to a faster utilisation of thyroxine relative to secretion.

The incidence of pale watery muscle (PSE) is greater in leaner breeds of pigs and thus the underlying physiological mechanisms to explain this might be expected to be those endocrinological systems which are involved in the control of growth and leanness. Pituitary adrenal function (both cortex and medulla) and thyroid function have been implicated in the aetiology of the PSE condition. Early studies of the involvement of thyroid function showed that feeding iodocasein to pigs resulted in an improvement in meat quality (Ludvigsen 1957) and that PSE pigs might be considered hypothyroid. Although some of the later work showed that goitrogens increased the tendency to PSE (Ludvigsen 1968; Topel & Merkel 1966) other work on thyroid turnover (Sorensen 1961; Romack et al) supports the contention that PSE prone pigs are hyperthyroid. The terms hypo- and hyper-thyroid are clinical terms and the typical clinical signs of hyperthyroidism is an ectomorphic body type with an excitable nature which may be more applicable to a stress sensitive type of pig. Since hypo- and hyperthyroidism are clinical terms they should not be applied to 'PSE' or stress sensitive type of pigs unless the same clinical syndrome exists, the reasons for this will be discussed later.

One of the aims of this review is to consider whether a physiological explanation can be given to the apparent anomalies and conflict in the literature in relation to the role of thyroid function in the aetiology of the PSE condition. Several approaches have been used to study the role of thyroid function and other endocrine systems in relation to the PSE condition. These approaches can be classified under 6 broad headings (1) genetic (2) correlation (3) pharmacological (4) clinical (5) biochemical (6) physiological. Many workers have used more than one approach.

In the genetic approach the main criterion for assessing thyroid function has been the measurement of circulating hormone levels, in earlier studies as protein bound iodine (PBI) and later directly as T_4 and T_3. Judge et al (1968) found higher PBI levels in stress susceptible pigs, and lower thyroid uptake 24 h post administration of radioiodine. The difference in circulating T_4 levels depends on the stress involved in sampling (Moss 1981) and as will be discussed later thyroid uptake studies must be interpreted with caution. Marked differences in the T_4 level of Landrace and Large White pigs were only found when samples were taken from the live animal under restraint and not when taken at slaughter (see Table 1). In stress situations both decreases in circulating T_4 (Moss 1984) and increases (Spencer 1984) have been reported. In other species rapid increases in T_4 secretion have been noted in response to stress (Falconer 1972).

Larger thyroid glands have been noted in stress susceptible pigs (Ludvigsen 1968; Topel & Merkel 1966). This may reflect hypertrophy to overcome ineffective trapping and/or increased TSH secretion. Loss of inorganic iodide from the thyroid gland parallels the loss of organic iodide (Isaacs et al 1966; Falconer & Hetzel 1964) and thus is related to stimulation. From kinetic models proposed (see Fig 1) it is evident that such a response could give rise to lower thyroid uptakes although the pigs may have higher TSH/T_4 ratios (See Table 2).

TABLE 1   Thyroxine responses of Landrace and Large White pigs to the stressors of restraint and pre-slaughter handling

	Plasma Thyroxine ($\mu g\ dl^{-1}$)	
	Landrace	Large White
In pens[a]	4.0	4.9
At slaughter[b]	4.2	4.1
Significance	NS	*

[a]   The pigs were restrained by a noose round the upper jaw, and blood samples were collected from an ear vein within 5 min of restraining the pig at the progeny station before transport to slaughter.

[b]   The pigs from their individual pens were mixed on the lorry, transported for a period of 35 min to the abattoir and killed by electrical stunning after 1 h in lairage and blood was collected at exsanguination.

The sex and breed differences in thyroid function in Table 2 are also of interest in consideration of the relationship between thyroid function and leanness.   In both breeds the leaner sex has lower T_4, lower FTI, higher TSH and higher TSH/T_4 ratio.   Also by comparison the leaner more stress sensitive breed (Landrace) has lower T_4, lower FTI and higher TSH/T_4 ratio.   Campbell (1984) suggested from breed/sex interactions on meat quality that the pigs could be ranked in the following order of decreasing post mortem glycolytic rate Landrace (LR) and Large White (LW) boars > LR gilts > LR barrows > LW gilts and barrows.   A similar ranking order would be obtained if the ranking was done on the basis of increasing FTI, or decreasing TSH ie lowest FTI or highest TSH in pigs of fastest post mortem glycolytic rate.   Thus it appears from these preliminary sex/breed studies that thyroid function may be involved in both the control of leanness and meat quality.   The low within litter variation and high intraclass correlation, 0.64 for T_4 and 0.77 for FTI, (calculated as in Snedecor 1965) indicates that thyroid function may have a high heritability coefficient and may thus be useful in selection schemes.

Table 2 Thyroid status[a] of Landrace and Large White boars, barrows and gilts

	T_3 Uptake Ratio	T_4 μg dl^{-1}	FTI[c]	TSH[b] μU ml^{-1}	TSH/T_4 Ratio
Landrace:-					
Boars	0.68	3.37	5.32	5.68	1.75
Barrows	0.57	3.60	6.58	4.03	1.16
Gilts	0.60	3.85	6.47	4.69	1.26
Significance of Difference:-					
Litters (8)	NS	***	***	NS	NS
Sex	NS	NS	*	NS	NS
Large White:-					
Boars	0.72	4.12	6.02	6.08	1.50
Gilts	0.65	4.59	7.13	5.17	1.14
Significance of Difference:-					
Litters (4)	NS	NS	NS	NS	NS
Sex	NS	NS	NS	NS	NS

a Measured in blood obtained from the ear vein after restraining the pig.

b TSH units based on hTSH (Phadebas[R] TSH Test, Pharmacia Diagnostics)

c FTI - Free Thyroxine Index

The correlation approach where specific parameters eg T_4 levels are related to meat quality parameters using statistical regression techniques suffer from several problems, particularly the influence of the environmental component. If correlations are attempted within a breed which does not contain halothane positive pigs then the assumption made is that there is an association between the level of the parameter and the rate of post mortem glycolysis or other related parameters. Debate still remains as to whether the parameter of thyroid function should be measured in the basal (ie 'unstressed') state or in the stressed state. Both measurements are important, but what is more crucial is the interpretation of the mechanism at the biochemical level. Problems in the control of the environmental influence have led to conflicting results. In Large White boars in two trials out of three correlations between thyroxine levels at slaughter and meat quality parameters were low and not statistically significant, whilst in the third trial thyroxine levels at slaughter were negatively correlated with LD pH 5 min post mortem (r = 0.55, p<0.05) and positively correlated with drip loss (r = 0.63, p<0.01). Judge et al (1968) reported significant positive correlations (p<0.05) between thyroid uptake and the time to rigor onset in the Poland China but not the Chester White breed. In both breeds radiobound iodine was positively correlated (p<0.05) with muscle lactic acid immediately post mortem. Thyroxine levels in slaughter blood were higher in pigs killed after an overnight lairage (Moss & Robb 1978) and may be explained by decreased peripheral conversion of T_4 to T_3 during fasting (Merimee & Fineberg 1976). Fasting pigs for 24 h, however, did not significantly affect the circulating T_4 levels in blood samples obtained by biopsy (Moss 1984).

Literature reports indicate a diurnal rythmn in thyroid function in humans related to activity patterns (Nicoloff 1970a,b). A similar relationship appears in young pigs in that a pre-feeding rise in thyroxine excretion occurred whether or not pigs were fed (Moss & Jordan 1980). Thus thyroid function may alter in relation to some conditioned response even when the stimulus for that response is removed. Such conditioning will have a marked effect on the measured relationship between thyroid function parameters and meat quality.

In the clinical approach parameters which are used to diagnose clinical hypo- and hyper-thyroid states have been applied to study thyroid function in pigs. Thyroxine circulates in the blood stream bound to protein, thyroid binding globulin (TBG), and it is only the unbound or free thyroxine (FT_4) which is considered to be physiologically active. Techniques have been made available recently to measure FT_4 levels directly on a routine basis in clinical diagnostics (Bordoux et al 1982), prior to this an indirect method to measure the free thyroxine index (FTI) was used.

The indirect method is based on the principle that triiodothyronine (T_3) will bind to the unoccupied binding sites on TBG without displacing bound T_4 (T_4-TBG). From the equation below it can be seen that the number of unoccupied binding sites, ie those available for T_3 binding, depends on (1) the amount of TBG and (2) the equilibrium between T_4 and TBG. The results of T_3 binding are expressed as a ratio and confusion may arise as some methods use the proportion of radiolabel bound to the serum, whilst others use the proportion bound to the separating medium (eg charcoal, resin, sephadex). The T_3 binding methodology and free thyroxine index has been reviewed by Evered et al (1978). The 'T_3 uptake ratios' given in table 2 were determined by the method of Herbert et al (1965) in which lower 'T_3 uptake ratios' indicate less unoccupied sites, the free thyroxine index was calculated as total T_4/T_3 uptake ratio.

$$T_4 + TBG \rightleftharpoons T_4\text{-}TBG$$
$$(\text{bound T4})$$

$$\text{at equilibrium} \quad k = \frac{[T_4][TBG]}{[T_4 - TBG]}$$

where k = equilibrium binding constant

Although an indirect measure of FT_4, the FTI was designed essentially to assess thyroid function in pregnancy when total T_4 levels could be abnormally high but due to increased production of TBG there was no increase in the physiological activity at the target area (ie cellular or subcellular level).

Problems exist in the use of clinical tests in animal studies and careful evaluation should be made to test the relevance of such tests. A prime example of misuse being the application of an FTI type diagnostic kit in poultry where T_4 is only loosely bound to albumin (Falconer 1971) and also in the same species a possible overestimation of T_4 if a competitive protein binding technique is used with cross reaction for T_3 (Moss & Balnave 1978) since T_4/T_3 ratios are lower in poultry than in humans (Sadovsky & Bensadoun 1971). In the studies of

Moss & Robb (1978) and Hall et al (1975) the same technique was used to measure FTI with higher values indicating greater free thyroxine, however, other workers have used the inverse relationship (Evered et al 1976). Hall, Lucke & Lister (1975) found that both total T_4 and FTI increased as a response to halothane anaesthesia, whilst Moss & Robb (1978) found increases in T_4 and FTI after overnight lairage which, associated with decreased cortisol levels, was considered a less stressful situation. In both of these cases it seems unlikely that TBG levels could change markedly in the time scale of response, however, part of the increase in FTI could be accounted for by the displacement of T_4 from binding proteins by free fatty acids which may also increase under the stress conditions.

In preliminary studies Pietrain pigs were found to have low thyroid uptakes (Moss 1975) and is in agreement with similar reports by Judge et al (1968). The time of measurement of radioactivity in the thyroid gland after the administration of the dose is a critical factor (Moss 1975). In clinical diagnostics an index (T) based on the excretion of radioiodine during uptake studies has been used. This 'T' index has not been used extensively on pigs but Lister (1976) suggests that the major difference between Pietrain and Large White pigs may be in the greater excretion of radioiodine in uptake studies. This could be explained by lowered uptakes and/or greater secretion rates accompanied by lowered recycling on the basis of the kinetic model in Fig 1.

Several methods may be used in the pharmacological approach eg use of goitrogens, thyroidectomy with or without replacement therapy, destruction of the thyroid gland with high doses of radioiodine administration. Topel & Merkel (1966) found that goitrogens produced hypertrophy of the thyroid gland but only a proportion of those pigs fed methylthiouracil developed PSE. Since the goitrogens effectively reduce iodide trapping and conversion to T_4, adaptive hypertrophy may occur in such situations and the resulting thyroid status may be one in which T_4 levels after goitrogen treatment are similar to pretreatment levels but there is increased TSH secretion. Fischer (1974) comments on the feedback mechanisms involved in thyroid function. Ludvigsen (1968) found goitrogens to increase PSE and the administration of iodocasein and thyroxine to reduce PSE. Ineffective trapping of iodide may be the explanation of the results.

58

Fig. 1. Compartmental model for the study of thyroidal iodine
metabolism

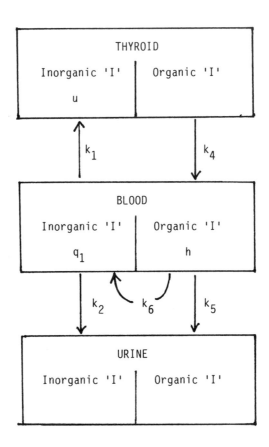

K_1 - rate constant for uptake into thyroid gland
K_2 - rate constant for excretion of inorganic iodide
K_4 - rate constant for secretion of hormone
K_5 - rate constant for excretion of organic iodide
K_6 - rate constant for breakdown of organic iodide
u - amount of radioiodide in thyroid gland
q - amount of inorganic radioiodide in plasma
h - amount of organic radioiodide in plasma

The biochemical actions of thyroid hormones have been established at the cellular level (Tata 1967) where the sequence of events appears to be stimulation of synthesis of RNA, stimulation of synthesis of mitochondrial proteins, synthesis of key enzymes and corresponding increase in metabolic rate. The short term effects of thyroid hormone appear not to be related to synthesis of mitochondrial proteins (Bronk 1967; Chen & Hoch 1977) but to activation of the bc_1 complex and possibly stimulation of succinate dehydrogenase (Verhoeven et al 1985). In laboratory animals thyroxine has also been shown to have an effect on the fibre type distribution, with greater activity of myosin ATPase in hyperthyroid animals (Ianuzzo et al 1978). Hyperthyroid animals had greater succinic dehydrogenase and glycerol phosphate dehydrogenase activity in the muscle fibres. Differences in fibre type distribution occur in stress susceptible breeds of pigs (Didley et al 1970; de Bruin 1971; Sair et al 1970). It is therefore possible that these fibre type differences may reflect differences in thyroid function.

An excitable nature is a typical feature in clinical hyperthyroidism and therefore it seems relevant to consider the role of thyroid function in relation to behaviour and how this might be related to stress sensitivity. Rats have been selected for emotionality on the basis of the open field test (Hall 1954). Such selected strains of emotionally disturbed rats have been observed to have hyperthophied thyroid glands, lower thyroid uptake values and higher pituitary TSH activity, and treatment with propylthiouracil produced typical signs of increased emotionality in these animals (Feuer & Broadhurst 1962a,b,c). Nevertheless these workers considered the emotionally disturbed rats to be hypothyroid relative to non emotional animals. These differences in thyroid function between emotional and non emotional rats are similar to those reported between stress susceptible and stress resistant pigs where the suggestion of hypothyroidism in stress susceptible strains has also been made (Ludvigsen 1968; Judge et al 1968). It has previously been suggested that stress sensitivity in pigs may be equated with emotionality in laboratory species (Moss & Robb 1978). Further support for this is given by the fact that the male laboratory species are more emotional than females, and that the differences in T_4 level between boars and gilts, is in the same order as between Landrace and Large White pigs (Moss & Robb 1978). Preliminary

Fig 2   Urinary ratio in Pietrain and Large White pigs.

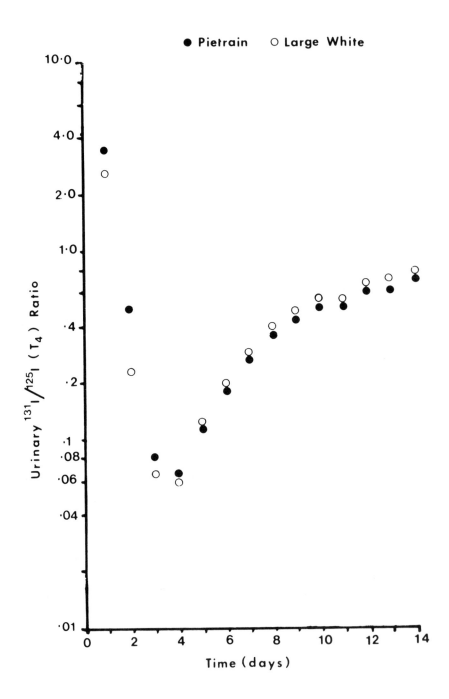

investigations suggest that sex differences in other parameters of thyroid function may also occur (see Table 2).

Problems relating to the application of clinical or pharmacological approaches to thyroid function may be explained if detailed physiological studies of thyroid function are undertaken. Such studies are time consuming but are essential to the valid application of more rapid or simple tests. The major disadvantage of the physiological approach is that animals may have to be catheterised and restrained in metabolism cages, a situation which may be difficult to relate to production practices.

Most physiological studies involve the measurement of turnover rates using isotope techniques. Thyroxine secretion rate may be measured by external scintillation counting of the thyroid gland. Such an approach by Bodart & Francois (1972) showed that stress sensitive breeds had faster thyroxine secretion rates. Thyroxine utilisation rate may also be measured directly by the rate of loss of isotopically labelled thyroxine (usually ^{125}I-T_4) from the blood circulation, half lives of T_4 using this technique have been calculated in the range 1 to 2 days for pigs, which is much shorter than those in human studies. Thyroxine secretion and utilisation rates may also be measured indirectly using a dual labelling technique which in human studies had been shown to be a sensitive indicator of thyroid function (Nicoloff 1970). The technique as described by Nicoloff (1970a,b) involves simultaneous administration of radiolabelled sodium iodide (eg ^{131}I) and radiolabelled T_4 (eg ^{125}I) and determining the ratio of the two isotopes in the urine. Initial studies using this approach (Moss & Lister - unpublished observations) showed that although the shape of the resulting curve in pigs was similar to humans the time scale was different (see Fig 2). According to Nicoloff (1970b) the gradient of the release slope can be increased by TSH administration and is greater in the hyperthyroid state. In a study involving 6 LW boars a similar approach to Nicoloff was used except that the administration of ^{125}I-T_4 was delayed until 4 days after $Na^{131}I$ administration. In this situation the minimum ratio was reached earlier, gradients of the release slope were inversely related to EEL (colour) values ($r = 0.89$ $p<0.05$) and drip loss ($r = -0.64$ NS) as shown in Fig 3.

To interpret these above results further it is necessary to consider the kinetics of thyroidal iodine metabolism. Several

62

Fig.3 Relationship between Urinary Release Slope and EEL values

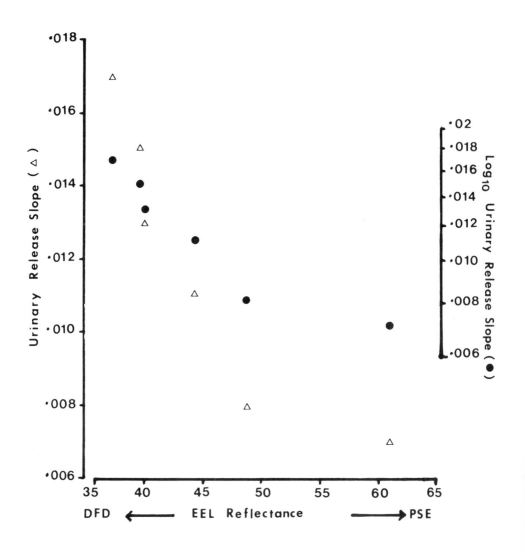

compartmental models have been proposed (Oddie 1949; Riggs 1952) the one shown in Fig 1 is based on that by Brownell (1951). Using the compartmental model of Fig 1 an equation relating the amount of radioiodine in the thyroid gland after intravenous injection of $Na^{131}I$ can be obtained.

$$U = \frac{k_1 Q^0}{k_1 + k_2 - k_4} [e^{-k_4 t} - e^{-(k_1+k_2)t}]$$

or if u is expressed as the percentage of dose administered

$$U = \frac{100\ k_1}{k_1 + k_2 - k_4} [e^{-k_4 t} - e^{-(k_1+k_2)t}]$$

Theoretical uptake curves drawn from these equations are shown in Fig 4a and 4b and it is evident that only in the case when the difference between individuals is due to plasma clearance rate (k_1+k_2) will thyroid uptake measurements discriminate between them in the early period post mortem. At later times (10 to 40 h) the measured uptake (U%) decreases with increased T_4 secretion rate and the time to peak decreases and U% increases with increased rate of thyroidal iodine trapping.

If recycling of radioiodine occurs then the slope after the maximum will be less than the true secretion rate. Brownell (1951) proposed the following relationships to account for this:-

$$U = U^0 e^{-k'_4 t}$$

where U^0 = theoretical maximum uptake

k'_4 = apparent secretion rate

from this the following relationships can be derived

$$k'_4 = k_4 (1-Er) \quad \text{where } Er = \frac{k_1}{k_1 + k_2}$$

provided the rate of organic iodide excretion is small

64

Theoretical Thyroid Uptake calculated from Compartmental Model

Fig.4a Alterations in Secretion Rate (K_4)

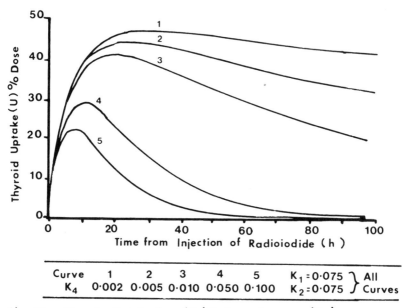

Curve	1	2	3	4	5	$K_1 = 0.075$	All
K_4	0.002	0.005	0.010	0.050	0.100	$K_2 = 0.075$	Curves

Fig.4b Alterations in Uptake Rate (K_1) & Excretion Rate (K_2)

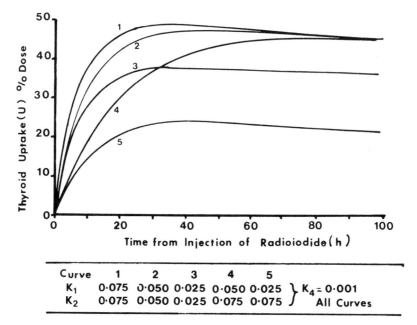

Curve	1	2	3	4	5	
K_1	0.075	0.050	0.025	0.050	0.025	$K_4 = 0.001$
K_2	0.075	0.050	0.025	0.075	0.075	All Curves

Using the above approach various kinetic parameters relating to thyroidal iodine metabolism were calculated. Pietrain pigs appeared to have a rapid initial release phase in the first 24 h after radioiodine administration and thus two secretion rates were calculated the first for the period up to 24 h and the second for the period after 24 h from the time of administration of radioiodine. The main difference between the breeds was the higher plasma clearance rate (k_1+k_2) lower uptake rate (k_1) and higher excretion rate (k_2) in Pietrain pigs.

This review has not covered all aspects of thyroid function in relation to PSE, the relationships between thyroid function and adrenal medulla (Harrison 1964; Melander & Sundler 1972) and adrenal cortical function (Peterson 1958; Beale, Croft & Powell 1973) must also be considered. There are differences in thyroid function between stress sensitive and stress resistant pigs but the descriptions hypo- or hyperthyroid should not be used. Differences in thyroid function can be explained however by consideration of the kinetic parameters where it appears that in stress sensitive pigs the rate of excretion of iodine (k_2) is greater and the rate of uptake lower (k_1). Thus if a stress situation requires increased thyroxine secretion or increases utilisation these demands cannot be met. Such an hypothesis could explain the larger thyroid glands of stress sensitive pigs and the lower T_4 values reported for stress sensitive pigs. Further studies on the kinetics of thyroidal iodine metabolism and how this is affected by other endocrine systems and both long term and short term stressors needs to be undertaken to fully understand the role of thyroid function in the aetiology of the PSE condition. It is also important to understand the adaptation in thyroid function that may occur due to long term stressors.

REFERENCES

Beale, R.N., Croft, D. and Powell, D. 1973. Some effects of thyroid disease on neutral steroid metabolism. J. Endoc., 57, 317-324.
Bodart, C. and Francois, E. 1972. A comparative study of thyroid metabolism in Belgian and Pietrain pig breeds. I.A.E.A. Symposium, Vienna. Use of Isotopes in Animal Physiology.
Bordoux, P., Verelst, J., Branders, C. and Ermans, A.M. 1982. Clinical evaluation of free thyroxine in serum by direct radioimmunoassay. Proc. Int. symp. Radioimmunoassay and Related Procedures in Medicine, Vienna, June 1982. IAEA pp 241-251.

66

Bronk, J.R. 1966. Thyroid hormone: Effects on electron transport. Science 153, 638-639.

Brownell, G.I. 1951. Analysis of techniques for the determination of thyroid function with radioiodine. J. Clin. Endoc. 11, 1095-1105.

Brown-Grant, K., Harris, G.W. and Reichlin, S. 1954. The influence of the adrenal cortex on thyroid activity in the rabbit. J. Physiol., 126, 41-51.

Chen, Y.L. and Hoch, F.L. 1977. Thyroid control over biomembranes: Rat liver mitochondial inner membranes. Arch. Bioch. Biophys. 181, 470-483.

de Bruin, A. 1971. Proc. 2nd Int. Symp. Condition and Meat Quality of pigs p. 86-89 (Centre for Agricultural Publishing and Documentation, Wageningen)

Didley, D.D., Aberle, E.D., Forrest, J.C. and Judge, M.D. 1970. Porcine muscularity and properties associated with Pale, Soft, Exudative Muscle. J. Anim. Sci., 31, 681-685.

Evered, D., Vice, P.A., and Clarke, F. 1976 The Investigation of Thyroid Disease, Medical Monograph 9, Radiochemical Centre, Amersham, England.

Falconer, I.R. 1971 The Thyroid Gland in 'Physiology and Biochemistry of the Domestic Fowl'. (Ed. Bell, D.J. and Freeman, D.B., Academic Press, New York) pp 459-471.

Falconer, I.R. 1972. Use of sheep with exteriorized thyroid glands to measure transient changes in thyroid hormone secretion during restraint. J. Physiol. 222, (1) 3 P.

Falconer, I.R. and Hetzel, B.S. 1964. Effect of emotional stress and TSH on thyroid vein hormone level in sheep with exteriorized thyroids. Endocrinology, 75, 42-48.

Feuer, G. and Broadhurst, P.L. 1962a. Thyroid function in rats selectively bred for emotional elimination. I. Differences in thyroid hormones. J. Endocrin., 24, 127-136.

Feuer, G. and Broadhurst, P.L. 1962b. Thyroid function in rats selectively bred for emotional elimination. II. Differences in thyroid activity. J. Endocrin., 24, 253-262.

Feuer, G. and Broadhurst, P.L. 1962c. Thyroid function in rats selectively bred for emotional elimination. III. Behavioural and Physiological changes after treatment with drugs acting on the thyroid. J. Endocrin., 24, 385-396.

Fischer, K. 1974. Die Bedeutung endoknner storugen fur die Entwicklung Von Qualitotsa bweichungen bei Schweinefleisch. Die Fleischwirschaft, 54, 1212-1231.

Hall, C.S. 1954. In 'Readings in Child Development'. (Ed. Martin, W.F. and Stendler, C.B.)(Harcourt Brace, New York) pp 58-60.

Hall, G.M., Lucke, J.N. and Lister, D. 1975. Treatment of porcine malignant hyperthermia. Anaesthesia, 30, 308-317.

Harrison, T.S. 1964. Adrenal Medullory and Thyroid Relationships. Physiol. Rev., 44, 161-185.

Ianuzzo, D., Patel, P., Chen, V., O'Brien, P. and Williams, C. 1977. Thyroidal trophic influence on skeletal muscle myosin. Nature, 270, 74-76.

Isaacs, G.H., Athans, J.C. and Rosenburg, I.N. 1966. Effects of Thyrotropin upon thyroidal iodide; studies using thyroid venous cannulation and two radioiodine isotopes. J. Clin. Invest., 45, (5), 758-767.

Judge, M.D., Briskey, E.J., Cassens, R.G., Forrest, J.C. and Meyer, R.K. 1968a. Adrenal and thyroid function in 'stress susceptible' animals (sus domesticus). Am. J. Physiol. <u>214</u>, 146-151.

Lister, D. 1976. Hormonal influences on the growth, metabolism and body composition of pigs. In 'Meat Animals; growth and productivity (Ed. Lister, D., Rhodes, D.N., Fowler, V.R. and Fuller, M.F.)(Plenum Press, New York) pp 355-372.

Ludvigsen, J. 1957. On the hormonal regulation of vasomotor reactions during exercise with special reference to the action of adrenal cortical steroids. Acta. Endoc., 26, 406-416.

Ludvigsen, J. 1968. Some thyroid and adrenal breed characteristics and their possible relation to pale, exudative muscles in pigs. In 'Recent Points of View on the Condition and Meat Quality of Pigs for Slaughter' (Ed. W. Sybesma, P.G. van der Wal and P. Walstra) (IVO, Zeist, Netherlands) pp 113-116.

Melander, A. and Sundler, F. 1972. Interactions between catecholamines, 5-hydroxytryptamine and TSH on the secretion of thyroid hormone. Endocrinology, <u>90</u>, 188-193.

Merimee, T.J. and Fineberg, E.S. 1976. Starvation-induced alterations of circulating thyroid hormone concentrations in man. Metabolism <u>25</u>, (1), 79-83.

Moss, B.W. 1975. PhD Thesis, University of Bristol.

Moss, B.W. 1981. The Development of a Blood Profile for Stress Assessment. In 'The Welfare of Pigs'. Current Topics in Veterinary Medicine and Animal Science, No 11. (Ed. Sybesma W.) (Martinus Nijhoff, The Hague, Boston, London) pp 112-125.

Moss, B.W. 1984. The effect of preslaughter stressors on the blood profiles of pigs. Proc. Eur. Meat. Res. Workers, Bristol 1984. pp 20-21.

Moss, B.W. and Balnave, D. 1978. The influence of elevated environmental temperature and nutrient intake on thyroid status and hepatic enzyme activities in immature male chicks. Comp. Biochem. Physiol. <u>60B</u>, 157-161.

Moss, B.W. and Jordan J.W. 1980. Preliminary investigations of urinary thyroxine excretion by young pigs. Res. Vet. Sci. <u>28</u>, 1-5.

Moss, B.W. and Robb J.D. 1978. The effect of pre-slaughter lairage on serum thyroxine and cortisol levels at slaughter and meat quality of boars, hogs and gilts. J. Sci. Fd. Agric. <u>29</u>, 689-696.

Nicoloff, J.T. 1970a. A new method for the measurement of acute alterations in thyroxine deiodination rate in man. J. Clin. Invest., <u>49</u>, 267-273.

Nicoloff, J.T. 1970b. A new method for the measurement of thyroidal iodine release in man. J. Clin. Invest., <u>49</u>, 1912-1921.

Oddie, T.H. 1949. Analysis of radioiodine uptake and excretion curves. Brit. J. Radiol., <u>22</u>, (257) 261-267.

Peterson, R.E. 1958. THe influence of the thyroid on adrenal cortical function. J. Clin. Invest., <u>37</u>, 763-773.

Riggs, D.S. 1952. Quantitative aspects of iodine metabolism in man. Pharm. Rev., <u>4</u>, 284-369.

Romack, F.E., Turner, C.W., Laskley, J.F. and Day, B.N. 1964. Thyroid secretion rate in swine. J. Anim. Sci., <u>23</u>, 1143-1147.

Sadovsky, R. and Bensadoun, A. 1971. Thyroid iodohormones in the plasma of the rooster (Gallus Domesticus). Gen. Comp. Endoc. <u>17</u>, 268-274.

68

Sair, R.A., Lister, D., Moody, W.G., Cassens, R.G., Hoekstra, W.G. and Briskey, E.J. 1970. Action of curare and magnesium sulphate on striated muscle of stress susceptible pigs. Amer. J. Physiol. 218, 108-114.

Sorensen, P.H. 1961. Use of radioactive iodine in studies of thyroid function in cattle and pigs. Conference on the use of radioisotopes in animal biology and medical sciences. Academic Press Inc. New York.

Snedecor, G.W. 1965. Statistical Methods, 4th Edition (Iowa State University Press) p 283.

Spencer, G.S.G., Wilkins, L.J. and Hallet, K.G. 1984. Hormonal and metabolite changes in the blood of pigs following loading and during transport and their possible relationship to subsequent meat quality. Proc. Eur. Meat. Res. Workers, Bristol, 1984. pp 15-17.

Tata, J.R. 1967. The formation and distribution of ribosomes during hormone induced growth and development. Biochem. J. 104, 1-16.

Topel, D.G. and Merkel, R.A. 1966. Effect of exogenous goitrogens upon some physical and biochemical properties of porcine muscle and adrenal gland. J. Anim. Sci. 25, 1154-1158.

Verhoeven, A.J., Kamer, P., Groen, A.K. and Tager, J.M. 1985. Effects of thyroid hormone on mitochondrial oxidative phosphorylation. Biochem. J. 226, 183-192.

HORMONAL IMPLICATIONS IN MALIGNANT HYPERTHERMIA AND THE PORCINE STRESS SYNDROME

J B Ludvigsen

National Institute of Animal Science
Rolighedsvej 25, DK-1958 Frederiksberg C, Denmark

ABSTRACT

The role of the endocrine system in stress-susceptibility in pigs is discussed against the background of seventy-five years of results of the Danish Landrace progeny testing system. Since 1910, selection of the bacon type of pig favoured heavy muscling and low feed cost. Feed intake was reduced from 3.77 feed units per kg of live weight gain in 1910 to 2.65 in 1985, and thickness of backfat was reduced from 4.5 to 1.7 cm.

Changes in body composition are mediated within wide limits through changes in the antagonistic interplay between hormones. There is substantial evidence that selection of heavy muscled pigs on feed conversion results in increasing activity of growth hormone (STH). This stimulates protein synthesis and deposition and the antagonistic effect of STH on the activity of TSH and ACTH, which are important hormones in the adaptive capacity to strain situations, such as transportation.

Investigations in the 1950s revealed that reduction of the death rate in transport and lairage caused an increase in the incidence of pigs condemned at slaughter for muscle degeneration (PSE). The effects of thyroid active substances, iodinated casein, iodinated histidine and thyroxine, on the quality of tinned ham indicated that specific compounds of thyroglobulin, not involving thyroxine, may control the buffering capacity of skeletal muscles, preventing the critical drop of pH during lactic acid formation. Changes in the electrophoretic pattern of serum proteins in stress susceptible pigs during handling or transport reflected the profound metabolic effects these animals are exposed to.

INTRODUCTION

Since the provisions for the improvement of the Danish Landrace bacon type pig were framed early this century, the overall criterion for the selection of breeding stock was improvement of feed conversion rates concomitantly with reduction of fat content of carcasses, as measured by the thickness of the backfat. The ongoing selection of breeding stock was based on improved performance of litter mates as compared to the average feed conversion rate of pigs at the progeny testing station in the same season of the year. The conversion rate improved from 3.77 feed units per kg live weight gain in 1910 to 2.65 in 1984, and the thickness of backfat decreased from 4.5 cm in 1930 to 1.7 cm in 1985. During the seventy-five years of organised selection, adjustments of the test meals have been inevitable with increasing knowledge of the nutritional requirements of the growing pig, especially the importance of vitamins and minerals. The basic test diet (3 kg acidified skim milk daily, barley according to live weight and supplementation with vitamins A and D) was not changed from 1928 to 1970, when skim milk was substituted by soybean meal, macro- and microminerals and vitamin supplementation adjusted to requirements. Systematic selection for increased protein deposition at the expense of fat is primarily expressed as an increase in skeletal muscle volume. Pietrain and Belgium Landrace pigs are typical examples of short pigs with voluminous fore- and hindquarters, whereas in Danish Landrace pigs the greater body length influences the size and appearance of the hindquarters, which do not look heavily muscled. As the ratio of protein to water in skeletal muscles is around 27/73 and the fat to water ratio in adipose tissue is 90/10, less food is required to produce 1 kg of muscle than 1 kg of fat. The decrease in feed units per kg of weight gain from 3.77 to 2.65 over seventy-five years justified the use of feed conversion rates for selection in heavy muscled pigs.

In 1954 meat quality was incorporated in the progeny testing programme using a subjective scale for the evaluation of colour and consistency (Table 1).

TABLE 1   The subjective scale used to evaluate pork colour and consistency that was incorporated in the Danish progeny testing programme in 1954

Score	Pork Colour and Consistency
0.5	Gray, same colour as boiled meat, very moist surface
1.0	Very pale, pinkish moist surface.
1.5	Pale, pink, slightly moist surface.
2.0	Slightly paler than desirable, almost dry surface.
2.5-3.0	Ideal red colour, dry surface
3.5-4.0	Slightly darker than desirable, dry surface.
4.5-5.0	Very dark and dry surface

A score of 0.5 to 1.5 indicated various degrees of PSE and 2.0 a slightly inferior quality. Instruments to measure meat quality directly did not exist, and although objective measurements of meat quality are preferable, the subjective scale became quite useful in the evaluation of the frequency of PSE in the progeny of boars. An example of colour and structure grading of a selected strain of stress susceptible pigs (LAB) used in the early PSE experiments (Ludvigsen 1954) and the offspring of boars from the same breeding centre is shown in Table 2.

TABLE 2   An example of colour and structure grading at progeny testing in Denmark. Pig strain LAB were stress susceptible

| | No. of pigs | Subjective scores | | Backfat | Live weight gain |
		0.5-2.0	2.5-4.0	(cm)	(FU/kg)
LAB	18	100%	0%	2.80	3.35
Boar I	16	62.5%	37.5%	3.15	3.08
Boar II	20	25%	75%	3.40	3.35
Boar III	124	53.2%	46.8%	–	3.25

Offspring from boar I were especially recommended for breeding because of the favourable feed conversion rate and backfat thickness. Pork quality was measured to comply with governmental meat inspection orders to condemn pigs suffering from 'muscular degeneration' of unknown origin, and pigs having a body temperature $1^{o}C$ above normal at slaughter.

The percentage of condemnations from 'muscular degeneration' of unknown origin, which is identical to the PSE condition, and death losses in transport and lairage from malignant hyperthermia was studied from 1948 to 1953 at a bacon factory (Ludvigsen, 1954).

TABLE 3   The percentage of carcasses condemned for 'muscular degeneration' (PSE), and the percentage of death losses (PSS) in transport and lairage, at a Danish bacon factory from 1948 to 1953

	1948	1949	1950	1951	1952	1953
Condemnations	0.19	0.20	0.25	0.34	0.52	0.57
Transport	0.10	0.06	0.04	0.02	0.02	0.03
Lairage	0.14	0.06	0.03	0.02	0.01	0.02
Total	0.43	0.32	0.32	0.38	0.55	0.62

Losses in transport and lairage dropped rapidly from 1948 to 1949, because of enforced restrictions on the loading density during transport, the death rate remained fairly constant from 1950 to 1953. Transport regulations had a marked effect on the survival of pigs in transport and lairage and thus contributed to an increase in condemnations for PSE, the increase in condemnations after 1950 exceeded the reduction in casualties. An obvious reason for the increase in PSE was the enhanced selection for meatiness, because of market demands for lean pork, which is still a major consumer requirement.

Endocrine pathophysiology in stress susceptible pigs

Metabolic, genetic and other aspects of the porcine stress syndrome were reviewed in a recent publication (Ludvigsen 1985a). The importance of the halothane test in future research was emphasised, as stress susceptibility and its close linkage with malignant hyperthermia will be of increasing importance to the pig industry, as long as the genetic trend runs towards leaner pigs. Unfortunately, the halothane test was not available to test stress susceptibility in the early fifties, when experiments to demonstrate the influence of the thyroid in the pathogeny of PSE were performed. It was, however, quite evident that the strain of pigs used in the experiments were prone to malignant hyperthermia, because of their clinical reactions to physical strain, such as running (Ludvigsen, 1957), being moved around in balance experiments ('cage shock'), and being transported. Very often pigs from that strain developed clinical symptoms of cyanosis and anoxia in the balance cages.

Previously unpublished observations show how serious physical strain may effect the percentage distribution of serum proteins in the circulation (Table 4).

TABLE 4 Percentage distribution of serum proteins in five normal pigs and in malignant hyperthermia (MH) type pigs under various strain conditions

		Albumin	$\alpha_1 + \alpha_2$	β_1	β_2	$\gamma_1 + \gamma_2$
Normal pigs	Mean	50.9	16.7	3.4	11.3	17.7
	SD	1.78	1.07	0.59	0.82	1.10
		MH type pigs after 'cage shock' and anorexia				
Pig No. 43	6/1	28.8	24.9	4.9	14.9	26.9
	13/2	35.6	22.0	4.5	12.2	25.8
	1/4	39.5	22.0	6.1	14.6	17.9
Pig No. 45	6/1	26.0	36.3	7.7	14.6	15.5
	13/2	48.8	18.8	4.5	13.6	14.4
	28/4	36.3	16.8	4.7	11.8	30.4
Pig No. 41	13/2	44.1	19.6	5.9	10.7	19.8
	28/4	38.9	21.4	6.1	13.0	20.8
		MH type pigs before and after transport				
Pig No. 42	Before	50.5	21.5	3.5	10.1	14.5
	After	37.5	34.4	3.7	11.2	13.4
Pig No. 44	Before	52.4	21.1	4.0	10.4	12.3
	After	47.2	21.6	4.6	10.3	16.5

In episodes of stress there was obviously a marked change in the electrophoretic pattern of serum proteins. As a percentage of total circulating proteins there was a marked drop in albumin, an increase in $\alpha_1 + \alpha_2$ and β_1, a variable increase in β_2 and $\gamma_1 + \gamma_2$, most pronounced in pigs after 'cage shock'.

The electrophoretic pattern in the drip from a PSE M. longissimus dorsi is shown in Table 5.

TABLE 5   Percentage distribution of extracellular proteins in the drip from a PSE M. longissimus dorsi

Albumin	$\alpha_1 + \alpha_2$	β_1	β_2	$\gamma_1 + \gamma_2$
5.3	8.8	15.3	20.0	50.9

About fifty percent of the extracellular protein belonged to the globulin groups. The variable increase in protein fractions in blood serum, apart from albumin, may reflect an absorption into the blood stream from the extracellular protein fraction present in transudates from muscle cells in acute stress reactions.

The role of the thyroid

Selection for heavily muscled pigs results in a gradual change in hormonal homeostasis in favour of the anabolic hormone systems at the expense of the catabolic systems. The beneficial effect of thyroid active substances on skeletal muscle colour and structure is obviously a result of a stimulating effect of the thyroid hormone on the buffering capacity of muscle tissue, preventing the critical drop in pH during the formation of lactic acid in the anaerobic phase of muscle contraction. The thyroid hormone may also be involved in the reaction to halothane, delaying the onset of muscle rigidity in halothane positive pigs (Ludvigsen, 1985b). The effect of various thyroid hormone substances on the quality of tinned ham is shown in Table 6 (Ludvigsen, 1954 and 1955).

The hams of PSE prone pigs fed iodinated casein preparations A and B had a significantly better colour and structure than the controls, which were of an unacceptable quality. The improved quality of the hams from pigs treated with iodinated casein was also evident in the reduced weight loss during curing and pasteurisation of the hams.

TABLE 6   Organoleptic evaluation[1] of tinned ham from PSE prone pigs fed compounds with thyroid hormone activity

Treatment	n	Colour	Structure	Taste	Weight loss during curing %
I 0 A_3^2	20	3.40[b]	3.50[b]	3.18[b]	13.4
I 0_4B^3	20	2.70[b]	2.50[b]	2.47[b]	13.2
I H[4]	10	2.00[a]	2.10[a]	1.87[a]	13.7
Thyroxine	10	1.20[a]	0.90[a]	0.60[a]	14.5
Controls	20	0.90[a]	0.70[a]	0.75[a]	15.8

[1] The evaluation scale used to assess the colour, structure and taste of the tinned hams was:

6, superb; 5, very good; 4, good; 3, acceptable;
2, pale or two-toned colour, dry structure, inferior taste;
1, mostly pale in colour, dry or stringy structure, inferior taste; 0, PSE.

[2] Iodinated casein preparation A

[3] Iodinated casein preparation B

[4] Iodinated histidine

a,b
    Mean values with different superscripts are significantly different ($P < 0.05$)

Because of the difference in their effects, the two iodinated casein preparations were examined by the carbon dioxide production of mice fed the compounds. Preparation A increased the metabolic rate by 0.4%, whereas preparation B increased the rate by 26.9%. The difference indicated that the effect on muscle structure of preparation A with the very low metabolic effect seemed not attributable to thyroxine, but to some other thyroglobulin compounds. This was supported by the results of feeding thyroxine, which had an insignificant effect on the quality of the ham. Iodinated histine improved colour and structure compared to the controls as well as the thyroxine treated pigs. Reviewing the literature, Mitchell and Heffron (1982) concluded that high as well as low thyroid activity could lead to PSE. In a study of nine patients to determine their susceptibility to malignant hyperthermia, Campbell et al (1981) found that the fasting concentrations of serum thyroxine of the control and experimental groups were very similar when resting, whereas when exercised the thyroxine concentration of the control group increased significantly while that of the MH group did not.

Conflicting evidence in the literature about the role of the thyroid may result from data produced where the physical conditions of the experimental animals are not comparable. Jensen and Barton-Gade (1985) showed that average daily live weight gain of pigs from 25 to 90 kg live weight was 699 g for halothane genotype NN (homozygous negative) and 659 and 660 g for genotypes Nn (heterozygous) and nn (homozygous positive) respectively, and Eikelenboom et al. (1976) found higher growth rates for non-susceptible than for susceptible pigs fed ad libitum.

Preliminary results in a study of the metabolic pattern in Danish Landrace halothane positive and halothane negative male pigs at 90 to 100 kg live weight are given in Table 7.

TABLE 7   Heat expenditure (KJ/hour) during heat stress in Danish Landrace halothane positive (H^+) and halothane negative (H^-) pigs

Time of day	9-10.00	10-11.00	11-12.00	12-13.00	13-14.00
Room temperature	18°C		40°C		18°C
H^+ (n=4)	904	1035	1284	1150	816
H^- (n=6)	847	871	1054	786	676

The pigs were brought into the respiratory chambers at 07.00 hours. Although rested, the H^+ pigs had a higher heat expenditure than their H^- littermates. With increasing room temperature up to 40°C, heat expenditure of the H^+ pigs increased more rapidly than that of the H^- pigs, and heat expenditure dropped faster in the H^- pigs. At 13-14.00 hours the heat expenditure of the H^+ pigs was still higher than that of the H^- pigs, although below that at 9-10.00 hours. That H^+ pigs had a higher heat expenditure than H^- pigs explains the observations of Jensen et al. (1985) that halothane positive pigs have a lower daily weight gain on the same amount of feed than halothane negative pigs because of a higher heat loss. However, the higher heat expenditure of H^+ pigs does not harmonize with a reduced thyroid activity. The H^+ pigs in the respiratory trials were more exitable and nervous than the H^- pigs, and muscle tremor, especially in the tail muscles, was a fairly constant symptom. The higher heat expenditure of H^+ pigs may well be a symptom of latent excitability and muscle tremor which inevitably leads to increased heat loss.

The role of the adrenals

The hormones of the adrenals are also important regulators of adaptation. Optimal activity of the adrenal cortical hormones is indispensible, especially in the acute phase of a strain situation, to prevent a circulatory collapse which may be fatal. Pigs that are prone to PSE and especially MH are highly susceptible to circulatory collapse and may respond positively to adrenal cortical steroid treatment in acute circulatory failure. Heffron and Mitchell (1982) concluded that it is likely that cortisol is deficient in PSE and MH. An adrenal cortical insufficiency, although not considered essential, may facilitate the onset of the syndrome.

Campbell et al. (1981) found that MH patients had a significantly lower serum cortisol than controls after food and exercise although not while resting and fasting. As well as conflicting evidence on the importance of the thyroid in stress susceptibility, the same obviously applies to the role of the cortical steroids. Concerning the measurement of circulating concentrations of hormones, it should be remembered that in episodes of PSS hormones may not reach the target organs because of circulatory barriers.

CONCLUSIONS

Selection of the Danish Landrace for increased feed efficiency and decreased backfat thickness produced a borderline change in the hormonal balance in favour of anabolic hormone systems, primarily STH, at the expense of the catabolic hormone systems, primarily the pituitary-thyroid and the pituitary-adrenal cortex axes.

As STH activity is antagonistic to TSH and ACTH activity, the mobilisation of the thyroid and the adrenal cortex may be insufficient to meet strain situations.

Although the results of measuring the activity of the thyroid and the adrenals may seem conflicting, the effect of feeding iodinated casein to stress prone pigs on meat quality indicates that thyroid active compounds of thyroglobin, other than thyroxine, are essential in skeletal muscle function. Future research should examine interactions between hormones, hormone release and mechanisms of action of hormones on tissue membranes and enzyme systems. Because of continuing selection of pigs for meatiness, it is essential for the benefit of both producers and consumers that the link between stress susceptibility and meatiness is broken.

REFERENCES

Campbell, I.T., Ellis, R.F., and Evans, R.T., 1981. Metabolic rate and blood hormone and metabolite levels of individuals susceptible to malignant hyperpyrexia at rest and in response to food and mild exercise. Anaethesiology, 55, 46-52.

Jensen, P. and Barton-Gade, P.A., 1985. Performance and carcass characteristics of pigs of known genotypes for halothane susceptibility. In: Stress Susceptibility and Meat Quality in Pigs (Ed. J. Ludvigsen) (EAAP Publication No. 33) pp 80-87.

Eikelenboom, G.D., Minkema, D. and vanEldik, P. 1976. The application of the halothane test. Differences in production characteristics between pigs qualified as reactors (MHS-susceptible) and non reactors. Proc. 3rd Int. Conf. on Production Diseases in Farm Animals, (PUDOC, Wageningen) pp 183-185.

Ludvigsen, J. 1954 and 1955. Investigations on the so-called 'muscular degeneration' in pigs. I. 272, 278, 279. Report from the National Institute of Animal Science, Copenhagen.

Ludvigsen, J. 1957. On the hormonal regulation of vasomotor reactions during exercise with special reference to the action of cortical steroids. Acta Endocrinol. XXVI, 406-416.

Ludvigsen, J. 1985a (Ed) Stress Susceptibility and Meat Quality in Pigs. (EAAP Publication No. 33, 1985).

Ludvigsen, J. 1985b. Intermediary calcium and magnesium. In: Stress Susceptibility and Meat Quality in Pigs. (Ed. J. Ludvigsen) (EAAP Publication No 33) pp 106-118.

Mitchell, G. and Heffron, J.J.A. 1982. Porcine stress syndromes. In: Advances in Food Research (Ed. C.O. Chichester) (Academic Press, New York) 28, 167-230.

FURTHER ASPECTS OF ELECTROLYTE METABOLISM IN RELATION TO MALIGNANT HYPERTHERMIA AND THE PORCINE STRESS SYNDROME

P. Fogd Jørgensen

Department of Veterinary Physiology and Biochemistry
The Royal Veterinary and Agricultural University
1870 Frederiksberg C, Copenhagen

ABSTRACT

Marginal potassium supply results in a highly modified MH syndrome and is followed by a reduced K/Na ratio in muscle tissue and a linear association between the K/Na ratio and the metabolic acidosis. The results indicate a reduced anaerobic metabolism during MH.

Retention of Na, K, Mg, and Ca is only slightly different in MHS and MHR pigs fed the same diet. During low K supply renal potassium conservation is maximized and the partial depletion is a likely result of marginal supply as well as increased excretion during repeated stress episodes. A negative result of the halothane test in pigs considered susceptible to partial K depletion should consequently be interpreted carefully.

INTRODUCTION

Stress-induced plasma electrolyte changes are common findings in several species and are a result of an electrolyte shift from the intra- to the extracellular compartment. In particular skeletal muscle, liver and erythrocytes should be considered as immediate sources for the electrolytes and the shift from equilibrium seems to reflect the degree of stress and stress-susceptibility of the animal to a high extent (Allen et al., 1970; Berman et al., 1970; Lucke et al., 1976; Jørgensen, 1981).

An extreme situation appears during malignant hyperthermia (Figure 1, average of three pigs) where P_i and K increase by a factor of two while the raised Ca and Mg levels may reflect a simultaneous haemoconcentration. Similar electrolyte changes are found in blood from pigs died during transportation to slaughter (Ludvigsen, 1954). After a reversible stress situation, equilibrium may be restored by an increased electrolyte excretion combined with a shift from the extra- to the intracelluar compartment.

Several investigations concerning the causal mechanisms of the porcine stress syndrome (PSS) and malignant hyperthermia susceptibility (MHS) have produced evidence that modification of transmembrane Ca transport, particularly in skeletal muscle mitochondria and sarcoplasmic reticulum fragments, is involved (cf. Cheah & Cheah, 1985).

Metabolism of magnesium (an antagonist to Ca in some Ca-Mg interactions

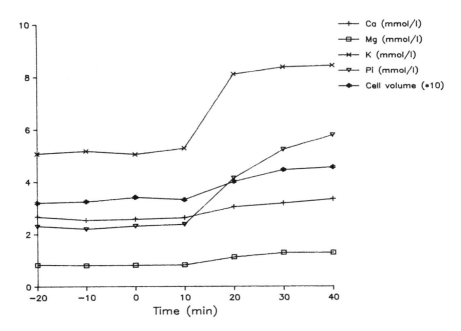

Fig. 1. Plasma electrolytes during malignant hyperthermia in three pigs. Zero indicates the time for start of halothane anaesthesia.

and a synergist in others) is also to some extent involved in stress reactions and PSS. Some interactions may be explained by the curarising effect of pharmacologically high concentrations of extracellular Mg (Sair et al., 1970; Lister & Ratcliff, 1971). Other interactions such as the variable influence of Mg-aspartate on stress in laboratory animals (Mesmer et al., 1981; Kaemmerer et al., 1984) and meat quality and stress-susceptibility in pigs (Schmid, 1981; Schumm, 1982, 1983; Ludvigsen, 1985) are less clear.

Normal muscle function requires intracellular homeostasis of Na and K as well as Ca and Mg. Experimental situations with low potassium supply and repeated stress episodes in MHS pigs result in significant modifications of the MH syndrome during halothane anaesthesia (Jørgensen, 1982a, 1983).

In this paper some aspects of the metabolism of the above mentioned cations is presented. The experimental procedures have been described in detail elsewhere (Jørgensen, 1982a,b, 1985a,b).

RESULTS AND DISCUSSION

1. Influence of different diets on MH in MHS pigs

MHS pigs were separated into control and experimental groups. The ex-

TABLE 1. Major feed components in control feeds and the casein mixture.

Feed mixture	FU$_s$per 100 kg	Crude protein (%)	Crude fat (%)	Crude fibre (%)	Crude ash (%)	Ca[1]	Mg[1]	Na[1]	K[1]
Control I	102	15.8	3.7	4,6	4.6	6.35	1.25	1.73	5.32
Control II	107	19.0	5.0	5.0	7.0	7.61	1.50	2.34	7.56
Casein	117	21.6	0.5	0.2	4.7	7.73	0.70	2.95	1.17

[1] g per FU$_s$

perimental groups were fed a purified casein-containing diet with an energy, protein and mineral content as summarised in Table 1. Experimental groups were further subdivided in up to three groups where one received supplemental potassium and one supplemental lipids to raise daily intake to that of the controls. The major results regarding MH initiation, the degree of acidosis during anaesthesia, and muscle tissue composition are given in Tables 2 and 3. High K comprises controls, and casein + K, while low K comprises casein, and casein + lipids.

TABLE 2. Summary of clinical chemical findings during halothane anaesthesia of MHS pigs in relation to high and low K supply.

Potassium supply	Initiation of MH	Degree of metabolic acidosis
High	Fast (< 2 min)	Marked
Low	Very slow (9-10 min, often > 20 min)	Insignificant

Animals fed low K diets had a significantly modified MH syndrome, sometimes delayed beyond the 20 min limit which was the maximum duration of anaesthesia employed. Furthermore the metabolic acidosis almost completely disappeared.

Nutritional modifications of MH have previously been reported by Marchsteiner et al. 1978. In later investigations Mg-aspartate in high doses led to a more or less delayed MH syndrome (Kammerlander, 1981; Schmid, 1981, Ludvigsen, 1985). The cited findings were, however, not as marked as those

TABLE 3. M. longissimus dorsi electrolytes and glycogen (mmol/ kg) in relation to low potassium supply of MHS pigs.

Potassium supply	Na	K	Ca	Mg	Glycogen
High (n = 10)	21.0	106	0.70	12.3	31.1 (n = 6)
Low (n = 6)	33.2**	80***	0.81	13.3*	76.8*(n = 3)
SD (75 df)	7.5	7.5	0.20	0.77	15.9 (14 df)

Significantly different from high K supply: * $P < 0.05$, ** $P < 0.01$, *** $P < 0.001$.

caused by protein deficiency or casein feeding (Jørgensen, 1982a) and it has not been established whether the effect of Mg-aspartate is a consequence of a very high Mg intake or a direct consequence of the aspartate.

In the present investigation tissue analyses showed an altered K/Na ratio probably as a result of an altered Na/K-ATPase activity, density or distribution in the sarcolemma. Furthermore the high glycogen levels among low K fed animals strongly suggest that partial K depletion led to protection of the animals from anaerobic glycogenolysis. These findings are opposite the ultimate composition of PSE meat where low K and high Na is caused by extreme stress susceptibility (Henry et al., 1958).

Severe K depletion leads to myopathy and muscle weakness (Welt et al., 1960; Knochel & Schlein, 1972). Severe anoxia during anaesthesia may also result in damage to skeletal muscle fibres in swine (Bader, 1985). It is, however, not likely that myopathic disturbances are responsible for the present findings as glycogen levels were high and plasma creatine kinase was not significantly different in low (293 ± 192, n = 3, sd) and high (430 ± 134, n = 6, sd) K fed animals.

2a. Electrolyte metabolism in MHS and MHR pigs

Four MHS and four MHR (presumed to be of the NN genotype) were subjected to 72 h balance studies before and after three weeks of low potassium supply. Feces and urine were sampled twice daily and the pigs were subjected to only one anaesthetic episode to identify their phenotype with respect to MHS. Daily intake, renal excretion, and apparent retention (intake − (renal + fecal excretion)) as well as plasma and urine concentrations of Ca, Mg, Na and K are presented in Tables 4 to 7.

a) Calcium and magnesium (Tables 4 and 5). Daily intake of these elements was similar in both MHS and MHR pigs fed standard diets. The apparent

TABLE 4. Calcium metabolism in 4 MHS and 4 MHR pigs before and after 3 weeks of low potassium supplementation. Plasma and urine concentrations, daily intake from feed, renal excretion, and apparent retention (intake - (fecal + renal excretion)).

		Concentration (mmol/1)		Intake (mmol/kg/24 h)	Renal excretion (mmol/kg/24 h)	Apparent retention (% of intake)
		Plasma	Urine			
MHS	Before	3.11 ± 0.08***	0.78 ± 0.21*	9.1 ± 0.34 ***	0.11 ± 0.04*	49.3 ± 8.9[a]
	After	3.01 ± 0.19***	0.53 ± 0.12*	7.0 ± 0.35*	0.10 ± 0.02	42.4 ± 5.8
MHR	Before	2.64 ± 0.06	0.37 ± 0.05	9.1 ± 0.81 *	0.04 ± 0.01	36.5 ± 9.3
	After	2.55 ± 0.09	0.31 ± 0.12	7.1 ± 0.24	0.05 ± 0.03	38.9 ± 7.6

Significance level below a figure indicates significance between time within genotype. Significance level after a figure indicates significance between genotype at the same time. Students t-test. a: $P < 0.1$, *: $P < 0.05$, **: $P < 0.01$, ***: $P < 0.001$.

TABLE 5. Magnesium metabolism in 4 MHS and 4 MHR pigs before and after 3 weeks of low potassium supplementation. Plasma and urine concentrations, daily intake from feed, renal excretion, and apparent retention (intake - (fecal + renal excretion)).

		Concentration (mmol/1)		Intake (mmol/kg/24 h)	Renal excretion (mmol/kg/24 h)	Apparent retention (% of intake)
		Plasma	Urine			
MHS	Before	1.05 ± 0.03***	2.32 ± 0.89* *	2.9 ± 0.11 ***	0.25 ± 0.10	28.0 ± 8.6[a]
	After	1.04 ± 0.07*	0.92 ± 0.30**	1.4 ± 0.07*	0.17 ± 0.04[a]	24.5 ± 4.2
MHR	Before	0.88 ± 0.02	1.70 ± 0.04 ***	3.0 ± 0.26 ***	0.33 ± 0.11	18.7 ± 4.7
	After	0.94 ± 0.09	1.25 ± 0.08	1.5 ± 0.04	0.26 ± 0.05	24.3 ± 7.3

Significance levels as indicated in Table 4.

retention, however, tended to be higher in MHS animals. Urinary excretion is of quantitatively minor importance. The higher plasma levels, urinary concentration, and renal excretion of especially Ca in MHS pigs suggest, however, that maintenance of the intracellular equilibrium of this element may be impaired in MHS animals, a fact pointing towards a more general alteration of transmembrane Ca transport than just in sarcoplasmic reticulum and mitochondria of skeletal muscles. During low potassium supplementation Mg supply was also low because of a low dietary Mg content. The apparent

retention, however, was similar to the initial levels in both groups because of a higher absorption of the magnesium compound given.

Reaccumulation of Ca in sarcoplasmic reticulum after muscle contraction is tightly coupled with the action of $(Ca^{2+}+Mg^{2+})$-activated ATPase in
SR (Stekhoven & Bonting,1981) and Mg is essential for the enzyme activity.
A similar $(Ca^{2+}+Mg^{2+})$-activated ATPase is present in the erythrocyte membrane from different species and is tightly coupled with Ca transport over
the erythrocyte membrane (Davis et al., 1982). As this enzyme has been
associated with malignant hyperthermia in dogs (O'Brien et al., 1984) the
opportunity exists to further characterise the Ca-Mg interaction with respect to the $(Ca^{2+}+Mg^{2+})$-activated ATPase in MHS and MHR pigs. Such investigations may contribute to clarify the reason why extra supply of magnesium preparations, especially as aspartate-HCl, in some situations seems
to influence mortality and meat quality in a beneficial way (Schumm, 1982;
Schmitten et al., 1984) while in other situations are of no importance.
As the activity of the ATPase is regulated by thyroid hormones,investiga-
tions of this enzyme may also contribute to a better understanding of the
role of thyroid hormones in MHS or PSS.

b) Sodium and potassium (Tables 6 and 7). On the standard diet potassium retention in MHS and MHR pigs is similar while Na rentention is high
in MHS animals. With low K supply the relative retention of K increases
while that of Na correspondingly decreases to almost zero in both groups.
The relatively small differences between groups does not seem to explain
the marked metabolic and tissue changes previously mentioned. During the
course of low K supply the kidneys gradually compensate for the low supply
by switching from the normal situation with K excretion and partial Na re-
absorption (aldosterone effect) to the opposite with maximum K retention
and a correspondingly high Na excretion. The electrolyte shift is accompanied by a doubling of diureses (not shown). The potassium saving mechanisms are reflected in the very low urine concentration of the element and
the Table clearly demonstrates that plasma K level is rather inaccurate
in predicting the K status of an individual. Urine gives a more reliable
measure as also found in man (Welt et al., 1960).

2b. Renal electrolyte excretion in MHS pigs at rest and after halothane
anaesthesia

To clarify whether the altered cation composition of skeletal muscle
is a combined consequence of marginal supply and increased demand of espe-

TABLE 6. Potassium metabolism in 4 MHS and 4 MHR pigs before and after 3 weeks of low potassium supplementation. Plasma and urine concentrations, daily intake from feed, renal excretion, and apparent retention (intake - (fecal + renal excretion)).

		Concentration (mmol/l)		Intake (mmol/kg/ 24 h)	Renal excretion (mmol/kg/24 h)	Apparent retention (% of intake)
		Plasma	Urine			
MHS	Before	6.3±0.3	57.7±23.8	9.3±0.34[a]	5.44±0.32	25.5±1.7*
		**	**	***	***	***
	After	5.3±0.5*	0.7±0.21*	1.4±0.07*	0.12±0.04	77.1±8.0
MHR	Before	6.2±0.6	64.6±20.0	10.4±0.92	5.67±0.67	18.6±3.1
		***	***	***	***	***
	After	4.5±0.4	0.9±0.11	1.6±0.05	0.14±0.03	76.4±9.4

Significance level below a figure indicates significance between time within genotype. Significance level after a figure indicates significance between genotype at the same time. Students t-test. a: $P < 0.1$, *: $P < 0.05$, **: $P < 0.01$, ***: $P < 0.001$.

TABLE 7. Sodium metabolism in 4 MHS and 4 MHR pigs before and after 3 weeks of low potassium supplementation. Plasma and urine concentrations, daily intake from feed, renal excretion and apparent retention.

		Concentration (mmol/l)		Intake (mmol/kg/ 24 h)	Renal excretion (mmol/kg/24 h)	Apparent retention (% of intake)
		Plasma	Urine			
MHS	Before	148±7	23.5±5.0	4.0±0.15	2.44±0.59**	21.5±7.0*
					**	*
	After	145±5	21.0±3.2	4.1±0.21*	3.95±0.31	-8.3±15.3
MHR	Before	148±1.4	28.2±5.7	3.9±0.35	3.15±0.37	0.1±12.7
				*	a	
	After	147±2	24.7±5.5	4.5±0.14	3.68±0.37	9.0±8.4

Significance levels as indicated in Table 6.

cially potassium in connection with acute stress situations, four MHS pigs were investigated. The renal electrolyte excretion was measured 0-5 h and 5-24 h after halothane anaesthesia leading to moderate malignant hypertermia. The pigs were fed the control diet or the casein diet. As potassium conservation was maximum on the casein diet only the results from control fed animals are shown in Figure 2.

Halothane testing resulted in an almost twice as high excretion of Na, K, Mg, and Cl during the first five hours after the test. It is therefore

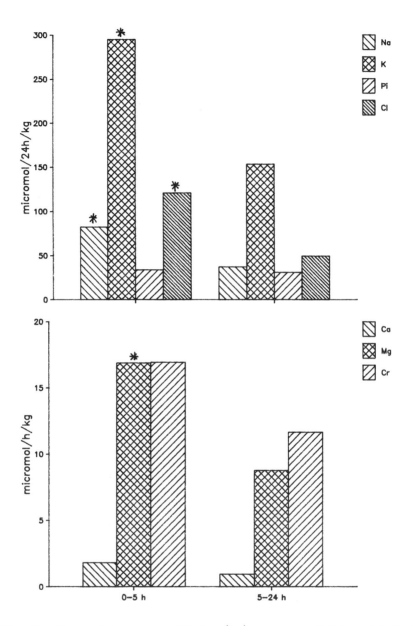

Fig. 2. Electrolyte and creatinine (Cr) excretion 0-5 h and 5-24 h after halothane anaesthesia of four MHS pigs. Sodium, potassium, phosphorous (Pi), chloride (Cl), calcium and magnesium. * indicates that 0-5 h excretion is significantly higher than 5-24 h (P < 0.05).

likely that MHS pigs - although their electrolyte metabolism during resting periods is not significantly different from MHR pigs - due to their responsiveness to different stressors, are susceptible to mineral depletion where there is a marginal supply of electrolytes of major intracellular importance. Such a situation may appear in unthrifty animals. The same situation may appear in very young animals as such animals have a high growth rate and an immature enzyme and electrolyte composition of skeletal muscles (Widdowson & Dickerson, 1964; Jørgensen, 1981). A negative result of a halothane test in such animals should consequently be interpreted with care.

The author wishes to express his gratitude to The Danish Agricultural and Veterinary Research Council and The Velux (1981) Foundation for the financial support to the investigations.

REFERENCES

Allen, W.M., Berrett, S., Harding, J.D.J. and Patterson, D.S.P. 1970. Experimentally induced acute stress syndrome in Pietrain pigs. Vet. Rec. 87, 64-69.

Bader, R. 1985. Histological investigations of skeletal muscles in pigs. 36th Ann. Meeting EAAP, 3 pages.

Berman, M.C., Harrison, G.G., Bull, A.B. and Kench, J.E. 1970. Changes underlying halothane-induced malignant hyperthermia in Landrace pigs. Nature 225, 653-655.

Cheah, K.S. and Cheah, A.M. 1985. Malignant hyperthermia: Molecular defects in membrane permeability. Experientia 41, 656-661.

Davis, P.J., Davis, F.B. and Blas, S.D. 1982. Studies on the mechanism of thyroid hormone stimulation in vitro of human red cell Ca^{2+}-ATPase activity. Life Sci. 30, 675-682.

Henry, M., Romani, J.-D. and Joubert, L. 1985. La myopathie exsudative depigmentaire du porc. Maladie de l'adaptation. Essai pathogénique et conséquences practiques. Rev. Path. Gen. Physiol.Clin. 58, 355-395.

Jørgensen, P.F. 1981. Muskelfunktion hos svin. Et enzymatisk-biokemisk-genetisk studium over muskeludvikling, muskelkomposition og stress-følsomhed. (Carl Fr. Mortensen, København).

Jørgensen, P.F. 1982a. Malign hyperthermifølsomhed hos svin. Nogle årsager til reduceret penetrans. Årsberetn. Inst. Sterilitetsforsk. 25, 53-67.

Jørgensen, P.F. 1982b. Influence of nutrition on malignant hyperthermia in pigs. Acta vet. scand. 23, 539-549.

Jørgensen, P.F. 1983. Malignant hyperthermia in pigs modified by low-potassium diets. Acta vet. scand. 24, 512-514.

Jørgensen, P.F. 1985a.Kaliumomsætning i relation til malign hyperthermifølsomhed hos svin. Effekt af lav kaliumtilførsel på reaktionstype, kationsammensætning i væv og renal kationekskretion. Årsberetn. Inst. Sterilitetsforsk. 28, 99-111.

Jørgensen, P.F. 1985b. Some aspects of electrolyte metabolism related to malignant hyperthermia and stress susceptibility in pigs. In "Stress susceptibility and meat quality in pigs" (Ed. J.B. Ludvigsen), (EAAP Publication no. 33, Denmark). pp. 88-97.

Kaemmerer, K., Kietzmann, M. and Kreisner, M. 1984. Untersuchungen mit Magnesium. 2. Die Wirkung von Magnesiumchlorid und Magnesiumaspartat-hydrochlorid auf Stressreaktionen. Zentralbl. Vet. Med. A. 31, 321-333.

Kammerlander, P. 1981. Über die Wirkung des Insulins auf den Glukosestoff-
wechsel isolierter Schweinefettzellen in abhängigkeit von der Halothanreaktion der Spendertiere und von oralen K- und Mg-Aspartat-Gaben.
Inaug. Diss. (Veterinärmedizinische Universität, Wien).

Knochel, J.P. and Schlein, E.M. 1972. On the mechanism of rhabdomyolysis in
potassium depletion. J. clin. Invest. 51, 1750-1758.

Lister, D. and Ratcliff, P.W. 1971. The effect of the pre-slaughter injec-
tion of magnesium sulphate on glycolysis and meat quality in the pig.
In "Proc. 2nd int. Symp. Condition Meat Quality Pigs", Zeist (Eds. J.
C.M. Hessel-de Heer, G.R. Schmidt, W. Sybesma, P.G. van der Waal).
(Pudoc, Wageningen). pp. 139-144.

Lucke, J.N., Hall, G.M. and Lister, D. 1976. Porcine malignant hyperthermia.
I. Metabolic and physiological changes. Br. J. Anaesth. 48, 297-304.

Ludvigsen, J. 1954. Undersøgelser over den såkaldte "muskeldegeneration"
hos svin. I 272 Beretn. Forsøgslaboratoriet, Copenhagen

Ludvigsen, J.B. 1985. Intermediary magnesium- und calcium metabolism in
stress-susceptible pigs. In "Stress susceptibility and meat quality
in pigs" (ed. J.B. Ludvigsen). (EAAP Publication no. 33, Denmark). pp.
106-118.

Marchsteiner, J., Waginger, H., Schmid, S. and Leibetseder, J. 1978. Ernäh-
rungsabhängige Beziehungen zwischen der Insulin- and ATP-ase-Aktivität
und der malignen Hyperthermie beim Schwein. Z. Tierphysiol. Tierer-
nähr. Futtermittelk. 40, 114-115.

Mesmer, M., Fischer, G. and Classen, H.G. 1981. Stress-Abschirmung durch
Magnesium. Günstige Wirkungen einer parenteralen Magnesium-Therapie
auf die Entstehung von Stress-Ulzera bei Ratten. Arzneimittel-Forsch-
ung 31 (I), 389-391.

O'Brien, P.J., Forsyth, G.W., Olexon, D.W., Thatte, H.S. and Addis, P.B.
1984. Canine malignant hyperthermia susceptibility: Erythrocytic defects - Osmotic fragility, glucose-6-phosphate dehydrogenase deficiency and abnormal Ca^{2+} homeostasis. Can. J. comp. Med. 48, 381-389.

Sair, R.A., Lister, D.L., Moody, W.G., Cassens, R.G., Hoekstra, W.G. and
Briskey, E.J. 1970. Action of curare and magnesium on striated muscle
of stress-susceptible pigs. Am. J. Physiol. 218, 108-114.

Schmid, E. 1981. Über die Abhängigkeit der β-adrenergen Stimulation der
Glucoseutilisation isolierter Schweinefettzellen von der Halothan-Re-
aktion sowie oralen K- und Mg-Aspartat-Gaben. Inaug. Diss. (Veterinärmedizinische Universität, Wien).

Schmitten, F., Jüngst, H., Schepers, K.-H. and Festerling, A. 1984. Untersuchungen über die Wirkung des magnesiumhaltigen Ergänzungsfuttermittels Cytran auf die Fleischbeschaffenheit von stressresistenten und
stressanfälligen Schweinen. Dtsch. tierärztl. Wschr. 91, 149-151.

Schumm, H. 1982. Transportverluste bei Schlachtschweinen nach Verabreichung des magnesiumhaltigen Ergänzungs-Futtermittels Cytran vor dem
Transport zur Schlachtung. Tierzüchter 34, 515-518.

Schumm, H. 1983. Fleischqualität bei Schweinen nach Verabreichung des mag-
nesium-haltigen Ergänzungsfuttermittels Cytran. Schweinewelt 8, 264-
270.

Stekhoven, F.S. and Bonting, S.L. 1981. Transport Adenosine triphosphatases:
Properties and functions. Physiol. Rev. 61, 1-76.

Welt, L.G., Hollander, W. and Blythe, W.B. 1960. The consequences of potassium depletion. J. chr. Dis. 11, 213-254.

Widdowson, E.M. and Dickerson, J.W.T. 1964. Chemical composition of the body. In "Mineral Metabolism" Vol. 2A (Eds. C.L. Comar and F. Bronner).
(Academic Press, New York). pp. 1-247.

IMPLICATIONS OF THE PORCINE STRESS SYNDROME
FOR ANIMAL WELFARE

Pierre MORMEDE and Robert DANTZER

Institut National de la Recherche Agronomique
Psychobiologie des Comportements Adaptatifs
INSERM U.259, Rue Camille Saint-Saëns
33077 BORDEAUX CEDEX, FRANCE

ABSTRACT
     As a pathological condition resulting in higher morta-
lity rates from exposure to environmental stressors, the
porcine stress syndrome (PSS) has obvious implications for
animal welfare in intensive pig husbandry. There is now good
evidence to suggest that this condition appears as a consequence of an excessive environmental pressure on a sensitive
animal. Malignant hyperthermia is an extreme exemple of
individual sensitivity. Although the control of PSS can be
achieved by breeding programs that consider stress-susceptibility as a selection character, this should not deter the
identification and correction of stressful environmental
factors in intensive pig husbandry.

INTRODUCTION

     An animal can be said to be in a state of well-being if
it is both physically and mentally healthy. Although there is
still some debate about how to assess the mental experiences
of animals if they can be studied at all (Dawkins, 1980),
nobody would argue against the necessity of preventing disease, injury and above all mortality in order to ensure animal
welfare. This means that any pathological condition arising
from the way farm animals are produced, kept and treated must
be considered as detrimental to their well-being, independently of its possible association with reduced productivity.
The case of the Porcine Stress Syndrome (PSS) is interesting
to consider in this respect since this often lethal condition, that has both idiopathic and environmental determi-
nants, has a strong influence on productivity, because of
increased mortality rates and alterations of meat quality.

In this paper, we will first concentrate on the historical aspects of PSS to follow the progressive simplication of the concept, before considering how this pathological condition is related to stress. With this information, we will be able to determine whether PSS summarizes all the stress problems encountered in swine production, and how, by getting rid of this condition, we would be able to improve both productivity and welfare.

HISTORICAL ASPECTS OF PSS

Nearly three decades ago, the attention of pig producers was drawn to increasing numbers of unexplained deaths in market-size pigs. These deaths occurred in otherwise healthy animals, in the absence of any apparent malpractice in management and housing. The condition was characterized by an acute shock-like syndrome involving increased heart rate, hyperventilation, hyperthermia, rigidity of muscles and blood acidosis. Death occurred in most cases within a few minutes or a maximum of 30-60 min after the beginning of the symptoms. Autopsies did not reveal any pathognomonic sign besides the rapid development of rigor mortis and the paleness and oedema occurring in specific muscles, although cases of acute myocardial necrosis were sometimes noted. Since the symptoms usually developed in pigs subjected to stress associated with mixing of animals, transport, physical exercise and exposure to high temperature, this condition was termed the porcine stress syndrome (Topel et al., 1968).

PSS soon became the subject of an abundant literature because of its close relation with the pale soft exsudative (PSE) condition affecting meat quality (see Cassens et al., 1975 for a review). The main impetus for this research was economic. In order to understand the causes of both death and alterations in meat quality and to discover preventive or curative therapeutics, animal scientists concentrated on attempts to find a way to identify animals at risk before they developed the pathology. Two main approaches were used for differentiating stress-susceptible (SS) from stress-resistant (SR) pigs. One was based on the study of physiological and metabolic reactions to heat stress while the other was more concerned with responses to physical exercise. In the absence of a high enough correlation between either thermal or exercise challenge and proneness to PSS and PSE, such research revealed itself to be of no more use than comparison of breeds immune or liable to the disease.

The situation improved considerably when, at the end of the sixties, it was observed that pigs submitted to neuromuscular blocking agents and/or potent anesthetic inhalation agents such as halothane, displayed symptoms very similar to those of PSS. Susceptible animals developed a syndrome of Malignant Hyperthermia (MH) characterized by muscular rigidity and hyperthermia and they died within a few minutes if the anesthetic was not withdrawn. Since halothane-positive animals were found to be more likely to develop PSS and to give PSE meat, halothane challenge soon became a reliable and easy way to assess the stress susceptibility status of individuals and to further study their biochemical, physiological and metabolic characteristics (Sybesma and Eikelenboom, 1969).

Malignant hyperthermia was subsequently shown to be the result of a defect in muscle calcium metabolism (see J.J.A. Heffron, this volume). This defect is genetically transmissible and is controlled by a single recessive autosomal gene. Finally, the so-called "stress syndrome" turned out to be a very specific metabolic problem triggered by any one of a large number of environmental stimuli and this outcome raises the question of the place of PSS in the general concept of stress.

PSS AND THE CONCEPT OF STRESS

Since Selye's definition of stress as a non-specific reaction elicited by nocuous environmental factors (Selye, 1936), the main breakthrough has been the recognition of a pivotal role for the central nervous system in this reaction (Dantzer and Mormède, 1979, 1983).

The response to a given threatening stimulus has several
components which can be classified under two principal headings : the behavioural response involving the motor system
and the neuroendocrine response triggering metabolic adaptation. The hypophyso-adrenocortical axis (ACTH-glucocorti-
coids) and the catecholaminergic systems have been more
extensively studied but other hormones such as sex steroids
also play an important role, although we still lack basic
knowledge about them. The responses to environmental stimuli
are under control of the brain which, through sensory inputs,
information processing and body control, is the interface
between the organism and its environment.

Behavioural responses to aversive stimuli are tightly
linked to concurrent neuro-endocrine activation, and this
shows that the two sides of the response are integrated at
the highest level. We have termed "brain state" the particular reactivity of an animal, and it can be evaluated by the
response to a non-specific psychological stimulus such as
exposure to a novel environment. The "brain state" of an
animal is strongly dependent on genetic factors. For
instance, we have observed that in several breeds of Chinese
and crossbred pigs, basal serum ACTH, glucocorticoids and
adrenaline levels were highly correlated with behavioural
reactivity (Mormède et al., 1984). However, the precise
genetic analysis of the phenomenon is still open to investigation. The "brain state" can also be influenced by early
experience but this aspect is less well understood in farm
animals (see for instance Dantzer et al., 1983, in calves).

The stress response is therefore a multi-component
response and is strongly dependent upon many factors related
to the animal itself but also to the threatening situation
(nature and intensity of the stimulus, availability of
behavioural coping...). The porcine stress syndrome can be
triggered by many unrelated stimuli such as transportation,
mixing or even sexual activity. Mere psychological factors
are potent enough to induce PSS. This "syndrome" is therefore
non specific as far as the initiating stimulus is concerned.
However, PSS is highly specific if we consider that its

occurence is dependent upon the existence of a single gene probably controlling the synthesis of a discrete protein in the striated muscle membrane. In this respect PSS has certainly little to do with stress as a psychophysiological concept, but we can still question whether the general concept of stress can help us understand PSS and MH.

PSS AND GENERAL STRESS REACTIVITY

Many attempts have been made to relate PSS to the neuroendocrine and behavioural characteristics of the animals. Stress-sensitive strains of pigs have been shown to differ from stress-resistant strains in many aspects of their hypothalamo-adrenocortical function, such as a low cortisol/ ACTH ratio, inability of dexamethasone to inhibit ACTH secretion and a high metabolic clearance rate of cortisol (Marple et al., 1972, 1974 ; Marple and Cassens, 1973 ; Sebranek et al., 1973). However, no difference could be found when animals from the same litter and differing only in their sensitivity to halothane were compared (Aberle et al., 1976 ; Dantzer and Hatey, 1981 ; Mormède and Dantzer, 1978). In these experiments interbreed differences were by far the most important in endocrine characteristics as well as in beha- vioural reactivity (Dantzer and Mormède, 1978).

Do these results mean that meat quality, PSS or malig- nant hyperthermia are not at all related to stress or, more exactly, to the way stress is now conceived ? We must first emphasize the fact that PSS and MH are probably two overlap- ping but different entities. American Hampshire pigs for instance, considered to be stress-susceptible (Marple et al., 1972), have a low incidence of halothane-sensitive subjects (Webb, 1980). This may explain the discrepancies described above : although various endocrine differences have been described between stress-susceptible and stress-resistant pigs, MH sensitivity seems to be related neither to endocrine nor to behavioural reactivity. However, it is still conceivable that the biobehavioural characteristics of the animals may influence the expression of the malignant hyperthermia gene.

The possible influence of sympathetic nervous activity is a good example. The importance of catecholamines in the development of MH is still debatable (cf. D. Lister, this volume). Although a huge increase of plasma catecholamines can be measured during the MH reaction to halothane (Lucke et al., 1976 ; Gronert et al., 1977), this release does not appear to be necessary for the development of the syndrome (Gronert et al., 1977) but the catecholaminergic reactivity of the animal can modulate the intensity of the syndrome in pigs (Lucke et al., 1978), as well as the development of experimental myopathy in rats (Kuncl and Meltzer, 1977). Furthermore, PSS-prone animals were shown to have different catecholamine levels in plasma and brain when compared to stress-resistant pigs (Altrogge et al., 1980 ; Hallberg et al., 1983). Considering the importance of catecholamines and glucocorticoids in muscle metabolism and their correlation with behavioural reactivity of pigs, it would be worthwhile investigating further the importance of basal psychoneuro-endocrine characteristics of the animals of the expression of PSS and MH. Genetic engineering could give us such an opportunity in the next few years by introducing the halothane sensitivity gene in different biobehavioural backgrounds (A. Archibald, this volume).

CONCLUSION

The concept of stress in pigs has been initially defined on the basis of an increased susceptibility to various stressors such as forced exercise or warm exposure inducing shock-like symptoms (porcine stress syndrome) and low quality meat (pale, soft and ex udative pork). The halothane test, inducing malignant hyperthermia in susceptible animals proved to be a very successful approach to PSS but turned out to introduce a bias in the study of the stress response. This pathological condition has been attributed to a primary dysfunction in the regulation of calcium ion movements in muscle cells. There is little doubt that the physical health of animals plagued with this condition is at great risk, but there is no evidence that they are suffering mentally as long as the syndrome is not triggered. The important genetic component in this syndrome has provided a strong impetus for the development of breeding programs incorporating an assessment of stress-susceptibility, and they have been generally successful indecreasing its incidence in the field. However, it must be pointed out that such means should not be considered as sufficient and totally satisfactory from the point of view of welfare. The improvement in mortality rates and in meat quality brought about by selection programs can be also achieved by modifying the environment of the animals and by decreasing the amount of physical and mental load they are subjected to. For example, mortality rates during transport can be significantly decreased by improving loading procedures, e.g. decreasing the slope of ramps or using elevators (see Dantzer and Mormède, 1979, for a review). In addition, it is quite likely that the condition is exacerbated by certain aspects of management, such as confinement and absence of physical exercise, and also by increased biobehavioural reactivity in a number of breeds. Within this context, PSS can be viewed as a sensitive indicator of a management problem in intensive pig husbandry. However, to imagine that eliminating halothane-sensitive animals is a "cure for stress" (Winstanley, 1979) is at best very naive as the search for farm animal welfare must take into account the whole range of interactions between the animals and their environment.

REFERENCES

Aberle, E.D., Riggs, B.L., Alliston, C.W., Wilson, S.P., 1976. Effects of thermal stress, breed and stress susceptibility on corticosteroid binding globulin in swine. J. Anim. Sci., 43, 816-820.

Altrogge, D.M., Topel, D.G., Cooper, M.A., Hallberg, J.W. and Draper, D.D., 1980. Urinary and caudate nuclei catecholamine levels in stress-susceptible and normal swine. J. Anim. Sci., 51, 74-77.

Cassens, R.G., Marple, D.N. and Eikelenboom, G., 1975. Animal physiology and meat quality. Adv.Food Res., 21, 71-155.

Dantzer, R. and Hatey, F., 1981. Plasma dopamine-beta-hydroxylase and platelet monoamine oxidase activities in pigs with different susceptibility to the malignant hyperthermia syndrome induced by halothane. Reprod. Nutr. Dévelop., 21, 103-108.

Dantzer, R. and Mormède, P., 1978. Behavioural and pituitaryadrenal characteristics of pigs differing by their susceptibility to the malignant hyperthermia syndrome induced by halothane anesthesia. I. Behavioural measures. Ann. Rech. Vét., 9, 559-567.

Dantzer, R. and Mormède, P., 1979. Le stress en élevage intensif, Masson, Paris.

Dantzer, R. and Mormède, P., 1983. Stress in farm animals. A need for a reevaluation. J. Anim. Sci., 57, 6-18.

Dantzer, R., Mormède, P., Bluthé, R.M., Soissons, J., 1983. The effect of different housing conditions on behavioral and adrenocortical reaction in veal calves. Reprod. Nutr. Dévelop., 23, 501-508.

Dawkins, M., 1980. Animal suffering : the science of animal welfare. Chapman and Hall, London.

Gronert, G.A., Milde, J.H., Theye, R.A., 1977. Role of sympathetic activity in porcine malignant hyperthermia. Anesthesiology, 47, 411-415.

Hallberg, J.W., Draper, D.D., Topel, D.G., and Altrogge, D.M., 1983. Neural catecholamine deficiencies in the porcine stress syndrome. Amer. J. Vet. Res., 44, 368-371.

Kuncl, R.W., Meltzer, H.Y., 1977. Role of the adrenal in an experimental myopathy. Exp. Neurol., 57, 322-330.

Lucke, J.N., Denny, H., Hall, G.M., Lovell, R., Lister, D., 1978. Porcine malignant hyperthermia. VI : the effects of bilateral adrenalectomy and pretreatment with bretylium on the halothane-induced response. Br. J. Anaesth., 50, 241-246.

Lucke, J.N., Hall, G.M., Lister, D., 1976. Porcine malignant hyperthermia. I : metabolic and physiological changes. Br. J. Anaesth., 48, 297-304.

Marple, D.N., Judge, M.D. and Aberle, E.D., 1972. Pituitary and adrenocortical function of stress-susceptible swine. J. Anim. Sci., 35, 995-1000.

Marple, D.N., Cassens, R.G., 1973. Increased metabolic clearance of cortisol by stress-susceptible swine. J. Anim. Sci., 36, 1139-1142.

Marple, D.N., Cassens, R.G., Topel, D.G., Christian, L.L., 1974. Porcine corticosteroid-binding globulin : binding properties and levels in stress-susceptible swine. J. Anim. Sci., 38, 1224-1228.

Mormède, P. and Dantzer, R., 1978. Behavioural and pituitary-adrenal characteristics of pigs differing by their susceptibility to the malignant hyperthermia syndrome induced by halothane anesthesia. 2. Pituitary-adrenal function. Ann. Rech. Vét., 9, 569-576.

Mormède, P., Dantzer, R., Bluthé, R.M., Caritez, J.C., 1984. Differences in adaptive abilities of three breeds of Chinese pigs. Behavioral and neuroendocrine studies. Génétique, Sélection, Evolution, 16 : 85-102.

Sebranek, J.G., Marple, D.N., Cassens, R.G., Briskey, E.J., Kastenschmidt, L.L., 1973. Adrenal response to ACTH in the pig. J. Anim. Sci., 36, 41-44.

Selye, H., 1936. A syndrome produced by diverse nocuous agents. Nature, 138, 32-33.

98

Sybesma, W. and Eikelenboom, G., 1969. Malignant hyperthermia syndrome in pigs. Neth. J. Vet. Sci., 2, 155-160.

Topel, D.G., Bicknell, E.J., Preston, K.S., Christian, L.L. and Matsushima, C.Y., 1968. Porcine stress syndrome. Modern Vet. Pract., 49, 40-60.

Webb, A.J., 1980. The halothane test : a practical method of eliminating porcine stress syndrome. The Veterinary Record, 410-412.

Winstanley, M., 1979. A cure for stress. New Scientist, 84, 594-596.

DISCUSSION - Session I

D. Lister   The papers in this session have covered a wide spectrum of
science and philosophy. We have heard about cellular, neurophysiological,
endocrinological and finally behavioural mechanisms. The problem in
porcine stress is that we can never identify which of these characteristics is a cause and which is a symptom of the condition. Cellular
function measurements are affected by chemical changes that occur in the
samples of muscle or other tissues taken from animals as biopsies; they
may change during the early post sampling procedure and can lead to some
quite unexplainable findings. So far as neuroendocrine enquiries are
concerned we have been persuaded by clinicians of the value of measurements
of circulating levels of hormones. In the science of PSS we now recognise
that clinical judgements are quite inadequate and that we ought to be
looking at receptor concentrations, turnover of hormones and the like.
It is not enough to rely on measurements of circulating levels of hormones
and metabolites.

So far as the behavioural studies are concerned, we can demonstrate a
behavioural component in PSS, but this cannot be the entire explanation for
it is impossible to control meat quality in stress-susceptible animals
simply by eliminating behavioural mechanisms. For example, if you
completely anesthetise and curarize a Pietrain or a Poland China pig,
thereby eliminating all behavioural reactions, you still get PSE meat. We
cannot say, therefore, that all PSE and porcine stress problems are
attributable to behavioural reactions.

These, I think, are some of the important issues that we might consider in
our discussion.

P Mormède   It is important to differentiate between the interest in the
general behaviour of the animal as reflecting its psychobiological
characteristics and the particular role of behaviour during the development
of MH. Of course, in the latter case the animal need not behave at all,
as MH can be developed in curarized animals.

J J A Heffron   What relevance has the anaerobically-induced Ca^{2+} efflux
in porcine MH mitochondria (noted by K S Cheah and coworkers and referred
to in Dr Lister's paper) to the primary pathogenesis of MH? I hope the
answer will not omit the fact that the mitochondria isolated by Cheah and
coworkers were from muscle obtained 15 to 20 minutes post mortem, and
derived from electrically stunned pigs, to my knowledge.

D Lister   Anaerobically induced Ca^{2+} efflux has been proposed by
Dr Cheah as a means of identifying PSS potential in pigs. But others
(e.g. I.G. Scott and R E Jeacocke at the Meat Research Institute, Langford,
Bristol) showed that the results for rate of Ca^{2+} efflux from mitochondria
could be influenced substantially by the 'condition' of the muscle at the
time the mitochondria were extracted. Thus my own conclusion is that
mitochondrial Ca^{2+} efflux is better at describing what has happened to
muscle at the time of slaughter and the changes occurring in muscle
thereafter than predicting the likely properties of the living muscle
which are pertinent to the pathogenesis of MH. I think there is a serious
problem somewhere along the line with mitochondrial function but I do not
think it is simply a question of the anaerobic release mechanism for Ca^{2+}.
Perhaps phospholipase A_2 activity and calmodulin are involved?

<u>J J A Heffron</u>   In 1973 G A Brooks and R G Cassens (J. Anim. Sci. <u>37</u>, 688-691) showed that muscle mitochondria isolated post mortem had reduced respiratory characteristics and poor respiratory coupling which, in present day terminology means that they would release calcium rapidly. Nowadays the important thing to do is to look at mitochondrial Ca^{2+} cycling ability. In fact, Dick Denton and colleagues at the Biochemistry Department of the University of Bristol have shown that mitochondrial Ca^{2+} cycling under aerobic conditions is extremely important for regulating the Krebs cycle, something that ths not really been acknowledged by many people working in the PSE or MH areas. Concerning the phospholipase A_2 activity, we can explain that in other ways, and that enzyme has now been shown not to be calmodulin dependent in a very recent paper. It depends on how you estimate the calmodulin.

<u>G Eikelenboom</u>   Concerning the primary defect in MH, I would like to hear comments on the proposition that it lies in the muscle cell, beyond the neuromuscular junction, and that calcium ions play a key role in it.

<u>J J A Heffron</u>   That the defect is distal to the motor endplate is indicated by: (a) experiments showing that acetylcholine does not induce contracture in MH pig muscle in vitro; (b) tubocurarine does not block halothane-induced MH but does block the suxamethonium-induced syndrome; (c) dantrolene, a muscle relaxant acting distal to the endplate, blocks the development of MH; (d) clinical concentrations of halothane cause contracture of MH muscle in vitro (see references in Mitchell, G. & Heffron, J.J.A. 1982, Adv. Food Res., <u>28</u>, pp 194-195); (e) intracellular ionized calcium is raised in human MH muscle fibres (Lopez, J.R., Alamo, L., Caputo, C., Wininski, J. & Ledezma, D. 1985. Muscle & Nerve, <u>8</u>, 355-358). The evidence that halothane does act beyond the neuro-muscular junction is quite good, blocking studies have shown that the probable site of action would appear to be at a sarcolemmal level. The calcium induced calcium release system of the sarcoplasmic reticulum does not appear to have the sensitivity to calcium that is required. You need up to 10 micromolar calcium to trigger it. Secondly, it is not blocked by dantrolene. I admit that blocking by dantrolene alone would not be sufficient evidence for that being as it were, sensitive component. So you have to move out to the sarcolemma and halothane does have effects on the sarcolemma which may serve as the initial stimulus for the sequence of events which manifests itself as MH.

<u>D Lister</u>   Dantrolene does seem to have a protective effect and a curative effect, where do you think its action lies.

<u>J J A Heffron</u>   I do not think anybody knows at the moment. The early report of it blocking calcium release from sarcoplasmic reticulum by W B Van Winkle (1976, Science, <u>193</u>, 1130-1131) has not been shown by others. So it's site of action must be more distal as it were.

I would say that dantrolene probably relies on an action on some general membrane perturbation that is caused by the anaesthetic.

<u>N G Gregory</u>   In an attempt to answer Dr Eikelenboom's question on whether the defect in the porcine stress syndrome lies in the muscle or in the stimulus which the muscle is presented with, I would like to add an observation. We have found that there is no precise relationship between the lactacidotic responses to intravenous isoprenaline infusion and

whether a pig reacts or fails to react to halothane. The premise was that halothane reactors have a defect in their muscle and would be expected to produce more acid in response to β-adrenergic agonists than halothane negative pigs. The lack of a clear relationship in our experiments, which were performed in barbiturate-anaesthetised pigs, leads us to suggest that the nature or extent of the agonist is critical, and that it is an over simplification to consider that the defect lies solely in muscle.

B W Moss   I believe that the PSS condition can be explained by a general membrane dysfunction. Several speakers have implicated changes at the membrane level. Both protein and lipid may be involved. Would Prof Heffron like to comment on the relative contribution of protein conformation changes or changes in the lipid component, for example in the content of cholesterol or the ratio of unsaturated to saturated fatty acids in the membrane?

J J A Heffron   Only very minor differences in neutral lipid and phospholipid composition of skeletal muscle of stress susceptible pigs were reported by D B McIntosh and M C Berman (1974, S. Afr. Med. J., $\underline{48}$, 1221). Cholesterol levels in erythrocyte membranes of normal and MH humans were not significantly different (Adams, M., Heffron, J.J.A. and Lehane, M. 1984. J. Physiol, $\underline{355}$, 15p). These findings suggest that it is the protein composition of the membranes that is changed in MH and PSS.

These findings would be consistent with recent studies (Lieb, W.R., Kovalycsik, M. and Mendelsohn, R. 1982. Biochim Biophys. Acta, $\underline{688}$, 388-398) on the site of action of anaesthetics in membranes. It is probably the integral proteins of the plasma membrane that provide the major focus of interaction with a particular anaesthetic, not merely lipid solubility which was the earlier view point.

P Mormède   Concerning the broader issue of the porcine stress syndrome let's go up to the brain to look for its origin. If the 'autonomic' nervous system controls the autonomous muscles, it is not itself autonomous but strongly dependent on brain control. Stress is primarily the product of brain function and manifests itself through behaviour and various endocrine changes, the autonomic nervous system being only one output among many.

P Sellier   As regards the aetiology of PSS and research on the primary effect of the halothane gene, it could be better to look at differences between halothane genotypes and not only at differences between the two halothane phenotypes. We have now some available methods for genotyping pigs at the Hal locus. Normal heterozygotes carry one halothane sensitivity (Hal^n) gene and it could be of interest to compare them to normal homozygotes (zero Hal^n gene) and halothane positive homozygotes (2 Hal^n genes) as a possibly more successful approach to finding the 'product' of the gene.

B W Moss   I'd like to comment on the study of MAO activity in the brain. The activity of this enzyme may very well depend on uptake of transmitter at the synaptic membrane. Thus, MAO activity in vivo could be altered by a general membrane dysfunction. This would not be detected using in vitro systems of measurement.

J B Ludvigsen   The difference observed by Dr Moss in plasma thyroxine response in Landrace and Large White during restraint, of 4.0 versus 4.9 µg dl^{-1}, may reflect slower hormone release in the Landrace under that condition whereas the Large White responded rapidly to the stress. Furthermore, the slight increase in circulating thyroxine at slaughter in the Landrace, to 4.2 ug dl^{-1}, might indicate difficulties, e.g. circulatory barriers, in thyroxine reaching the target, whereas the circulatory circumstances of the Large White are much better.

B W Moss   The difference in the thyroxine level between the ear vein restraint sample and the slaughter sample may be due to a difference in circulatory barriers as you suggest but also due to the different types of stressors under which these samples were taken. Results from experiments carried out on the turnover of thyroidal iodine indicate that the clearance of plasma iodide (K_2 in my compartmental model) is increased in stress susceptible breeds and the uptake (K_1) is reduced. I suggest that the Landrace pigs respond to stress by an increased utilisation of thyroxine and this is seen as a maximal response at restraint and slaughter. In Large White pigs, because their sensitivity is lower the response may not be maximal until the additional stress of slaughter. This hypothesis is based on thyroid turnover studies and urinary ratio studies as outlined in my paper.

C P Ahern   Would you agree with the suggestion the catecholamines may not be absolutely essential for initiating MH when triggered by halothane, but these hormones may be necessary in triggering the porcine stress syndrome.

D Lister   I am absolutely certain that catecholamines and the sympathetic nervous system have a role to play in PSE, because I do not see how glycolysis could be stimulated other than by a reduction in ATP which might be prompted by β-adrenergic agonists. I am not quite so clear about the role of catecholamines and the sympathetic system in the initiation of the MH reaction. It may be possible to get an attenuated form of MH in the absence of catecholamines when the syndrome is triggered by halothane. Catecholamines are probably essential for triggering MH by succinylcholine.

There is one point that I think we have missed in our discussion. We have concentrated on the triggering effects to start the reaction off, that really is not a major problem. The problem in MH and certainly in PSE, is that the stress susceptible animals are quite incapable of switching off the reaction and we have not had any comment at all about the mechanisms that control switching off. All our efforts have been devoted to the mechanisms controlling switching on. I think Dr Kozak-Reiss is one of the first people to identify the importance of the inter-relationship between the acidosis and the changes in muscle stimulation. Do you think that your explanation is similar to the cascade effect where once the acidosis starts to develop the reaction goes on and on.

G Kozak-Reiss   I think that acidosis produces a cycling phenomenon between metabolism, uncoupling of mitochondria and increased calcium.

J J Heffron   Do you think that phosphocreatine levels are lower in MH muscle in vivo? (As they are in your isolated specimens in Krebs solution monitored by ^{31}P-NMR).

G Kozak-Reiss   I believe the difference in phosphocreatine content that we observed between halothane negative and positive animals is an artefact due to the biopsy procedure we used.

D Lister   If you are extremely careful, and freeze the muscle in situ instantaneously with liquid nitrogen or a similar technique, then you can get extremely high levels of phosphocreatine. It is entirely dependent on the success of your getting close to the resting stage. I would imagine that all pigs have roughly the same level of phosphocreatine in the true resting condition and most of the differences we observe are attributable to sampling technique and the state of the animal during the period of sampling.

G Kozak-Reiss   A comment on Fig. 7 of the paper by Prof Ludvigsen. Overall, the heat expenditure of HN and HP animals was quite similar throughout the whole experiment. The differences observed at particular times could be related to the heat production of muscle tissue. In fact, a slight increase in free Ca^{2+} in the cytoplasm, as is assumed to occur in MH, activates metabolism. Experiments performed with isolated muscle in thermostatic chambers demonstrated this point. When a little reduction of membrane potential is produced by increasing extracellular K^+, below the threshold for activation of contraction, an increase in heat production is recorded. In animals at rest, this may perhaps be sufficient to account for the increased heat expenditure in HP animals.

F O Ödberg   This is a rather theoritical remark elicited by the findings mentioned by Dr McLoughlin.

The current hypothesis in ethology about some pathological behaviours and the so-called displacement activities is that they are homeostatic mechanisms which regulate arousal, they help to cope with a conflicting or difficult situation. Mormède and Dantzer published data recently on pigs which support this: the possibility to perform stereotypies decreased the plasma glucocorticoid level. Behaviour can alter physiological parameters. It is important to realise this when one measures these, especially those related to CNS activity (neurotransmitters) and the adreno-hypophyseal axis . There are also adaptations at the receptor level: chronic amphetamine administration decreases the amount of active DA receptors.

SESSION II

THE EVALUATION OF MEAT QUALITY

Chairman: P.V. Tarrant

METHODS FOR EVALUATION OF MEAT QUALITY IN RESEARCH AND PRACTICE,
USED IN MEMBER STATES OF THE EUROPEAN COMMUNITY

E. Kallweit

Institut für Tierzucht und Tierverhalten (FAL), Mariensee,
3057 Neustadt 1, Fed. Rep. of Germany

ABSTRACT

Pork quality is determined in many different ways in EC countries and the information obtained is used in breeding programmes, for research and for meat marketing and processing. To examine differences in methodology a questionnaire was sent to nineteen institutes and the answers are summarised and discussed here. A few methods are standardised and give comparable results, but others differ in the technique used or time of measurement, making comparisons of results between laboratories difficult or impossible. So far, meat quality criteria are not used in pig carcass grading systems in the EC.

INTRODUCTION

In discussing the PSE and DFD problems of pork it is not always clear how these extremes are defined or on what measurements they are based. Sometimes differences between individuals and/or countries are obvious. The aim of this survey was to obtain information on the techniques used for determination of meat quality in the EC and how these evaluations are utilised.

Questionnaires were sent to nineteen institutes in the following member states: Belgium (B), Federal Republic of Germany (D), France (F), Ireland (IR), Italy (I), Netherlands (NL), Denmark (DK) and the United Kingdom (UK). Thus the majority of the institutes in the most important pork-producing countries of the EC were included. It is, however, likely that the list of addresses was not complete. In some instances the questionnaire may not have been clear enough or some questions - that now seem important having reviewed the answers - were not included. The questionnaire is reproduced in Fig. 1.

In answer to the first question (Why is the measurement performed?) only one purpose should have been ticked off, and a new form(s) used for any other purpose(s). This was handled correctly by most of the respondents. In some replies, however, several purposes were marked, and for the rest of the form it was not possible to distinguish which method applied to each particular purpose. With this limitation, a summary of

108

Figure 1:

<u>Q U E S T I O N N A I R E</u>

<u>METHODS FOR EVALUATION OF MEAT QUALITY IN RESEARCH AND PRACTICE</u>,
<u>USED IN THE EEC MEMBER COUNTRIES</u>

Why is the measurement performed?

    Progeny-, sib-, performance testing for selection   O

    Quality control in trade - buying   O

                    - selling   O

    Research   O

        O

Where are the measurements taken?

    Commercial slaughter house   O

    Research slaughter house   O

    Laboratory   O

        O

When are the measurements taken?

    Day of slaughter
      h, min. p.m.: _____   O

    Cold Carcass
      h, min. p.m.: _____   O

        O

Which tests are employed?

    pH   O

      brand, type of instrument, electrode

    Colour brightness
          method: _____   O

    Conductivity   O

        O

Water holding capacity   O

Structure, firmness, texture, tenderness[1]   O

Other methods[2]   O

_____

_____

References to the above (add reprints if available)

_____

_____

At what location of the carcass are the measurements taken?
(e.g.: LD, 13th/14th rib, homogenate of muscle .....)

_____

_____

How are the results of these tests utilised?

(e.g.: screening test, part of a selection programme, part of a meat
        quality or selection index, weighing factor, commercial
        purpose etc.)

_____

_____

_____

_____

Remarks:

[1] Underline;   [2] take extra sheet if necessary

the answers received is presented here.

Answers to the first question indicate that in several countries meat quality is measured in the context of breeding programmes, in others meat quality evaluation is used for market surveys or in research programmes only. In no case was meat quality used regularly as a part of the marketing of carcasses or for price determination.

Breeding programmes

The meat quality criteria that are applied in test stations are summarised in Table 1. Test station pigs are killed in commercial abattoirs; in addition in Denmark and West Germany special slaughterhouses are available at the station. Preslaughter treatment is clearly standardised only in Denmark. In three out of six countries no measurements are taken on the day of slaughter (pH_1), whereas measurements at 24 h post mortem are the rule.

The most frequently used trait is the pH value, but colour brightness (reflectance, light scattering) is used in several national programmes. Water binding capacity is determined only in France. Visual scoring is used in Ireland and the Netherlands. The location of measurement is also variable. Measurements may be taken in M. longissimus dorsi at the 13th/14th rib, last rib or first lumbar vertebra. In addition, one or more ham muscles may be used, and in Denmark a neck muscle is also used.

The use of the results in breeding programmes varies. In France, the Netherlands and West Germany an index is calculated which, for example in West Germany, is also part of a selection index. Sometimes meat quality is recorded but is not considered in the breeding programmes as yet (UK). In Ireland, boars for nucleus herds are discriminated against if their progeny is predominantly PSE.

Research and market surveys

Similar methods for measuring meat quality are used in research and marketing surveys (Table 2). Where available, information is presented for several institutes within a country. Meat quality measurements and assessments were performed in abattoirs in all countries surveyed but also in one or more special meat laboratories. Measurements are frequently made on the day of slaughter and always on the day after slaughter although the time of measurement is variable on that day. In

TABLE 1: Methods used in the European Community for meat quality evaluation in breeding programmes using progeny, sib and/or performance testing

Country	F	UK	DK	IR	NL	D
Where?						
Commercial abattoir	x	x	x	x	x	x
Special abattoir			x	x		x
Laboratory			x	x		
When?						
Day of slaughter	x	x		x		x
20-24 h postmortem	x		x		x	x
Measurements						
pH	x	x	x	x		
Colour	reflectance	FOP,EEL	reflectance			x Göfo
Water binding capacity	imbibition					
Visual score			x	x	x	x
Location						
LD 13/14 rib		x	x			x
LD last rib		x	x			
Ham	2 muscles		x	x	x[1]	
Neck			x			x
Utilization	index	-	KK value	PSE[2]	index	index

[1]First lumbar vertebra. [2]Discrimination against boars for nucleus herds if their progeny is predominantly PSE.

TABLE 2: Methods used in the European Community for the measurement of meat quality in research programmes and in screening tests.

	F	UK	DK	B	NL	IR	D
Commercial abattoir	x x	x x	x	x	x	x	x x
special abattoir	x	x	x	x			x
laboratory	x	x	x	x	x	x	x
day of slaughter	x	x	x	x	x	x	x x
20-24 h p.m.	x x	x	x	x	x	x x	x x
later	x	x	48		76		48
pH	x x	x	x	x	x	x	x x
colour	x	x	x	x	x	x	x x
water-b.-cap.	x	x	x		x	x	x x
visual score	x	x			x	x	x
structure		x	x				x
others	x		x	x			x x
LD 13/14 rib	x			x	x	x	x
LD last rib		x x	x				
ham	x x	x	x	x	x	x	x
others			x				
conductivity				x		x	x x

some laboratories additional measurements are made between 48 and 176 hr post mortem.

Of the methods listed (Table 2) pH measurement is most common. The other traits are determined by a variety of methods and instruments and results cannot be compared directly. The location of measurements within the carcass also varies, in M. longissimus dorsi (LD) between the 13th/14th rib (D,B,F), last rib (UK, DK) or first lumbar vertebra (NL). In the ham several muscles are used.

The utilization of the results is variable, depending upon the objectives of the work and cannot be described here in detail. Applications include questions of the genetic origin, technical aspects and marketing problems including consumer acceptance.

Methods

In the case of pH measurement different manufacturers' instruments are used, but there was in common the use of combined glass electrodes and calibration with buffer solutions of approximately pH 7.0 and 4.0. These facts should allow comparable pH values for all laboratories. However certain differences need to be taken into account, including whether the pH value is obtained in a homogenate or in intact muscle. Also the exact time of measurement of pH_1 is important because this value drops relatively rapidly at this time compared to later stages.

Some other traits and the methods of measurement used are listed in Table 3. Colour, for example, is determined by reflectance, light scattering, or by visual appearance with or without the application of standards. Therefore, hardly any results can be compared between research groups in terms of direct values. This is a common problem for all traits listed in Table 3.

DISCUSSION

Originally, visual assessment was the only way of assessing meat quality. Even then different schemes were used. With the Danish five point scale (0.5 PSE - 5.0 DFD) the optimum meat quality was 2.5 - 3.0, while in the German system a score of 5 indicated best meat quality, and very pale as well as very dark pork both received low scores.

When more objective measurements were developed, again a large number of different approaches were made. This may be of some advantage

TABLE 3: Methods used for the evaluation of various meat quality traits in laboratories in the European Community

	F	UK	DK	B	NL	I	IR	D
Colour	Retrolux	Fibre optic probe EEL	Elrepho Hunterlab Visual score	FOP Labscan Göfo	FOP Japanese colour scale Labscan Reflectance	CIELAB Hunter CIE	Fibre optic probe EEL	Fat O Meater Fibre optic probe Elrepho, Göfo Hunterlab, reflectance visual score
Water binding capacity	Imbibition	Drip loss Visual score	Biuret	Grau/Hamm	Drip loss Visual score	Grau/Hamm Drip loss Cooking loss	Drip loss	Drip loss Grau/Hamm Cooking loss
Structure Firmness Tenderness		Taste panel Visual score	Taste panel Instron Frank		Visual score Warner/Bratzler shear	Taste panel Instron Warner/Bratzler shear	Visual score	Instron W/Bratzler shear Volodkevitch Rigor Visual score Histology
Conductivity				MS tester				MS tester, WTW LF Digi 550 LF 191
Others	Yield of processed ham		Chemical analysis protein, fat, pigment	Temperature Transmission	Transmission	Chemical composition Histology		Transmission Taste panel Chemical analysis

in providing a choice of methods under various conditions.

Definitions of certain characteristics, however, should be more standardised. The terms PSE and DFD for example were originally descriptive in character. Nowadays, several attempts have been made to define these conditions by pH_1 and pH_{24} values but the limits used are quite different between laboratories and countries. Therefore, the proportion of PSE meat found in separate investigations may not be comparable. Unfortunately, definitions of this kind were not asked for in the questionnaire. Another very important point missing in this context is the preslaughter treatment of the pigs. Except in the Danish progeny testing scheme, this point is not well defined elsewhere. These two points could be the subject of a more general discussion.

The early postmortem time when meat quality is developing is also of great importance. Since meat is never consumed while it is still warm and processing usually starts after cooling, it would be acceptable if meat quality was determined rather late. Early pH values, however, allow conclusions on the final meat quality. If it was possible to measure meat quality directly at later stages, early measurements could be abandoned, in particular since they can be incorrect if time is not strictly standardised. Standardisation of time may be difficult in a commercial abattoir, where many animals are slaughtered per hour. The exception, where early meat quality evaluation is needed, is where it is used on a commercial basis as part of the grading system. In that particular case the present pH measurement seems to be too sensitive. Light scattering or conductivity may be better measures; new developments in these areas will occur in the near future.

ACKNOWLEDGEMENTS

The author wishes to thank all who have contributed to this survey by sending back the questionnaires. Their names are listed below.

- Dr P Barton-Gade, Danish Meat Research Institute, Roskilde, Denmark.
- Dr M Casteels, Rijjksstation voor Veetvoeding, Gontrode, Netherlands.
- Dr J P Chadwick, Meat and Livestock Commission, Milton Keynes, UK.
- Dr E Cosentino, Instituto di Produzione Animale, Universita di Napoli, Facolta di Agraria, 80055 Portici, Italia.
- Dr G Eikelenboom, Research Institute for Animal Husbandry 'Schoonoord', Zeist, Netherlands.
- Dr K O Honikel, Bundesanstalt für Fleischforschung, Kulmbach, FRG.

- Dr Ph Lampo, Rijsuniversiteit Gent, Diersgenesskunde, Merelbeke, Belgium.

- Dr G Monin, Station de Recherches sur la Viande (INRA), Theix, France.

- Dr J Scheper, Bundesanstalt für Fleischforschung, Kulmbach, FRG.

- Dr D Seidler, Fachhochschule Lippe, Lemgo, FRG.

- Dr F Schmitten, Institut für Tierzuchtwissenschaft, Bonn, FRG.

- Dr V Tarrant, The Agricultural Institute, Dunsinea Research Centre, Castleknock, Dublin 15, Ireland.

- Dr P D Warriss, AFRC Food Research Institute Bristol, Langford, Bristol, UK.

EXPERIENCE IN MEASURING THE MEAT QUALITY OF
STRESS-SUSCEPTIBLE PIGS

Patricia A. Barton-Gade, Eli V. Olsen
Danish Meat Research Institute, Maglegårdsvej 2, DK-4000 Roskilde, Denmark

ABSTRACT

Measurements with the automatic Danish meat quality probe allow a rapid, direct determination of the water holding capacity of pig meat. The aim of the work reported in this paper was to use a material of pigs with known or presumed halothane status (positive or negative) to develop methods of combining probe and pH_2 values to give the most correct estimate of the meat quality in a pig carcass.

The results showed that halothane positive Landrace pigs often showed both PSE and DFD meat within the same carcass and that it was necessary to combine probe and pH_2 measurements if all positive animals with a poor meat quality were to be identified. 88-100% of the halothane positive animals had an unacceptable meat quality in this material.

Halothane positive Large White pigs showed mixtures of PSE and DFD meat within the same carcass to a much lesser extent, and in fact were nearly always PSE in at least 2 of the 3 probe muscles measured. Thus, for this breed, probe values alone could identify all halothane positive carcasses and the further information gained from pH_2 measurements was minimal. In this material 89-100% of the halothane positive pigs had an unacceptable meat quality.

INTRODUCTION

The meat quality estimation used in Danish pig breeding for a number of years, the KK-index, was based on a combination of reflectance and pH_2 values (Barton-Gade, 1979).

Reflectance values are affected by other factors than water holding capacity, i.e. pigment content and % intramuscular fat, and as meat quality improved these factors have become increasingly important in explaining the variation found. Work to replace the KK-index with a method which was less sensitive to changes in pigment and intramuscular fat content, resulted in the development of a probe which gave a direct indication of water holding capacity (Andersen, 1984). Probe values were found to be independent of pigment content, and when the measurements were carried out continuously throughout a muscle, a method of calculation could be devised which eliminated to a great extent the effect of intramuscular fat (Barton-Gade & Olsen, 1984). Probe values in 3 muscles, longissimus dorsi, biceps femoris and semimembranosus have replaced the previous KK-index from July 1985, the results being given as a single index value (see below).

The short, considerate, standardised pre-slaughter treatment which Danish test pigs receive (Barton-Gade, 1974 - now modified, as all pigs are stunned in the CO_2-compact equipment), was chosen so that meat quality variation showed up mainly as PSE meat. However, a certain (small) percentage of the pigs - those presumed most sensitive - still show higher pH_2 values than normal. Reflectance values were corrected for higher than normal pH_2 values in the original KK-index, as indeed is the case in meat quality estimations based on reflectance values used in other countries (Lundström et al. 1979, Schwörer, 1980). For practical reasons the newly introduced method does nct contain any correction for higher than normal pH_2 values. The aim of the work described in this paper was to use a material of pigs with known halothane status (positive or negative) and other material (presumed positive) to develop methods of combining probe and pH_2 values to give the most correct estimate of a pig's meat quality.

MATERIALS AND METHODS

In the period March-June 1983 1179 pigs from two of the testing stations were subjected to a halothane test (5% halothane in oxygen for 3 minutes). Each litter group consisted of a boar, a castrate and a gilt, and was subjected to the normal procedure for Danish progeny test pigs, i.e. ad lib feeding to approximately 90 kg live weight.

The castrate and gilt were then slaughtered after the standardised pre-slaughter treatment, and the day after slaughter probe values were measured in three muscles, bicep femoris, semimembranosus and longissimus dorsi as well as ultimate pH values in these three muscles and rectus femoris, semispinalis capitis, serratus ventralis and triceps brachii.

This material was supplemented with further values obtained on progeny test pigs partially or wholly rejected for human consumption because of PSE development on the slaughter line. These pigs are almost certainly halothane positive (Barton-Gade, 1984, Jensen and Barton-Gade, 1985).

RESULTS

Experimental material - known halothane status

The frequency of halothane positive animals in the experimental material was low, especially for the coloured breeds (Table 1). Duroc and Hampshire pigs were therefore not suitable for the purposes of this experiment and were not considered further.

TABLE 1   Frequencies of halothane positive pigs for the four breeds

	Landrace	Large White	Duroc	Hampshire
Number tested	360	462	282	75
% positive	4.7	4.3	0.7	0

The litters containing halothane positive pigs had greater losses during fattening then litters without positive pigs, so that the actual numbers of castrates and gilts tested for meat quality was lower than the numbers tested for halothane sensitivity. Thus, only 8 of the 12 positive Landrace pigs reached slaughter (1 died under the halothane test), while 9 of the 12 positive Large White pigs reached slaughter. The average probe and pH_2 values of these pigs is compared with the average values of non-reacting pigs in Table 2.

TABLE 2 Average values for the Danish automatic meat quality probe and for pH_2 in relation to halothane reaction and breed

Description		Landrace			Large White		
		Hal+	Hal-	Signifi-cance	Hal+	Hal-	Signifi-cance
Number of pigs		8	147		9	249	
Probe value	longissimus dorsi	110	68	***	114	65	***
	biceps femoris	73	75		89	75	*
	semimembranosus	85	75		102	77	***
pH_2-value	longissimus dorsi	5.48	5.46		5.49	5.48	
	biceps femoris	5.76	5.61	**	5.57	5.57	
	semimembranosus	5.63	5.51	**	5.52	5.50	
	rectus femoris	6.05	5.77	***	5.74	5.71	
	semispinalis cap.	6.19	5.98	*	6.01	5.90	
	serratus ventralis	6.29	6.03	***	6.21	6.10	
	triceps brachii	5.72	5.67		5.66	5.64	

Halothane positive Landrace pigs had higher probe values than halothane negative pigs for longissimus dorsi only. However, pH_2 values were highest in the halothane positive group for 5 of the 7 muscles studied. Halothane positive Large White pigs had higher probe values in all three muscles compared to halothane negative pigs, but pH_2 values were not significantly different. Thus, these results confirmed that stress-susceptible Landrace pigs show both PSE and DFD meat, whereas stress-susceptible Large White pigs seem to show mainly PSE meat.

The actual frequencies of PSE and DFD meat in the experimental material are shown in Table 3. In this Table probe values of 80 and above were considered to be indicative of PSE meat, whereas pH_2 values higher than the average value +2s for l. dorsi, biceps femoris, semimembranosus and triceps brachii and + 1s for the other muscles were considered to be indicative of DFD meat (Barton-Gade, 1981).

TABLE 3 PSE and DFD frequencies in relation to halothane reaction and breed

Description		Landrace		Large White	
		Hal+	Hal-	Hal+	Hal-
% PSE	longissimus dorsi	75	22	100	15
	bicep femoris	25	22	56	20
	semimembranosus	38	26	89	38
% DFD	longissimus dorsi	0	2	0	2
	biceps femoris	38	10	0	2
	semimembranosus	25	5	0	2
	rectus femoris	25	5	0	2
	semispinalis cap.	50	19	33	6
	serratus ventralis	63	10	44	18
	triceps brachii	25	6	0	4

Halothane positive Landrace pigs had a much higher PSE frequency in longissimus dorsi compared to halothane negative pigs, whereas the PSE frequency in biceps femoris and semimembranosus were only slightly increased in the halothane positive group. Halothane positive Landrace pigs showed however much higher DFD frequencies than halothane negative Landrace pigs, so that when the total percentages of PSE and DFD meat were considered, the percentages of deviating animals were more or less similar for all muscles i.e. 75, 63 and 63% for respectively longissimus dorsi, bicep femoris and semimembranosus.

Looking at individual animals, one of the halothane positive Landrace pigs was normal in meat quality in all muscles measured, two were PSE with normal pH values, four were mixtures of PSE and DFD, while one was normal in probe values but DFD in so-called "red" muscles (see below).

The picture was different for Large White pigs, where halothane positive animals had much higher PSE frequencies than halothane negative animals for all three muscles measured and DFD frequencies were only higher in two muscles, semispinalis capitis and serratus ventralis. Looking at individual animals for this breed showed none with a normal meat quality. One was slightly PSE with normal pH-values, five were PSE with normal pH values and three were mixtures of PSE and DFD.

This result was somewhat unexpected because other work reported in the literature (Eikelenboom et al. 1978, Schwörer et al. 1980) seems to show that halothane positive Large White pigs do not have a poorer meat quality than halothane negative pigs.

Models describing the meat quality of a whole carcass

The available information to describe the meat quality of a whole carcass consists of probe- and pH_2-values in three so-called "white" muscles, longissimus dorsi, biceps femoris and semimembranosus and pH_2-values in a further four so-called "red" muscles. This information must be reduced to one or two figures, if it is to be used in practice. There are several possible methods of doing this.

Probe values without pH

It is of interest to investigate the effect of dispensing with pH_2 values entirely, as the practical work involved in measuring the meat quality in progeny test pigs could be reduced considerably. Probe values can either be weighted equally or unequally. The present meat quality index used in Denmark uses a weighting, where a substantial deviation in one muscle only has a disproportionately large effect on the result obtained:

$$\text{New meat quality index} = 10 - \sqrt{\frac{(\text{Probe}_{bf}-40)^2+(\text{Probe}_s-40)^2+(\text{Probe}_{ld}-40)^2}{192}}$$

If values within the square root sign come to more than 100, they are set to 100.

This index is compared with an equal weighting, scaled as described under "Probe values corrected for pH", in Table 4. Values below 5.0 are considered unacceptable.

Table 4 Comparison of two methods of combining three probe values

		Landrace		Large White	
		Hal+	Hal-	Hal+	Hal-
Equal weighting	average	4.42	6.36	2.41	6.30
	% < 5.0	38	15	89	20
Present index	average	3.37	5.78	2.12	5.79
	% < 5.0	75	22	89	23

The difference between positive and negative animals was larger with the present index than with an equal weighting for Landrace, whereas it was more or less the same for Large White. All differences were however highly significant ($p < 0.001$). The percentage of unacceptable values showed that an equal weighting of probe values could not be used for Landrace, as only 38% of the halothane positive animals would be identified as having PSE meat. The reason for this, of course, is that halothane positive Landrace pigs have a higher DFD frequency in "white" muscles, and muscles with higher than normal pH_2 values will have low probe values. The unequal weighting used in the present meat quality index seems to be much better, and in fact at least one of the three muscles showed higher than normal probe values in 6 of the 8 halothane positive Landrace pigs.

For Large White pigs the percentages considered unacceptable were the same irrespective of weighting, which is expected as most halothane positive Large White pigs were PSE in at least two of the muscles measured.

Probe values corrected for pH_2

Probe values will always be low, when pH_2 values are higher than normal. Thus, some of the halothane positive animals (25% in this material) will have an acceptable index, when this is based on probe values alone. Probe values were therefore corrected for pH_2, as follows:

The probe value in any given muscle was first normalised, centred and standardised, and then transformed to a scale from 0 to 10. The transformed value corresponding to probe value 80 (just PSE-meat) was set at scale value 5; transformed values corresponding to about 150 (extreme PSE-meat) are set at scale value 0 and transformed values corresponding to about 40 (excellent meat quality) to scale value 10. The hyperbolic function tan h was used to make a continuous connection between the three points.

The mean pH_2 value in the muscle considered was also transformed to a scale from 0 to 10. pH_2 values less than the overall mean value for the muscle concerned were transformed to scale value 0. pH_2 values higher than normal, i.e. the standard deviation multiplied by 2 higher than the overall mean, were transformed to scale value 5, and very high pH_2 values corresponding to 8 standard deviation units above the mean were transformed to scale value 10. The difference between probe scale value and pH_2 scale value yield the final scale value for that muscle (W_{muscle}). If the pH_2 value is not essentially greater than the overall mean the probe scale value is decreased very little, whereas a very high pH_2-value will result in a large correction and probe scale values will be decreased a lot.

Finally, the data from all three white muscles are combined in the mean scale value:

$$W \text{ index} = \text{mean of } W_{bf}, W_s \text{ and } W_{ld}$$

This value is compared to the present meat quality index, i.e. probe values not corrected for pH, in Table 5. Again values below 5.0 are considered as indicative of an unacceptable meat quality.

Table 5   Comparison of probe index with and without pH correction

		Landrace		Large White	
		Hal+	Hal-	Hal+	Hal-
W-index	average	2.99	5.73	1.95	5.46
	% < 5.0	63	26	89	39
Present index	average	3.37	5.78	2.12	5.78
	% < 5.0	75	21	89	23

The difference between positive and negative Landrace pigs increased when probe values were corrected for pH, but the percentage of animals indicated as having an unacceptable meat quality was slightly less than with the present index.

For Large White the difference between positive and negative pigs and the percentage of animals with an unacceptable meat quality was not affected by correcting probe values for pH.

pH-index for "red" muscles

Correcting probe values for higher than normal pH_2 values does not seem to be sufficient for Landrace. It was therefore investigated whether addition of a pH-index in "red" muscles could improve matters. There are four mainly "red" muscles with pH-values only. The pH_2 value in each muscle was first transformed to a 0-10 scale in such a way that the average value + 1 standard deviation corresponds to 5 and + 4 standard deviations to 10. All pH_2 values less than the average were set at scale value 0.

$$pH\text{-index } R = 10 - \sqrt{\frac{scale_{rf}^2 + scale_{sc}^2 + scale_{sv}^2 + scale_{tb}^2}{4}}$$

Any values less than 0 are set to 0.

With this method very high pH_2 values in all four muscles will result in an R-value of 0, while normal or low pH_2 values in all muscles will give an R-valueof 10. pH_2 values slightly higher than normal in all four muscles, or two normal pH_2 values combined with two moderately high pH_2 values, will give an R-value of 5. The values obtained are shown in Table 6.

Table 6   pH-index R in relation to halothane genotype and breed

		Landrace		Large White	
		Hal+	Hal-	Hal+	Hal-
R-value	average	4.76	7.99	6.45	8.00
	% < 5.0	50	12	33	9

The difference between positive and negative animals was, as expected, highly significant for Landrace (p <0.001) and 50% of the positive animals had unacceptably high pH_2 values as compared to only 12% in the negative animals. Even for Large White the difference between positive and negative animals was significant (p < 0.05) and 33% of the positive animals had unacceptably high pH_2 values.

Combined probe/pH$_2$-index

It is now possible to express the information in the whole carcass by combining the two values W and R:

$$\text{Combined index} = (k_1 \times W + k_2) \times (k_3 \times R + k_4)$$

The constants are chosen, so that the combined index has a scale from 0-10, where 10 is best and 5 is the limit between an acceptable and unacceptable meat quality. The values obtained are shown in Table 7.

Table 7   Combined index in relation to halothane status and breed

		Landrace		Large White	
		Hal+	Hal-	Hal+	Hal-
Combined index	average	2.20	5.04	1.78	4.87
	% < 5.0	88	44	89	50

When pH$_2$ values in red muscles were taken into account together with probe values corrected for pH, the best separation of positive and negative Landrace animals yet seen was obtained, and all positive animals with a poor meat quality were identified as such. For Large White pigs the separation between positive and negative animals was not improved and the percentage of animals with an unacceptable meat quality was the same as before.

Pigs presumed halothane positive

The number of halothane positive pigs in the above material was not large and to supplement this, pigs with a very rapid PSE development in the period 1983-84 were considered similarly. These pigs are perhaps not completely appropriate for the purposes of this experiment as they can be expected to mainly PSE, but previous experience has shown that pigs judged to be PSE on the slaughter line can reverse the process and be normal or slightly DFD the day after slaughter. The results are shown in Table 8.

Table 8. Average index values and percentages with unacceptable meat quality in pigs presumed halothane sensitive

	Landrace		Large White	
	Average	% <5.0	Average	% <5.0
Number of animals	51	-	23	-
Present index	1.52	90	0.53	100
W-index	1.22	100	0.50	100
R-index	8.03	8	8.06	4
Combined index	1.33	100	0.81	100

The results confirmed that the pigs were highly PSE but even so about 10% of the Landrace animals showed DFD meat, both in "white" and "red" muscles. Although the differences between the various index values were not large, indexes taking pH into account identified all Landrace animals as having an unacceptable meat quality as compared to 90% with the present index. Large White pigs were identified as unacceptable whatever the method used.

DISCUSSION

Previous work has shown that Landrace pigs are more prone to become DFD than Large White pigs, even under the considerate conditions of the standardised pre-slaughter treatment (Barton-Gade, 1977). The results of this work confirm that Landrace pigs genetically pre-disposed to poor meat quality become exhausted more quickly for the same pre-slaughter treatment. Thus, methods which only measure PSE status risk missing a certain proportion of the most sensitive animals for Landrace - in this work 25%. It was necessary to take into account pH values in both "white" and "red" muscles, if all animals with a poor meat quality were to be identified. Large White pigs on the other hand, even when genetically pre-disposed to poor meat quality, did not show higher DFD frequencies, at least in "white" muscles, and methods measuring PSE status were sufficient to identify the pigs concerned.

When halothane negative pigs have higher than normal pH_2 values, there are some difficulties in assessing their true genetic potential for meat quality. All pigs irrespective of genetic pre-disposition for meat quality, can become exhausted during the pre-slaughter period and thus show DFD meat. However, it must be assumed that most Landrace pigs showing DFD meat with the short, standardised pre-slaughter treatment have become DFD via PSE, and thus halothane negative animals with higher than normal pH_2 values must be penalised. For Large White on the other hand, higher than normal pH_2 values are most likely to be due to a combination of genetic and environmental factors, and it is debatable whether halothane negative Large White animals should be penalised for DFD meat.

The combined index reported in this work needs to be further refined and among other things the number of muscles necessary to define the R-index needs to be investigated in more detail. If the number can be reduced, then there will be less work involved.

REFERENCES

Andersen, I.E. 1984. Some experience with the portable Danish probe for the measurement of pig meat quality. Proc. Scient. Meet. 'Biophysical PSE-muscle analysis' (Ed. H. Pfützner) (Technical University, Vienna) pp 173-191.

Barton, P.A. 1974. A standardised procedure for the pre-slaughter treatment of pigs to be tested for meat quality. Proc. 20th Europ. Meet Meat Res. Workers, Dublin, Ireland, pp 52-54.

Barton, P.A. 1977. Comparison of Danish Landrace and Large White with respect to meat quality. Proc. 23rd Europ. Meet Meat Res. Workers, Moscow, USSR.

Barton-Gade, P.A. 1979. Some experience on measuring the meat quality of pig carcasses. Acta agric. scand., Suppl. 21, 61-70.

Barton-Gade, P.A. 1981. The measurement of meat quality in pigs post mortem. Proc. Symp. 'Porcine Stress and Meat Quality', (Ed. T. Froystein, E. Slinde, N. Standal) (Agric. Food Res. Society, As) pp 205-218.

Barton-Gade, P.A. 1984. Influence of halothane genotype on meat quality in pigs subjected to various pre-slaughter treatments. Proc. 30th Europ. Meet Meat Res. Workers, Bristol, England, pp 8-9.

Barton-Gade, P.A. and Olsen, E.V. 1984. The relationship between water holding capacity and measurements carried out with the automatic Danish meat quality probe. Proc. Scient. Meet. 'Biophysical PSE-muscle analysis' (Ed. H. Pfützner) (Technical University, Vienna) pp 192-202.

Eikelenboom, G., Minkema, D., Van Eldik, P. and Sybesma, W. 1978. Production characteristics of Dutch Landrace and Dutch Yorkshire pigs as related to their susceptibility for the halothane-induced malignant hyperthermia syndrome. Livestock Prod. Sci. 5, 277-284.

Jensen, P. and Barton-Gade, P.A. 1985. Performance and carcass characteristics of pigs with known genotypes for halothane sensitivity. EAAP Publication No. 33 (Ed. J.B. Lvdvigsen) pp 80-87.

Lundström, K., Nilsson, H. and Malmfors, B. 1979. Interrelations between meat quality characteristics in pigs. Acta agric. scand. Suppl. 21, 71-80.

Schwörer, D. 1980. Selection of meat quality and stress resistance with pigs in Switzerland, Schweiz. Verein. Tierzücht 56, 44-59.

Schwörer, D., Blum, J. and Rebsamen, A. 1980. Parameters of meat quality and stress resistance of pigs. Livestock Prod. Sci. 7, 337-348.

HOW TO MEASURE THE WATER-HOLDING CAPACITY OF MEAT?
RECOMMENDATION OF STANDARDIZED METHODS

K.O. Honikel

Federal Centre for Meat Research, Kulmbach
Federal Republic of Germany

ABSTRACT

In principle, water-holding capacity (WHC) is defined as the ability of meat to hold all or part of its own water. There exist, however, no reference unit nor reference procedures for measuring WHC which have been adopted in general in meat science or technology. Therefore a wide variety of methods are used, due to the fact that meat is handled and processed in a variety of ways. Also the meaning of WHC may vary. People who slaughter, chill, transport and sell fresh meat understand by WHC the weight or drip loss of carcasses or cuts. Consumers and processors of "ready to eat" meat understand by WHC its cooking loss. The present work illustrates that the WHC measured as drip loss does not allow conclusions about the cooking loss of fresh meat. Two standardized procedures for drip loss and cooking loss determination will be recommended after the factors that influence these WHC determinations are evaluated.

INTRODUCTION

Muscles of live animals contain 70 - 75 % water which is bound primarily to the muscle proteins within the muscle cell. The high pH of about 7.0 in the muscle cell and its physiological salt concentration allows the muscle proteins to bind about 90 % of the water intracellularly. This ability of muscles we call water-holding capacity (WHC). After the death of the animal the pH of normal beef and pork muscle starts to fall to its ultimate value of about pH 5.5. This pH fall reduces the ability of the muscle proteins to hold the water tightly. The WHC of the muscles decreases (Hamm, 1972). Additionally the velocity of the pH fall in combination with the temperature of the muscle during this time influences WHC. Slow pH fall and rapid temperature decrease induces cold shortening with an enhanced drip loss, whereas slow pH fall at very low chilling rates causes rigor shortening again with an increased drip loss (Honikel et al., 1986). Fast glycolyzing muscles at prevailing high temperatures result in PSE muscles with a rapid release of exudate from the meat (Honikel and Kim, 1985; Honikel, 1986).

So besides pH itself, the temperature/time/pH conditions in muscles in the first hours post mortem influence WHC. Drip loss of meat is effected by all these factors, the cooking loss, however, is effected primarily by the pH of the meat. Also,cooking loss is greatly effected by the conditions of cooking (Bendall and Restall, 1983; Kopp and Bonnet, 1985).

As different factors influence drip and cooking loss it cannot be expected that drip loss results allow reliable conclusions about cooking loss and vice versa. Therefore separate methods must be used.

In this paper the factors that influence drip and cooking loss will be described and taking these into account two standardized procedures will be recommended.

FACTORS WHICH INFLUENCE DRIP AND COOKING LOSS

As mentioned above, the pH of the meat influences WHC. Furthermore the WHC depends on the muscle type and the degree of marbling. Also the species of animal influences the WHC of the meat due to variations in the muscle composition and structure. In addition to these factors the drip loss depends on

1) size and shape of the sample. The surface to weight ratio is important for the amount of drip released. A larger surface area per weight unit increases the drip loss per unit time.

2) treatment during conditioning period. As mentioned above, rapid chilling of prerigor muscles may lead to cold shortening and very slow chilling may lead to rigor shortening as is shown in Fig. 1.

The shortening of sarcomeres was at a minimum when muscles were chilled to 10 - 15°C within the prerigor period i.e. at pH values above 6.0. The drip loss released from these muscles within 7 days of storage (Fig. 2) coincided, with respect to the minimum of drip loss and shape of curves, with the sarcomere shortening in Fig. 1. There was indeed a linear relationship between final sarcomere length and drip loss of pork and beef muscles as shown in Fig. 3.

As also mentioned above, PSE conditions are induced by a rapid pH fall at prevailing high temperatures. Results are presented in Fig. 4 for drip loss in slices of M. long. dorsi taken from a pig carcass with a pH_1 value of 6.0 and subjected to different chilling conditions. One pair of slices was stored and slowly chilled simulating PSE conditions, the other pair was chilled rather rapidly thus avoiding PSE conditions and simulating normal meat quality.

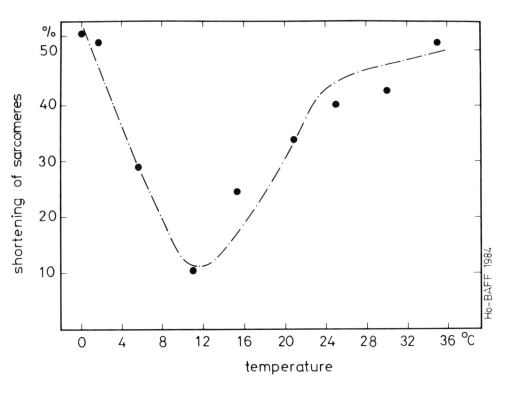

Fig. 1   Influence of temperature on the shortening of sarcomeres
in prerigor pork muscles

Pieces of slaughterfresh M. mastoideus were stored at 0° to 35°C
between 45 min and 24 hours post mortem. After 24 hours the sarcomere
length was measured according to Voyle (1971) and the degree of
shortening calculated by comparison with the sarcomere length of the
muscle before incubation. For each temperature a different muscle
sample was taken.

The "PSE" conditions lead (Fig. 4) to rapid release of juice within the
first 24 hours, whereas the "normal" muscle slices had little drip at the
first day increasing between day 1 and 5. After 17 days of storage the
difference in drip loss has been diminished. Therefore drip loss of meat
depends also on

3) the time of measurement post mortem and on

4) the length of time of measurement.

The drip loss also depends on

5) the chilling temperature of the meat after the ultimate pH is reached.
   Storage at 4°C results in a higher drip loss than storage at 0°C.

132

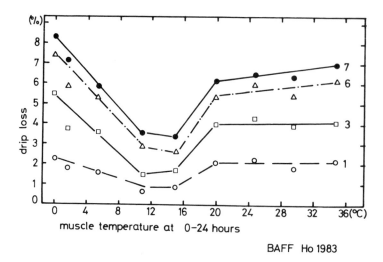

BAFF Ho 1983

Fig. 2 Drip loss of pork muscles stored at the various temperatures
indicated on the abcissa during the first day post mortem and subsequently stored at 0°C.
Cubes (about 50 g weight) of M. mastoideus obtained within 45 min
post mortem with pH values between 6.3 and 6.5 were stored during
the first day post mortem at 0° to 35°C. After 24 hours all samples
were kept in a chilling room at 0°C. At 1, 3, 6 and 7 days post
mortem (indicated on the right hand of the Figure) the drip loss
was determined. The exact procedure is described in the text.
For each temperature a sample from a different carcass was taken.

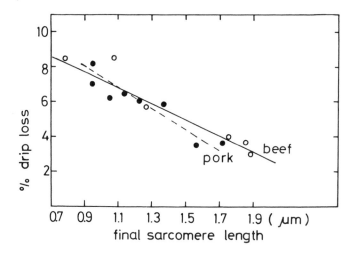

Fig. 3 Relationship between drip loss and final sarcomere length
in pork M. mastoideus and beef M. sternomandibularis.
Drip loss was determined between day 1 and 7 post mortem as described
in Fig. 2. Sarcomere length were determined according to Voyle (1971)

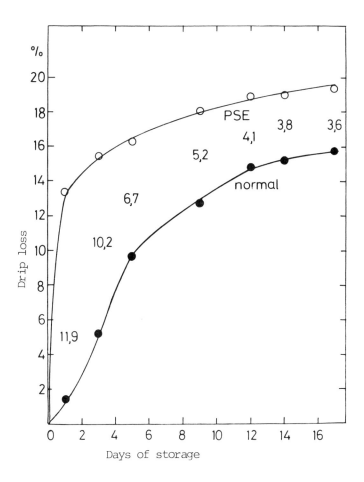

Fig. 4   Influence of temperature on the drip loss of slaughterfresh
M. long. dorsi of pork with a pH$_1$ of 6.0

Slices of M. long. dorsi were obtained within 45 min post mortem with
a pH of 6.0. Two slices were stored at 38°C between 45 min and 2 hours,
at 35°C between 2 and 3 hours and at 33°C between 3 and 4 hours post
mortem.They were then stored at 0°C until day 17. These were designated
as "PSE conditions". Two further adjacent slices of the same muscle
were incubated at 20°C between 45 min and 4 hours post mortem. Then
the temperature was reduced to 0°C and the slices stored until day 17.
These were designated as "normal conditions". Drip loss was measured
at the days indicated and was carried out as described below in the
recommended procedure. The figures between the curves are the
difference in drip loss in percent.

Finally the drip loss of meat samples depends on
6) the method of packaging used in the experiment.
Surface tight (heat shrunk) vacuum packaging releases less exudate from
the meat during storage than vacuum packaging with vacuum holes at the
corners of the meat sample. In both cases the amount of drip released differs
from measurements in plastic pouches or boxes under atmospheric pressure.

The above mentioned influence of pH on drip loss is shown in **Fig.** 5.

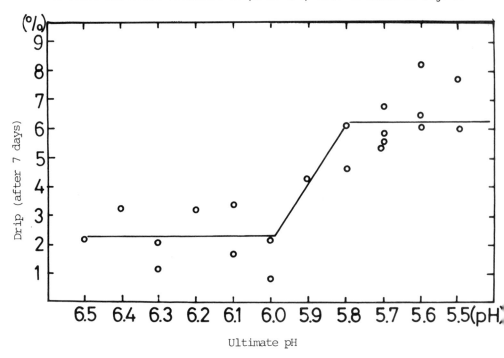

Fig. 5 Influence of ultimate pH value on the drip loss of
M. mastoideus of pork.

Samples were stored at 15°C to 25°C at the first day post mortem,
then at 0°C up to day 7, when the drip loss was measured. All muscle
samples had pH_1 values above 6.2. Drip loss was determined as
described in Fig. 2.

The relationship between ultimate pH of meat and drip loss was not
linear. In these experiments neither shortening nor conditions associated
with the development of PSE existed. There was little pH influence above
pH 6.0. Below this pH there was a sharp increase in drip loss.

The cooking loss of meat, however, had a linear relationship to pH not only in the prerigor (Honikel et al., 1981) but also in the postrigor state as shown in Fig. 6.

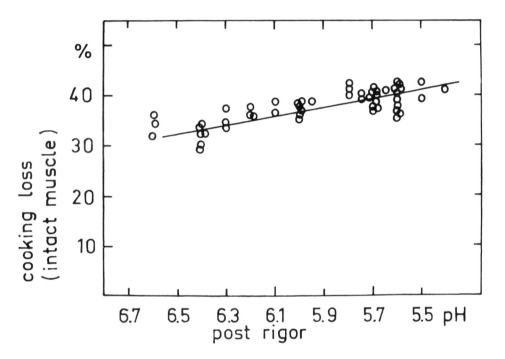

Fig. 6   Influence of the ultimate pH on cooking loss in M. mastoideus of pork.
Samples were taken from the carcass at the first day post mortem. Between day 2 and 7 the temperature of storage was 0°C. Cooking loss was measured 7 days post mortem.

Cooking loss was determined as described in the recommended procedures in the text but due to the size of M. mastoideus the slices weighed about 50 g.

Besides on pH the cooking loss depends on
1) the final temperature of heating.
In Fig. 7 the meat samples were heated to the indicated internal temperature and chilled immediately afterwards. In Fig. 7 again the influence of pH is shown. Furthermore Fig. 7 shows that in PSE and normal meat between 65°C and about 85°C the release of fluid was faster in PSE muscles, whereas at 55°C and 100°C the differences in cooking loss were virtually non existant.

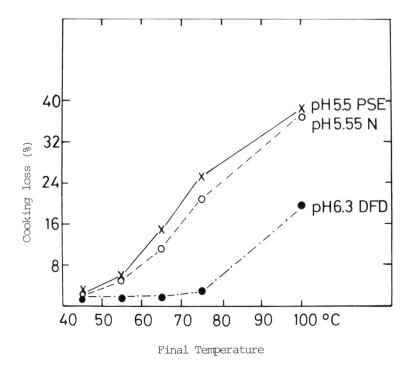

Fig. 7 Cooking loss of M. long. dorsi heated to different final
temperatures. Muscles of varying qualities and ultimate pH values
were studied.
Cooking loss was determined as described in the recommended
procedure in the text with the exception that the slices at 20°C
were put into a boiling water bath and taken off the bath
immediately after the prescribed final temperature was reached.

PSE muscles had a pH_1 of 5.6, normal muscles (N) had a pH_1 of 6.3
and DFD had a pH_1 of 6.5. Samples were taken from the carcass
after 24 hours and stored for a week at 0°C.

In the meantime we have confirmed this in a number of experiments. So
cooking loss depends on

2) the meat condition at certain final temperatures of heating;

the cooking loss also depends on

3) the velocity of heating (Fig. 8).

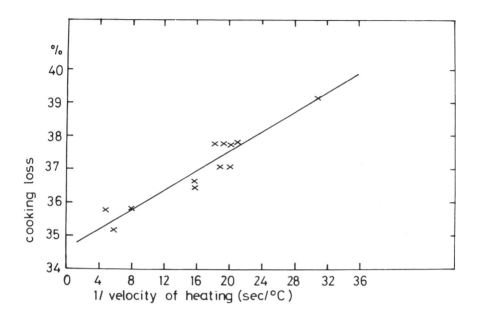

Fig. 8 Cooking loss of M. mastoideus of pork with an ultimate pH of 5.8 after 3 days post mortem, heated with different velocities.

Samples of 50 g weight as described in Fig. 6 were heated from 20° to 90°C. Between 20 and 55°C the heating rate was 5 sec/°C (or 12°C/min) in all cases. From 55° to 90°C the velocities indicated were chosen with a constant (linear) temperature increase. At the moment 90°C was reached in the centre, the samples were put in tap water.

The faster the velocity of heating the lower is the cooking loss (Kopp and Bonnet, 1985).

4) The length of heating after reaching the final temperature influences WHC. Longer heating time increases the loss of fluid (Bouton and Harris, 1972). Connective tissue is gelatinized and released from the meat.

5) Size, shape, marbling and composition of the meat also influence cooking loss. The surface/weight ratio is important. Meat with a considerable fat content releases fat as well as watery fluid which often increases total loss. Marbling of meat covering some muscle tissue may reduce cooking loss.

Contrary to drip loss which increases with time of storage (see Fig. 2 and 4) cooking loss does not change with the time of storage.

Fig. 9 shows that between day 1 and 8 post mortem there was no change in cooking loss. Fig. 9 also shows that neither cold (5°C during the first day) nor rigor shortening (35°C) had any influence on cooking loss.

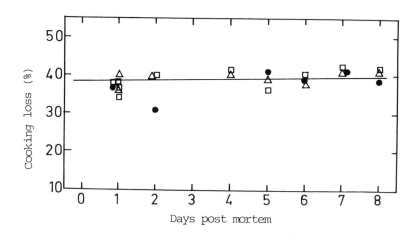

Fig. 9 Cooking loss of M. mastoideus of pork at different times post mortem.

The samples were taken from the carcass within 45 min post mortem and stored until 24 hours at 5°C (●), 20°C (□) and 35°C (Δ). Between day 1 and 8 they were stored at 0°C. The pH of all samples was 5.6 at the time of measurement. Cooking loss was determined as described in Fig. 6.

Due to the different factors that influence drip loss and cooking loss a good relationship between both methods of detecting WHC cannot be expected, when the pH of the samples is uniform. This is shown in Fig. 10. Whereas the drip loss in these 36 samples of M. long. dorsi of pork with a final pH of 5.5 - 5.6 varied from 1 to 17 percent i.e. a 17 fold increase, the cooking loss of these samples with few exceptions was between 35 and 45 percent i.e. a 1.3 fold increase.

Having considered the different factors that influence drip loss and cooking loss, and the poor relationship that exists between these properties of meat, we recommend that the following procedures are adapted as standard practice.

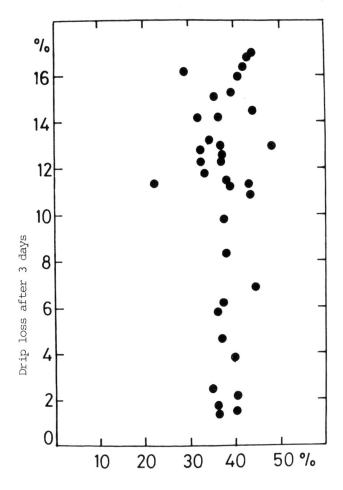

Cooking loss at 3 days post mortem

Fig. 10 Relationship between drip and cooking loss in M. long. dorsi
of pork.

The ultimate pH of all samples was 5.5 - 5.6.
The recommended procedures described in the text were applied.
Drip loss was measured between day 1 and 3. Cooking loss was
determined immediately afterwards.

RECOMMENDED PROCEDURES

Within the framework of the CEC Beef Production Programme, the
working group on Meat Quality published (Boccard et al., 1981) procedures

for measuring meat colour, tenderness by instrumental methods, sensory assessment, and chemical analysis. The present recommended procedures for measuring the drip loss and cooking loss of intact muscular tissue are an extension of that work. Both methods of measuring water-holding capacity are of paramount interest to meat scientists and technologists.

Where it is necessary to carry out both measurements on the same piece of muscle, cooking loss measurement will succeed drip loss determination.

The methods are described for transverse slices of M. long. dorsi as this is the most widely used muscle. Other muscles may be assessed in a similar manner but care should be taken in these cases to note the muscle fibre direction and to report it in the description of the method.

MEASUREMENT OF DRIP LOSS

Before carrying out the experiment the following data should be measured or known: pH of the sample, the time post mortem, and type and age of the animal. Knowledge of conditioning temperatures and/or sarcomere length could be helpful.

A slice, 2.5 cm thick, of M. long. dorsi sampled between the 8th thoracic and first lumbar vertebra with a freshly cut surface should be taken for measurement. Associated adipose tissue or parts of M. spinalis and M. multifidus dorsi should be removed. The facies should stay around the muscle. The temperature of the room during cutting should be similar to the temperature of the meat. The muscle should be weighed as accurately as appropriate in a plastic pouch (the weight is usually between 70 and 100 g) in which the slice may suspended by means of a net or a thread. The plastic pouch is sealed under atmospheric pressure. The samples are stored at 0° to 4°C for at least 48 hours. The time of storage must be stated in the method's description. The pouches should hang in such a way that the exudate dripping from the meat does not stay in contact with the meat. At the end of the experiment the muscle is taken from the pouch, dried gently with an absorbing tissue and reweighed. During the weighing care must be taken that no condensation of water vapour at the cold surface occurs. The drip loss is expressed as the weight loss in mg/g original weight of meat or percent of original weight.

Finally the pH of the muscle chould be measured again.

After drip loss evaluation the sample can be used immediately for cooking loss measurements. If there is a delay before measuring cooking loss the sample must be wrapped to avoid drying of the surface.

DETERMINATION OF COOKING LOSS

The sample remaining after drip loss measurement, of known weight and pH, is placed in a thin walled polyethylene bag (the bag must be waterproof and withstand 75°C) and sealed under moderate vacuum. The vacuum must be applied in order to remove air layers or air pockets between meat and wall of the bag. Furthermore during subsequent heating the meat must be completely immersed in water. With air pockets in the pouch part of the meat may be above the water level and that is not acceptable.

If the bag is not sealed, care must be taken that the wall of the bag touches the meat surface in order to allow optimum heat flow. The mouth of the bag must remain above the water level. Glass beads or similar devices should be put at the bottom of the bag in order to keep all meat surfaces immersed in the bath.

The properly sealed bag is placed in a water bath at preferably 75°C (other temperatures may be used, if necessary) where it will stay for 30 minutes. The pack is then placed for 40 min in running tap water (about 15°C) after which time the meat is taken from the bag, mopped dry and weighed. The heating loss will be expressed as g cooking loss/g initial weight (before cooking) or as percent heating loss.

In some experiments the measurement of drip loss may be unnecessary. In this case the sample may be used directly for cooking loss measurement after preparing it as described above for drip loss.

There are experiments in which the "history" of the meat is not known to the researcher. In this case it may be advantageous to measure the moisture content of meat and relate the loss to total moisture content or if appropriate to dry matter. Method of moisture determination is described by Boccard et al. (1981). Also the expression of g cooking loss/g protein is possible.

After measurement of cooking loss the sample can be used for tenderness measurements and sensory assessment according to the method described by Boccard et al. (1981).

CONCLUDING REMARKS

Adoption of these methods will mean that in the future the wide variety of ways for measuring WHC can be reduced to a necessary minimum. Drip loss and cooking loss measurements are of paramount interest to research in meat science and technology. Considering the distinct factors that influence the different methods, we recommend the use of these two methods as they are most closely related to the commercial and consumer uses of meat.

REFERENCES

Bendall, J.R. and Restall, D.J., 1983. The cooking of single myofibers, small myofibre bundles, and muscle strips of beef M. psoas and M. sternomandibularis muscles at varying heating rates and temperatures. Meat Sci. 8, 93-117.

Boccard, R., Buchter, L., Casteels, E., Cosentino, E., Dransfield, E., Hood, D.E., Joseph, R.L., MacDougall, D.B., Rhodes, D.N., Schon, I., Tinbergen, B.J. and Touraille, C., 1981. Procedures for measuring meat quality characteristics in beef production experiments. Livestock Production Sci. 8, 385-397.

Bouton, P.E. and Harris, P.V., 1972. A comparison of some objective methods used to assess meat tenderness. J. Food Sci. 37, 218-221.

Hamm, R., 1972. Kolloidchemie des Fleisches. P. Parey Verlag, Berlin und Hamburg.

Honikel, K.O., Hamid, A., Fischer, C. and Hamm, R., 1981. Influence of post mortem changes in bovine muscle on the water-holding capacity of beef. Post mortem storage of muscle at various temperatures between 0° and 30°C. J. Food Sci. 46, 23-25 and p 31.

Honikel, K.O., 1986. Chilling of pig muscles early post mortem and meat quality. In: Evaluation and Control of Meat Quality in Pigs, EC Seminar, Dublin, Nov, 1985.

Honikel, K.O. and Kim, C.J., 1985. Über die Ursachen der Entstehung von PSE-Schweinefleisch. Fleischwirtschaft 65, 1125-1131.

Honikel, K.O., Kim, C.J., Roncalés, P. and Hamm, R., 1986. Sarcomere shortening of prerigor muscles and its influence on drip loss. Meat Sci. in press.

Kopp, J. and Bonnet, M., 1985. Effects of cooking methods on meat quality. In: The long-term definition of meat quality: Controlling the variability of quality in beef, veal, pigmeat and lamb, (ed. G. Harrington). Commission of the European Communities, Luxembourg, p. 105.

Voyle, C.A., 1971. Sarcomere length and meat quality. Proc. 17th European Meeting of Meat Research Workers, Bristol, p 95-97.

REMOTE MONITORING OF POSTMORTEM METABOLISM
IN PORK CARCASSES

H.J. Swatland

Department of Food Science
University of Guelph
Guelph, Ontario N1G2W1, Canada

ABSTRACT

This review describes methods for the remote monitoring of postmortem metabolism in pork carcasses. Information on the degree of postmortem reflex activity was obtained from a load cell in the shackling chain and was confirmed by electromyography. The rate of depletion of adenosine triphosphate (ATP) was assessed by mechanically monitoring the loss of muscle extensibility and by measuring the loss of electrical capacitance of muscle fibre plasma membranes. These relationships were confirmed by biochemical assays for ATP and by nuclear magnetic resonance spectrometry. Fibre optics were used to measure the postmortem development of paleness in pork and to monitor glycogen depletion and pH. A photodiode array combined with an electronic flash unit provided a superior method for reflectance spectrophotometry of pork under commercial abattoir conditions.

INTRODUCTION

This review describes some of the methods that have been used at the University of Guelph to study various aspects of postmortem metabolism in pork carcasses. A major goal has been to find new ways with which to study the series of events that occur as the muscles of newly slaughtered pork carcasses are converted to meat. A few of the techniques that have been developed have been sufficiently robust to be used for problem solving in commercial abattoirs. Other methods that have been used are essentially the same as those used elsewhere and only the general results obtained are considered.

METHODS

1. EMG - electromyography to record reflex muscle activity during slaughter,

2. EEG - electroencephalography in attempts to measure the efficiency of stunning methods in producing anaesthesia,

3. ECoG - electrocorticography to measure neural involvement in reflex activity during slaughter, and

4. Spinal electrodes - to record and to induce extra spinal cord involvement in reflex muscle activity during slaughter.

5. Current measurements - of applied and transcortical electrical stunning voltages and amperages.

6. Shackling-chain load cell measurements to study the rate and extent of exsanguination and the degree of carcass bouncing caused by reflex activity postmortem,

7. Extensibility measurements of superficial muscle fasciculi to study the rate of rigor development, and

8. Distension measurements of deep regions of the carcass to study the development of carcass setting.

RESULTS

Physiological Techniques

The major uses of EMG, EEG, ECoG and spinal electrodes have been: (1) to quantify the amount of postmortem reflex activity, (2) to investigate the efficiency of stunning procedures, and (3) to search for sources of postmortem reflex activity. The basic premise is that postmortem reflex activity accelerates glycolysis and, thereby, exacerbates the development of the PSE condition of pork.

Postmortem reflex activity in pork carcasses involves the coordinated activity of major muscle groups such as those of the ham or the head and neck. With spinal electrodes it has been shown that the final common pathways from motor neurons to motor units remain viable for a number of minutes postmortem so that postmortem reflex activity appears to originate from the CNS or central nervous sytem (Swatland, 1976a). Termination of postmortem reflex activity, therefore, probably occurs by a cessation of CNS activity and not by a failure of the peripheral nervous system. Hence, the CNS is a major element in understanding the variability between pigs in their degree of postmortem reflex activity. However, it is difficult to ascertain where in the CNS that postmortem reflex activity arises. Waves of cortical activity occur in pigs that are still struggling after exsanguination (Swatland, 1976b) but whether this activity is motor or sensory feed-back in nature is difficult to determine. In terms of the remote monitoring of postmortem metabolism, the classical physiological techniques are difficult to apply under industrial conditions and much of the information that these methods yield may be obtained more easily by other methods.

EEG and ECoG have been used by several research groups to investigate the efficiency of various types of stunning methods for meat animals. However, with the various types of automated electrical stunning equipment that are of current interest to the pork industry in Ontario it is difficult to secure the electrodes onto the pig's head. Surgically-implanted ECoG electrodes require a substantial investment of time and effort for each animal and, all too often, the neural events that occur during slaughtering are obliterated by muscle activity in the head region (Swatland et al., 1984). The remote monitoring of blood pressure, particularly in the head region, would be of considerable interest since blood pressure appears to have an influence on the duration of unconsciousness after stunning. A likely mechanism for this is the effect of blood pressure on the reticular activating system of the brain.

Current Measurements

Surgically-implanted ECoG electrodes are needed to measure the small current that actually passes through the pig's brain during electrical stunning (Swatland et al., 1984). However, the overall current that passes across the animal is easily measured and may also yield interesting information. Voltages and amperages may change with time, perhaps as a result of changes in the distribution of blood in the major vessels and sinuses around the brain. A major difficulty is to be sure that the measured current is a realistic representation of the current that actually affects the nervous system. In some situations, part of the overall current may leak to a grounded part of the animal, such as a shoulder that touches a grounded metal surface, and may cause blood splashes in the region involved.

Mechanical Methods

Information about postmortem reflex activity in pork carcasses may be obtained from a load cell in the shackling chain or above the gambrel. This method was originally employed to study rates of exsanguination, essentially to confirm that lethal stunning currents would not interfere with subsequent exsanguination procedures (Swatland, 1983). It was soon found, however, that the bouncing of carcasses caused by postmortem reflex activity showed up quite clearly on the chart recorder and, rather suprisingly, that the activity of different parts of the carcass produced different recoil patterns (Figure 1).

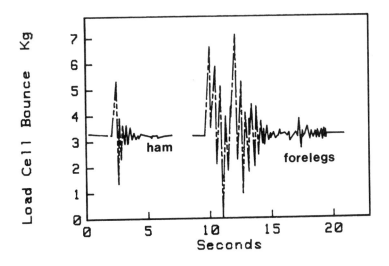

Fig. 1. Recoil patterns produced by postmortem reflex activity in different parts of the carcass.

Regional sources of reflex activity may be identified by electromyography. Figure 2, for example, shows integrated

148

EMG activity from the ventral muscles of the neck associated
with the head down reflex.

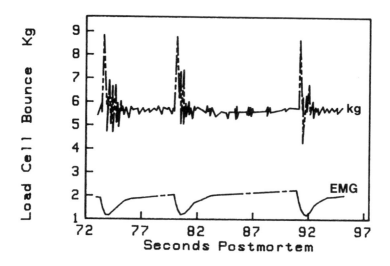

Fig. 2. Integrated EMG activity from ventral neck muscles associated with the head down reflex seen on the
load cell.

Another mechanical method that has recently been
adapted for use on pork carcasses is the rigorometer. The
rigorometer was originally developed by Bate-Smith (1939) as
a method for measuring the loss of extensibility of isolated
muscle strips during the development of rigor mortis. At
that time, the major interest was to demonstrate that the
development of rigor mortis was correlated with the disap-
pearance of adenosine triphosphate (ATP). Now that this
relationship is firmly established scientifically, the loss
of muscle extensibility in the carcass provides an indirect
method of measuring ATP depletion rates. The rigorometer

has been miniaturized and automated so that it can be attached to the surface of a carcass (Swatland, 1985a). A solenoid-operated lever pulls on a wire that is connected to a superficial muscle fasciculus of either the adductor or semimembranosus where these muscles are exposed at the mid-line of a split carcass. The primary sensor is a microdis-placement transducer that measures the amount of muscle extension produced by the rigorometer lever. The original prototype also included a strain gauge to measure the force with which the muscle fasciculus was loaded. This enabled the operational characteristics of the system to be examined – an important point since the force generated by a solenoid is a curvilinear function of the position of the core within the coil. The bias produced by this system was not a seri-ous problem in obtaining information on the loss of muscle extensibility. An important point is that different histo-chemical types of muscles fibres probably go into rigor mor-tis at different rates so that the output of a rigorometer system is merely the summation of changes in a heterogeneous population of muscle fibres (Swatland, 1984).

An alternative system operated pneumatically was developed to study the development of rigor mortis in deeper parts of the musculature. An expansion bulb of thin rubber is inserted inside the muscle mass and is inflated at a set rate. The amount of expansion of the bulb is measured from the amount of movement of a fluid-filled section of the inflation system. The key element of the design is to keep the internal volume of the inflation system as low and as constant as possible by using thick-walled, narrow-bore Teflon tubing. The interpretation of the results is aided by monitoring the relationship between pneumatic pressure and the amount of expansion of the bulb. The results produced by the system require some subjective analysis because of the microstructural damage produced by the bulb. Essentially what happens is that, when initial ATP levels are high, maximum expansion of the bulb occurs since the muscle mass offers little resistance to internal deformation. As rigor progresses and muscles slowly become more rigid, then the amount of expansion becomes limited. At this point, however, the expansion bulb begins to split the muscle mass in which it is embedded, presumably by tearing the connective tissues that bind the stiffening muscle fasciculi. This allows the expansion bulb to inflate to almost its original volume before the whole sequence is repeated again. Hence, with this system, the development of rigor appears as a saw-tooth pattern on the graphics display of the microcomputer.

Biophysical Methods

Meat is composed of essentially cylindrical muscle fibres separated by intercellular spaces containing connective tissues. The plasma membranes of living muscle fibres are extremely effective insulators that electrically isolate the electrolytes between and within cells. Within the muscle fibres the sarcoplasmic reticulum might also partition the internal electrolytes, although there is little evidence for this from electron microscopy. As far as can be ascertained structurally, the sarcoplasmic reticulum soon degenerates from a tightly ordered system into a loose jumble of distended vesicles. This may largely be a result of the tremendous increase in the volume of intermyofibrillar area that occurs postmortem (Figure 3).

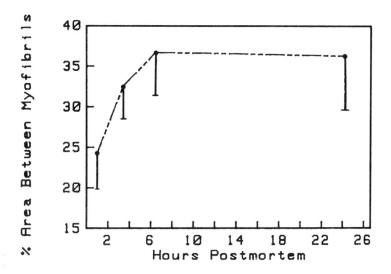

Fig. 3. Increase in area between myofibrils measured on electronmicrographs taken transversely through muscle fibres from M longissimus dorsi (bars = SD).

The fluid that fills the intermyofibrillar space originates from between the filaments within the myofibrils. Hence, the data shown in Figure 3 are matched by an increase in the packing of filaments per unit of cross section. Thus, when an alternating current is passed through a piece of muscle taken from a pork carcass soon after slaughter, the capacitance that is detected probably originates from the plasma membrane.

Three types of capacitance measurements have been used with food myosystems:

Parallel capacitance (Swatland, 1980)

Quality factor $Q = 2 \pi f Cp Rp$ (Jason and Lees, 1971)

Dissipation factor $D = 1 / Q$ (Kleibel et al., 1983)

where f = frequency, Cp = capacitance in parallel, and Rp = resistance in parallel.

Two major technical factors in biophysical measurements of pork quality are the electrical anisotropy of meat and temperature. Anisotropy exists when a measurement is affected by the direction in which it is made through the meat. Although anisotropy and temperature do affect capacitance measurements to some extent, their effect is far smaller on capacitance than on resistance. Thus, although capacitance measured in parallel with resistance is somewhat more difficult to measure than a compound ratio such as Q or D, capacitance measurements are not strongly biased by the direction in which the meat is measured or by meat temperature. All three types of measurement are affected by the frequency of the AC test current and by the geometry of the electrode configuration (for example, parallel needle electrodes versus parallel plate electrodes). With substances such as meat that are seldom microstructurally homogeneous

it may be difficult to standardize measuring systems so that
the data can be corrected to provide parameters that are
independent of the local measuring conditions. For example,
at certain times soon after slaughter, membrane capacitance
may be in parallel with a high resistance due to the transient flux of fluids into the intracellular compartment. In
this situation, a higher capacitance would be measured even
if the real capacitance of membranes was constant. Hence,
it is difficult to prove that capacitance per se has
increased in such situations.

At any time within one or two days of slaughter, PSE
pork has a lower capacitance and a lower resistance than
normal pork (Figure 4).

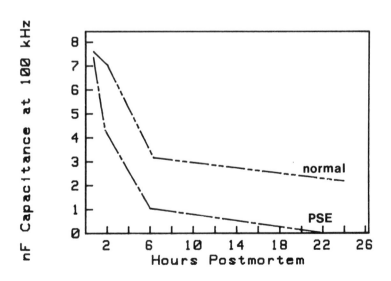

Fig. 4. Electrical capacitance of normal and PSE pork.

In PSE pork the membranes leak electrically because they are

154

damaged and there is more intercellular fluid. The
reduction of membrane capacitance was originally attributed
to lactate-induced membrane damage but recent research indicates that the loss of membrane capacitance is more closely
related to ATP depletion than to pH (Swatland and Dutson,
1984). A relationship between ATP depletion and loss of
capacitance can be demonstrated in real-time studies by NMR
(nuclear magnetic resonance), as shown in Figure 5.

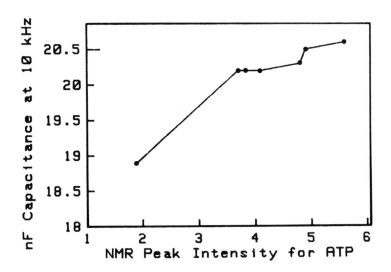

Fig. 5. Relationship between decline in electrical
capacitance and decrease in ATP concentration measured
in a parallel muscle strip by NMR spectrometry.

However, with both systems (electrodes in a piece of meat
versus a muscle core in an NMR tube) it is important to note
that the results obtained are for a heterogeneous population
of muscle fibres.

Biophysical methods such as the remote methods

described above provide a robust means of collecting data in a commercial environment. In research in Ontario, the correlations of capacitance with objective measurements of pork quality have always been slightly stronger than correlations of pH with objective measurements of pork quality. This still leaves a lot to be desired since pH correlations are not reliable enough for grading or for predictive uses. The advantage with biophysical measurements, however, is that they may be made much more easily than pH measurements. Glass pH electrodes are rather fragile and must be kept clean and constantly recalibrated. Stainless steel electrodes for biophysical measurements are extremely sturdy. Biophysical measurements may be taken in either of two modes. In a commercial environment measurements are made with portable apparatus with the investigator being able to keep pace with the line speed of the abattoir. In a research abattoir a group of carcasses may all be wired up to a central recording system that at programmed intervals measures the capacitance of the meat. This enables real-time longitudinal correlations of capacitance with a variety of other parameters such as ATP depletion assessed by NMR or loss of extensibility (Figure 6).

Fibre optics

Optical fibres that conduct light by means of total internal reflections at the junction between their core and their cladding are ideally suited to measuring the paleness or light-scattering that is usually associated with PSE.

156

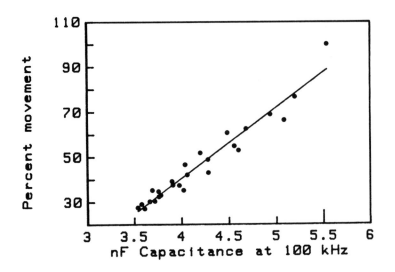

Fig. 6.   Relationship between decline in electrical
capacitance and the   onset of rigor mortis measured by
loss of muscle extensibility.

Reflectance from the meat is measured with a bifurcated
light guide.   One branch of the Y-shaped system takes light
into the meat, the other branch takes light reflected from
the meat to a photometer when the common trunk is pushed
into the meat.   An extra feature that may be added to the
basic Y-shaped light guide is to locate a multiple splice in
the main trunk so that it splits into a number of roots,
each with optical fibres connected to the main in- and out-
going branches.   A five-way split in the main trunk gives a
probe that integrates the reflectance from five different
areas simultaneously.

     There are some obvious advantages to dealing with a
parameter that is subjectively visible and of extreme judge-

mental importance. Although it is the loss of water holding or binding capacity in PSE pork that is the major commercial problem, subjective judgements of the severity of the PSE condition are strongly influenced by paleness.

A number of different fibre-optic configurations have been used for different purposes. A key factor in all configurations, however, is to use the strongest illumination possible. The ideal is to illuminate a large volume of tissue (> 25 cm^3) so that local variations are integrated and ambient illumination is unimportant. Since it is undesireable to keep switching high voltage illuminators on and off, some type of electronic shutter may be needed to protect the meat between measurements. For monochromatic measurements it is relatively easy to place both the illuminator and the photometer close to the carcass with a hard-wired link to the microcomputer that controls the shutter and data collection. With spectral scans it is usually necessary to have a long fibre-optic linkage to connect the carcass to the laboratory.

An alternative configuration, which is probably the most elegant solution, is to use a photodiode array. The photodiode array, as its name indicates, is a whole set of miniature photometers arranged in rows across the surface of a single chip. Thus, an image formed on the surface of the chip can be digitized almost instantaneously and converted to a television signal or loaded into a microcomputer data file. If a small prism is placed just in front of the photodiode array, incoming light from the fibre-optic light guide forms a spectral rainbow across the photodiode array so that the whole spectrum can be measured at once. High intensity illumination may then be obtained from an electronic xenon flash that is synchronized with the photodiode array. Thus, the full analytical power of an accurate reflectance spectrophotometer is now available for use under industrial conditions. The unit that we use at Guelph is called a Colormet and is manufactured by Instrumar (St. John's, Newfoundland). It is battery-operated and water-resistant and the unit is about the size of a domestic electric drill. The photodiode array system makes it possible to collect data at a number of wavelengths. Wavelengths between 600 and 700 nm are generally correlated most strongly with objective measurements of pork quality. It is also possible to compensate for intrinsic differences in myoglobin concentration in the pork by taking a reading at 555 nm. Figure 7 shows the results of typical experiment undertaken in a commercial abattoir. The objective was to determine if improved pre-slaughter handling methods would have a detectable effect on pork quality. The lower line shows the reduced reflectance (darker pork) that was obtained in the specially treated pigs.

Fibre optics may be used for more than simply measuring meat paleness. For example, glycogen has an absorbance peak at about 410 nm that can be measured with quartz optical fibres. In a typical biochemical experiment where a series

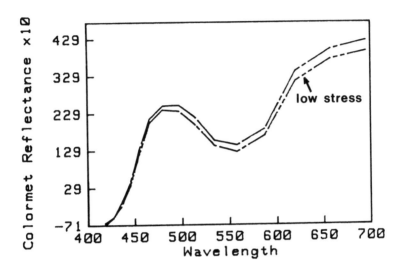

Fig. 7. Reduction of reflectance from pork as a result of reduced preslaughter stress.

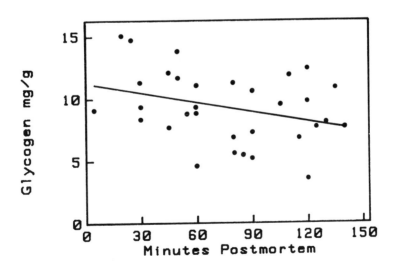

Fig. 8. Apparent variability in the rate of postmortem glycogenolysis in longissimus dorsi muscles, probably due to sampling and experimental errors (n = 8 pigs).

of samples is taken postmortem and assayed for glycogen, there is a considerable degree of variation between rates of glycogenolysis in different parts of the longissimus dorsi (Figure 8). Even the removal of the first sample may stimulate glycogenolysis in the surrounding part of the muscle. With fibre optics, a difference spectrum can be computed at any time postmortem (spectrum at zero time minus spectrum at time x) so that glycogenolysis can be followed in real-time at one exact location with a minimum of experimental interference (Figure 9). Fibre optics may also be used to measure pH by placing phenaphthazine indicator paper over the end of the common trunk of the light guide (Swatland, 1985b).

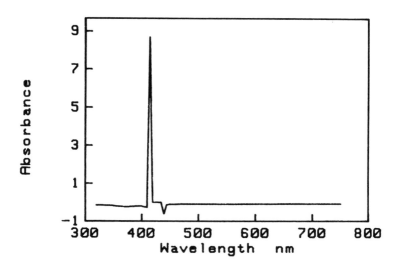

Fig. 9. Glycogenolysis measured remotely by fibre optics at a single site on the longissimus dorsi muscle. The absorbance spectrum at 165 minutes postmortem was subtracted from the absorbance spectrum at 15 minutes.

DISCUSSION

There are many possibilities for remotely monitoring postmortem metabolism in pork carcasses and thereby gaining new insights into the complex events that occur as muscles are converted to meat. Hopefully, one or more of these methods may be sufficiently robust to be incorporated into regular commercial use. Sooner or later the meat industry will make full use of the possibilities provided by computer-operated and robotic systems. At that time, there may well be a need for sensors that respond to pork quality and which can make management decisions. Even with the technology that is available at present it would be possible to sort carcasses automatically into groups with a high or low incidence of PSE.

Another topic of interest, at least in Canada, is the possibility of developing some type of grading system for pork quality. This would be relatively easy at 24 hours postmortem, but it may be an unworkable dream to do it 45 minutes postmortem as carcasses come off the kill floor. The major problems are essentially biological. Since both the rate and the final extent of lactate production appear to influence the development of the PSE condition, measurements made at 45 minutes may only catch those carcasses with a rapid PSE development. Thus, some carcasses that appear to be non-PSE at 45 minutes may subsequently develop PSE by 24 hours postmortem.

An idea that comes readily to mind in attempting to

improve the predictive power of 45-minute measurements is the possibility of combining two or more radically different systems such as capacitance measurements and fibre optics. At first sight it appears that a combination of two or more measurements might give a more accurate prediction. This idea, however, is fraught with problems that must first be solved. The main problem is that the cellular aspects of postmortem metabolism that occur as muscles are converted to meat are still only poorly understood. Nearly all the research that has been undertaken on postmortem metabolism has been undertaken by conventional biochemical methods in which meat samples are first reduced to a slurry or powder. Cellular integrity is, therefore, immediately sacrificed. To make the point more clear, let us consider a working hypothesis that is under investigation at Guelph.

Three different parameters that have been followed in their real-time changes postmortem are resistivity, capacitance and fibre-optic reflectance at 650 nm. Resistivity and capacitance progressively decline postmortem while reflectance increases. However, just before these progressive changes become apparent there is often a transient change in the opposite direction. For example, reflectance may show a transient decrease before it starts its final increase. Likewise, resistivity may show a transient increase before its final decline. A possible explanation for these changes is that muscle fibres in some carcasses may initially take up water by osmosis because of their glycogenolysis. Later on, the flux of fluid is reversed so that the muscle fibres lose much of their fluid to the intercellular space. The practical importance of this phenomenon, whatever its real explanation, is that these transient peaks are not all synchronous and that they are scattered through the first hour postmortem during the period in which we wish to make 45-minute measurements. The only hope of being able to overcome this problem is to understand more about it.

REFERENCES

Bate-Smith, E.C. 1939. Changes in elasticity of mammalian muscle undergoing rigor mortis. J. Physiol., 96, 176-193.

Jason, A.C. and Lees, A. 1971. Estimation of fish freshness by dielectric measurement. Department of Trade and Industry, Torry Research Station.

Kleibel, A., Pfutzner, H. and Krause, E. 1983. Measurement of dielectric loss factor. A routine method of recognizing PSE muscle. Fleischwirtschaft, 63, 1183-1185.

Swatland, H.J. 1976a. Electromyography of electrically stunned pigs and excised bovine muscle. J. Anim. Sci., 42, 838-844.

Swatland, H.J. 1976b. An electrocorticographic study of necrobiosis in the brains of electrically stunned and exsanguinated pigs. J. Anim. Sci., 43, 577-582.

Swatland, H.J. 1980. Postmortem changes in electrical capacitance and resistivity of pork. J. Anim. Sci., 51, 1108-1112.

Swatland, H.J. 1983. Measurement of electrical stunning, rate of exsanguination, and reflex activity of pigs in an abattoir. Can. Inst. Food Sci. Technol. J., 16, 35-38.

Swatland, H.J. 1984. Structure and Development of Meat Animals. (Prentice-Hall, New Jersey).

Swatland, H.J. 1985a. Development of rigor mortis in intact sides of pork measured with a portable rigorometer. J. Anim. Sci., 61, 882-886.

Swatland, H.J. 1985b. Optical and electronic methods of measuring pH and other predictors of meat quality in pork carcasses. J. Anim. Sci. 61, 887-891.

Swatland, H.J., Brogna, R.J. and Lutte, G.H. 1984. Electrical activity in the cerebral hemispheres of electrically stunned pigs. J. Anim. Sci., 58, 68-74.

Swatland, H.J. and Dutson, T.R. 1984. Postmortem changes in some optical, electrical and biochemical properties of electrically stimulated beef carcasses. Can. J. Anim. Sci., 64, 45-51.

THE RELATIONSHIP BETWEEN SLAUGHTERLINE AND 24-HOUR MEASUREMENT OF PIG MEAT COLOUR AND LIGHT SCATTERING BY THE USE OF HENNESSY GRADING AND FIBRE OPTIC PROBES

K. Lundström*, I. Hansson*, G. Bjärstorp**

*Department of Animal Breeding and Genetics
Swedish University of Agricultural Sciences
S-750 07 Uppsala, Sweden
**Swedish Meat Research Institute
P.O.B. 504, S-244 00 Kävlinge, Sweden

ABSTRACT

The Hennessy Grading System is now used in Sweden for the commercial grading of meat content. An experiment was performed to discover if the Hennessy Grading System could also be used to predict the final meat quality from slaughterline measurements.

The longissimus dorsi muscle was measured between the 3rd and 4th last ribs, immediately after the ordinary grading, about 30 min after exsanguination, using a Hennessy Grading Probe (GP) equipped with an extra software function to print the reflectance value. In addition, the Fibre Optic Probe (FOP) was used at the same incision as the GP. The day after slaughter, the carcasses suspected to be PSE were remeasured together with a sample of the remaining carcasses using both the GP and the FOP.

For recordings made at the same time the correlations between the GP and FOP values were high (r=0.85 and 0.88). A proportion of carcasses that were judged to be of normal quality at grading developed PSE the day after slaughter, causing non-linear relationships between measurements taken at different times.

Irrespective of the lack of accuracy the information about the meat quality obtained from grading may be useful, and the possible applications are discussed. The postponement of the ordinary grading for meat quantity would give a better measure of the meat quality which may also be a practical solution.

INTRODUCTION

Since April 1984 all pig carcasses in Sweden have been classified according to meat content by use of the Hennessy Grading System. The measuring probe uses the difference between reflected light from muscle and fat to identify the borders between the different tissues. In the last few years, it has become more desirable to improve also the quality of meat. Both the export and domestic markets have demands for a better meat quality. Some leading cutting plants want to have a classification also for meat quality and not only for meat quantity with a corresponding price differentiation. The most common method for meat quality estimations in the slaughterline, pH_1, can not routinely be performed with a speed of

120-200 carcasses per hour. With new equipment now available on the market, meat quality can also be evaluated by internal reflectance or light scatter without the need for a cut surface. However, several questions such as the time, anatomical site for measuring, measuring principle and equipment remain unsolved. If the instruments used for grading could also be used for meat quality assessment the cost would be reasonable. The reflectance function in the Fat-o-Meater grading system has been used at grading to assess the ultimate meat quality (Sack et al., 1984).

The ultimate meat quality has, however, not been achieved at 30 to 45 min. post mortem as pointed out by e.g. Barton-Gade (1981). This will cause non-linear relationships between slaughterline measurements and the final meat quality, with a certain proportion of carcasses developing PSE quality later. With a new Danish meat quality probe, about 50 per cent of carcasses with severe PSE after cooling were found at the slaughterline (Andersen, 1984). Seidler et al (1984a,b,c) have tried to solve the problem of late developing PSE by provoking the muscle to reach its final quality by using the Fat-o-Meater probe 5 min. before measuring with a modified Testron MS-tester in the same incision.

The purpose of this investigation was to determine if the Hennessy Grading Probe could be used for predicting the final meat quality from slaughterline measurements.

MATERIAL AND METHODS

Animals

The animals used in the study were commercial pigs slaughtered at one slaughterhouse in the southeast of Sweden during one day in May, 1985. The plant has two slaughterlines, with a line speed of 200 pigs per hour. Most of the carcasses from one of the lines were measured at the time of grading about 30 min. post mortem. The carcasses that from the measuring results were suspected to be PSE at the time of grading were sorted out and remeasured on the next day. In addition, a sample of carcasses not suspected to be PSE, were also measured the day after slaughter (see Table 1 for numbers measured).

Equipment

Commercial grading was performed with the Hennessy Grading System (Hennessy & Chong Ltd., Auckland, New Zealand). The grading includes two measurements of backfat thickness over M. longissimus dorsi; one at the tip of the last rib (f_1) and the second 12 cm cranially (f_2) between the 3rd and 4th (3/4) last ribs (Fig. 1). At this point also the thickness of the longissimus dorsi muscle is recorded.

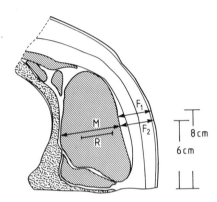

Fig. 1. Cross-section of the back at 3/4 last rib with measurements for grading indicated. The reflectance value is registered within the section denoted by R.

The meat percent of the carcasses is estimated by an equation, in which these three measurements are included. More details about the grading system as used in Sweden is given by Hansson and Andersson (1984). The carcasses were measured once more using a similar Grading Probe (GP) but with an extra software function included to get the reflectance value from the longissimus dorsi muscle at the 3/4 last rib displayed and printed out. The reflectance value (570 nm) was calculated as the mean of 5 recordings per mm when the GP passes the longissimus muscle (see Fig. 1).

Evaluation of meat quality was also made with the Fibre Optic Probe (FOP; TBL, Leeds, UK). This instrument determines the back-scatter of light transmitted into the meat. The recorded values for PSE-muscles are higher than for normal meat (McDougall and Jones, 1981). The same incision as for the GP was used, and only one recording was performed on each occasion.

RESULTS

The number of carcasses measured at the grading occasion and the day after slaughter with the instruments used is given in Table 1 together with overall means and standard deviations.

TABLE 1   General description of the meat quality measurements in *M. longissimus dorsi* at the time of grading and the day after slaughter (subscript $_1$ and $_2$, respectively).

Trait		Mean	Standard deviation
Slaughterline measurements	n=986		
GP_1		46.7	20.4
FOP_1		10.3	14.6
24-hour measurements	n=435		
GP_2		104.8	36.8
FOP_2		50.6	24.0

Due to the slaughterhouse routines, all carcasses were not available for measuring the day after slaughter. We, therefore, chose to measure the carcasses that were suspected to be PSE or almost PSE at the time of grading and only a sample of the rest. Of the carcasses measured the day after slaughter the proportion of PSE carcasses was thus too high.

The distributions for the instruments used at the time of grading were quite skewed independent of the instrument used, with only relatively few carcasses showing high values. The day after slaughter, the distributions show an overrepresentation of high values depending on the selection system.

The relationships between the variables are shown graphically in Figures 2 to 4 and as correlation coefficients in Table 2.

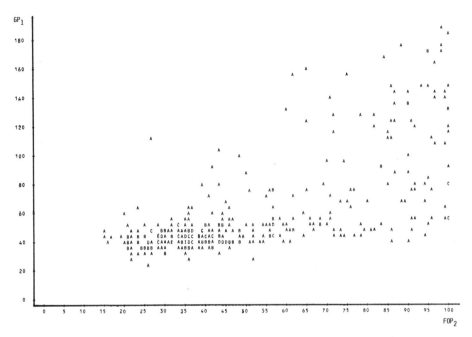

Fig. 2. Relationship between Hennessy Grading Probe values (GP) the day of slaughter and FOP values the day after slaughter measured at the 3/4 last rib in M.longissimus dorsi. A: 1 observation; B: 2 identical observations, etc.

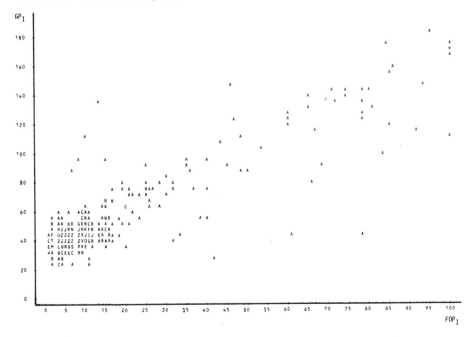

Fig. 3. Relationship between Hennessy Grading Probe values (GP) and FOP values the day of slaughter (see Fig. 2 for abbreviations).

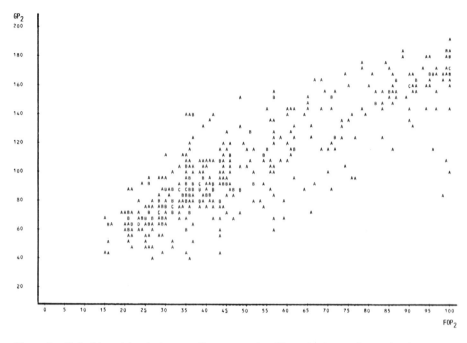

Fig. 4. Relationship between Hennessy Grading Probe values (GP) and FOP-values the day after slaughter (see Fig. 2 for abbreviations).

TABLE 2   Correlations between the meat quality measurements (all correlations are significant, P≤0.001).

	FOP$_1$	GP$_2$	FOP$_2$
GP$_1$	0.88	0.65	0.68
FOP$_1$	-	0.65	0.65
GP$_2$		-	0.85

The correlations were quite high between measurements on the same occasion, and somewhat lower between values recorded the day of slaughter and the day after slaughter. There were very few carcasses that had high reflectance or back scatter values at the grading occasion, without having high values also the day after slaughter as shown in Fig. 2. In contrast, quite many carcasses developed PSE quality after grading. As mentioned earlier, this will lead to non-linear relationships between measurements taken at different occasions. This must be considered when studying the correlations.

DISCUSSION

The variation along the longissimus dorsi muscle, as discussed by Lundström & Malmfors (1985), makes one spot measurements of meat quality uncertain. It is, however, possible to use the GP recordings both at the last rib and 3/4 last rib, to give a better accuracy. Other weak points with probe readings for the prediction of meat quality is the influence of muscle pigment as well as the amount of intramuscular fat (Barton-Gade & Olsen, 1984). The GP and FOP use both different wavelengths and different principles for measuring. The fairly good consistency between the GP and FOP values (Figures 3 and 4) obtained when measuring at the same time seems promising.

When meat quality assessments are made on the slaughter line at the time of grading to sort out the carcasses with less good quality, the thresholds for selection have to be discussed. The accuracy needed in the estimation of ultimate meat quality depends on how the information is used. We regard the following applications as the most essential.

1. A measurement of the level of meat quality at the slaughterhouse, used by the production manager for taking measures against the problem.
2. An estimate of the final meat quality of individual carcasses, used for sorting out carcasses for different ways of cutting and processing.
3. A measure of the final meat quality of the carcasses, used as base for differentiation in payment.

Even if not all carcasses with less good quality are identified the first step is applicable. The information can be used as a quick feed back on the stress situation on the farm and when transporting pigs to the slaughterhouse or during lairage, and steps can be taken to solve the problem.

The last two fields of application require a more accurate estimate than the first one. When used for price setting the estimate must be correct in most of the cases. When used for sorting, a resorting step can be performed. The consequences of false predictions of the ultimate meat quality then depend on the practical situation at the slaughterhouse and the above described fields of application. Carcasses identified as potentially of low quality at grading and that do not subsequently develop PSE (false positives) can be sorted out just prior to or during cutting, either subjectively or with the use of an instrument. The carcasses that are late in developing PSE characteristics are difficult to detect at the time of grading (false negatives).If 20 to 50 percent are not found at the time of grading (the number depending of the thresholds chosen) sorting without re-examination of the carcasses the day after slaughter will be too uncertain. Re-examination will increase the cost of quality control.

One solution to the problem of false predictions might be a local stimulation of the muscle either electrically or mechanically in order to selectively increase the rate of glycolysis. The approach was used by Seidler et al. (1984a,b,c) in order to increase the dielectric loss factor before measuring with a modified MS-tester, but has not been tested to provoke the development of the ultimate colour as far as we know. Another approach to the problem would be to postpone the ordinary grading for meat quantity to a time that is suitable for measuring meat quality. Two practical possibilities are available, either directly after the fast cooler (e.g. within 3 hours after killing) or next day at the time of delivering the carcasses from the cooling room. The first occasion suggested seems most interesting, but needs further investigation.

ACKNOWLEDGEMENTS

The Co-operative Slaughterhouse and Food Industry in Kristianstad is gratefully acknowledged for allowing us to measure carcasses irrespective of the trouble we caused. The kind cooperation by Mr. Sven Henningsson, and the technical assistance by Ms. Britt-Marie Möller and Mr. Göran Larsson are greatly appreciated. Financial support from Food Technology Company, Sundbyberg is gratefully acknowledged.

REFERENCES

Andersen, I.-L. 1984. Some experience with the portable Danish probe for the measurement of pig meat quality. In "Proc. Scientific Meeting on Biophysical PSE-Muscle Analysis" (Ed. H. Pfützner). (Vienna Technical University, Vienna). pp. 173-191.
Barton-Gade, P. 1981. The measurement of meat quality in pigs post-mortem. In "Proc. Porcine Stress and Meat Quality - causes and possible solutions to the problems" (Ed. T. Frøystein, E. Slinde and N. Standal). (Agricultural Food Research Society, Ås, Norway). pp. 205-218.
Barton-Gade, P. and Olsen, E. 1984. The relationship between water holding capacity and measurements carried out with the automatic Danish meat quality probe. In "Proc. Scientific Meeting on Biophysical PSE-Muscle Analysis" (Ed. H. Pfützner). (Vienna Technical University, Vienna). pp. 192-202.

Hansson, I. and Andersson, K. 1984. Pig carcase assessment in grading and breeding. Proc. 30th European Meeting of Meat Research Workers. (Bristol). pp. 31-32.

Lundström, K. and Malmfors, G. 1985. Variation in light scattering and water holding capacity along the porcine longissimus dorsi muscle. Meat Science. In press.

MacDougall, D.B. and Jones, S.J. 1981. Translucency and colour defects of dark-cutting meat and their detection. In "The problem of dark cutting in beef" (Ed. D.E. Hood and P.V. Tarrant). Current Topics in Veterinary Medicine and Animal Science 10, 328-339. (Martinus Nijhoff Publishers, The Hague.)

Sack, E., Fischer, K., von Canstein, B. and Scheper, J. 1984. Beziehungen des FOM-Reflexionswertes zu Merkmalen der Fleischbeschaffenheit. In "Proc. Scientific Meeting on Biophysical PSE-Muscle Analysis" (Ed. H. Pfützner). (Vienna Technical University, Vienna). pp. 304-310.

Seidler, D., Bartnick, E., Nowak, B., Gelhausen, A. and Kuhfuss, R. 1984a. PSE-Erfassung mit modifiziertem "MS-Tester" und FOM-Gerät in der Schlachtlinie. In "Proc. Scientific Meeting on Biophysical PSE-Muscle Analysis" (Ed. H. Pfützner). (Vienna Technical University, Vienna). pp. 203-220.

Seidler, D., Nowak, B., Bartnick, E. and Weisemann, M. 1984b. PSE-Diagnostik am Schlachtband. Teil 1: Untersuchungen über ausgewählte Parameter in der 1. Stunde post mortem. Fleischwirtschaft 64, 1379-1387.

Seidler, D., Nowak, B., Bartnick, E. and Huesmann, M. 1984c. PSE-diagnostik am Schlachtband. Teil 2: Gegenüberstellung der diagnostischen Möglichkeiten mit einem modifizierten MS-tester und dem FOM-Gerät auf der Basis biochemisch und strukturell orientierter Fleischbeschaffenheitsparameter. Fleischwirtschaft 64, 1499-1505.

PSE DETECTION USING A MODIFIED MS TESTER COMPARED WITH OTHER
MEASUREMENTS OF MEAT QUALITY ON THE SLAUGHTERLINE

D Seidler, E Bartnick, B Nowak

Fachbereich Lebensmitteltechnologie
Laboratorium Rohstoffkunde (Tier)
Fachhochschule Lippe
4920 Lemgo 1, Germany

ABSTRACT

Measurements of meat quality were made on 3028 pig carcasses. Data
were analysed to determine the relationship of the different measurements
to each other and their usefulness for the routine identification of meat
quality. Regressions showed that the relationships between the traits
measured were generally non linear. For routine measurements on the
slaughterline three methods were considered. Measurement of reflectance
values using the Fat-O-Meater (FOM) had the advantage that meat quality was
determined simultaneously with carcass grading. Measurement of electrical
conductivity in the muscle using the LF instrument type LF DIG 550 gave
the same degree of accuracy for assessment of meat quality as the FOM
reflectance value but required additional expenditure on equipment and
personnel. Measurement of the complex electrical conductivity using the
MS Tester (Lemgo modification) was clearly superior for discriminating
between PSE and non PSE material.

INTRODUCTION

The quality of lean meat is a very important factor in classifying

the quality of a carcass. This factor is of special significance when it

affects the subsequent uses of the meat. PSE muscle is the most

important quality defect in pork in the Federal Republic of Germany.

Genetically conditioned and activated by stress factors associated with the

transport and slaughtering procedure, it leads to a rapid accumulation of

acid products of metabolism in the musculature. The acidosis is associated

with a rise in temperature and may be attributed to anaerobic glycolysis.

Simultaneously, microstructural changes in the muscle tissue arise

that may be interpreted as an accelerated necrobiosis in the form of a

degenerative oedema. Degenerative changes are detectable in proteins,

membranes and more complex structures such as capillary beds. Due to the

nature of these changes there is a continuum between the normal and

definitely pathological conditions, giving rise to the difficulty of

finding an objective demarcation between PSE and meat of normal quality.

Under industrial conditions it is important to diagnose the defect

as early as possible, before the carcasses are transferred to the refrigerators. The two main approaches based on the pathogenesis of the condition are as follows:

measurement of the rapid postmortem biochemical changes associated with anaerobic glycolysis;

measurement of structural alterations in proteins and membranes, where swelling and osmosis must be taken into consideration.

MATERIALS AND METHODS

Three series of tests were performed on a total of 3028 left sides of carcasses.

Series A n = 1453 (December 83 to February 84)

Series B n = 1117 (July 84 to August 84)

Series C n = 458 (March 85 to April 85)

The objective was to develop a practical method of high precision for identification of PSE pork for use on the slaughterline.

Carcasses were taken from the slaughterline either individually or in groups of up to four, by random sampling. Measurements were taken unhurriedly on a parallel track because special importance was attributed to the exact collection of data. Further measurements were taken after 17 to 24 hours from some of the carcasses (series B, n = 1117) in addition to the data taken ca. 45 minutes post mortem.

The characteristics of meat was examined in M. longissimus dorsi by the following methods, listed in the sequence in which they were applied to the carcass.

1. Brightness (Rf value) was measured using the FOM reflectance value (Fa. SFK, Hvidovre, Denmark, FOM grading probe) at about   45 min post mortem.

2. Complex conductivity was measured using a modified MS Tester (Fa. Testron Meßgeräte GmbH, Vienna, Austria) in combination with a digital voltmeter, type DMN 6010 (Fa. BEWA, Holzkirchen, Germany). Measurements were made in pretreated tissue at about 45 min post mortem (D_1V) and in series B also at 17 to 24 h post mortem ($D_{24}V$).

3. Temperature (T_s) was measured using a Technotherm thermometer type 7600/7700 Pt 100 (Fa. Dewert, Bünde, Germany) at 45 minutes and 17 to 24 h post mortem.

4. pH was measured using a digital pH meter, Portamess type 651 (Fa. Knick, Berlin, Germany) and an electrode type 406-M6-KN (Fa. Ingold, Frankfurt, Germany) at about 45 minutes post mortem (pH_1) and in

series B also at 17 to 24 h post mortem (pH_{24}).

5. Water holding capacity was measured by the press method (Grau and Hamm 1953), with compressors model Braunschweig (Fa. Hauptner, Solingen Germany) tested for pressure. Samples were prepared on the slaughterline and measured in the laboratory according to Hofmann et al. (1982) at about 45 minutes post mortem (WHC_1) and in series B also at 17 to 24 h post mortem.

6. Measurement of electrical conductivity using an LF instrument type LF DIG 550 (Fa. WIW, Weilheim, Germany) was at 45 minutes post mortem (LF_1).

The data were statistically analysed to estimate the relative suitability of these methods and to determine limiting values for normal and PSE meat. The material was suitable for these analyses since PSE and non PSE carcasses were nearly equally represented in the total sample.

RESULTS AND DISCUSSION
Frequency polygons

Parameters differentiating between PSE and non PSE should form a bimodal curve in the frequency polygon. The polygon should be formed by a substantial amount of data. The frequency polygon for pH_1 values (n = 3028) complies with these demands (Fig 1). The pH_{24} value (n = 1117) does not meet the demand of bimodal distribution (Fig 2) and is therefore unsuitable for diagnosis of PSE.

The frequency polygon of the WHC_1 value approaches a normal distribution (Fig 3). Consequently discrimination by means of this meat structure parameter is less precise than by using the pH_1 value. The frequency polygon for WHC_{24} is nearly a normal distribution and therefore does not comply with the demand for biomodal distribution either (Fig 4). This also applied to other measurements made on the cold carcass, as is shown below.

The meat structure-related value, D_1 V, showed a bimodal distribution (Fig 5) similar to that of the pH_1 data. The FOM data (values of colour brightness, closely connected with the structural characteristics of the meat) were distributed in a totally different mode in the frequency polygon (Fig 6). At 45 minutes post mortem the curve was not bimodal. Presumably the structure characteristics of this value were not yet differentiated or were overlaid by other characteristics (Fig 6).

178

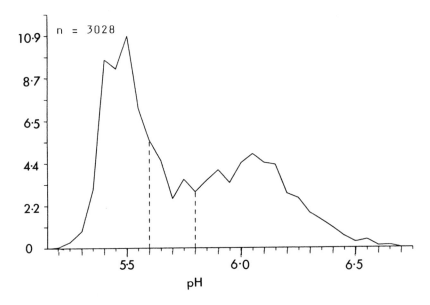

Fig. 1   Frequency polygon of the pH_1 values

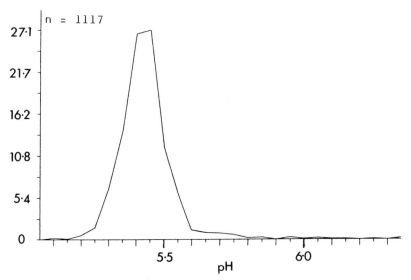

Fig. 2   Frequency polygon of the pH_{24} values

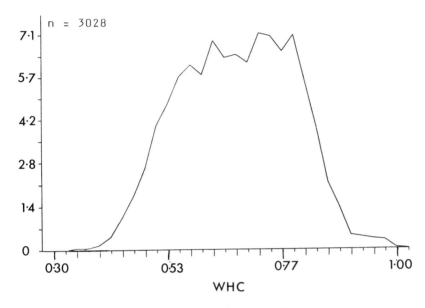

Fig. 3   Frequency polygon of WHC_1 values

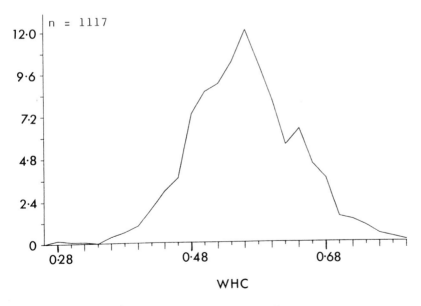

Fig. 4   Frequency polygon WHC_{24} values

180

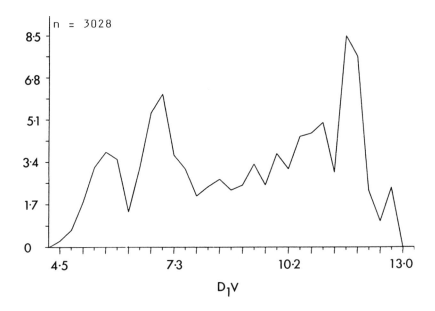

Fig. 5   Frequency polygon of the D_1V (complex conductivity) values

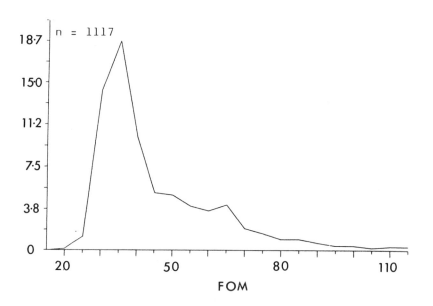

Fig. 6   Frequency polygon of the FOM reflectance values

Although the complex conductivity (including the dielectric loss factor) showed a distinct bimodality measured in pretreated tissue at 45 minutes post mortem with the MS Tester (Lemgo modification) (Fig. 5), the conductivity with the LF instrument, type LF DIG 550, was still considerably undifferentiated at that time. Without prealteration of muscle tissue and dielectric loss factor the frequency polygon demonstrated that simple conductivity measured using the LF instrument was insufficient for PSE identification (Fig. 7). An approximately normal distribution was found at 24 h post mortem. Therefore normal and PSE type material could not be differentiated by these parameters.

Relationship between meat quality measurements made at about 45 minutes post mortem

Firstly, the pH_1, a biochemical value, and WHC_1, a structure parameter, were compared. In addition, regressions of pH_1 and WHC_1 against the other other parameters are shown in Figures 8 to 12. Curvilinear functions describe the relationships more correctly than linear ones. Relationships described more correctly by a linear function are those between pH_1 and temperature (M. longissimus dorsi, M. adductor) ($r = 0.57$), WHC_1 and temperature ($r = 0.63$), WHC_1 and D_1V value ($r = 0.87$) and WHC_1 and LF_1 value ($r = 0.64$).

Multiple regression analysis

For multiple regression analysis of meat quality measurements showing curvilinear relationships it was necessary to prelinearise the data. This analysis (Table 1) showed that for the parameters pH_1, WHC_1 and temperature in M. longissimus dorsi the D_1V value measured using the modified MS Tester was superior ($R^2 = 0.83$) to the FOM reflectance value ($R^2 = 0.70$). A corresponding comparison with the LF instrument could not be made as these data belong to a different series of measurements. The distribution in the frequency polygon already showed that the LF_1 values were less useful than the complex conductivity (D_1V).

Limiting values for differentiating between PSE and non PSE material

Statistical analysis of the bimodal frequency polygon for pH_1 gave good agreement with the limiting values arrived at empirically:

Empirical	Mathematical/Statistical
$pH_1 \leq 5.60$ PSE	$pH_1 \leq 5.55$ PSE
$pH_1 > 5.80$ non PSE	$pH_1 > 5.90$ non PSE

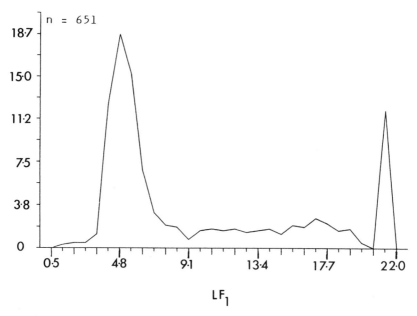

n = 651

LF_1

Fig. 7   Frequency polygon of the LF_1 (electrical conductivity) values

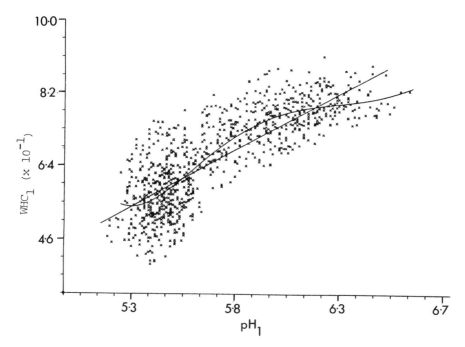

WHC_1 ($\times 10^{-1}$)

pH_1

Fig. 8   The relationship between pH_1 and WHC_1 was curvilinear (r = 0.76)

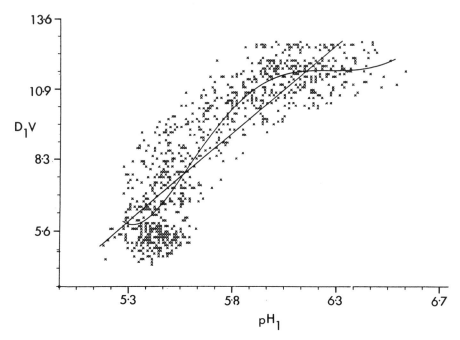

Fig. 9   The relationship between pH_1 and D_1V was curvilinear
(r = 0.89)

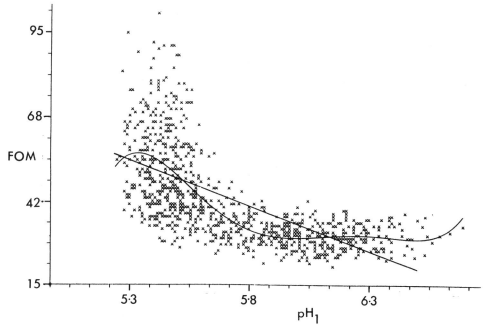

Fig. 10   The relationship between pH_1 and FOM reflectance
value was curvilinear (r = 0.72)

184

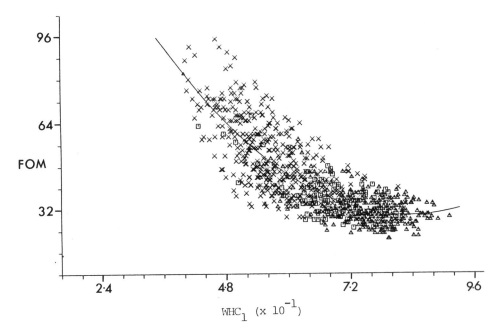

WHC$_1$ (x 10^{-1})

Fig. 11   The relationship between WHC$_1$ and FOM reflectance
value was curvilinear (r = 0.83)

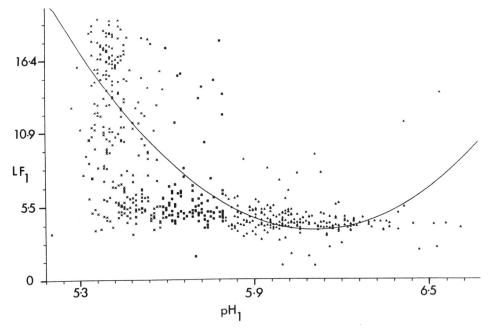

pH$_1$

Fig. 12   The relationship between pH$_1$ and LF$_1$ was curvilinear
          (r = 0.67)

As well as PSE and non PSE categories there exists a defined transition field.

TABLE 1   Multiple regression analysis of meat quality measurements

	Linearised	
	R	R^2
pH_1/WHC_1, T	0.77	0.59
WHC_1/pH_1, T	0.77	0.59
D_1V/pH_1, WHC_1, T	0.91	0.83
FOM Rf*/pH_1, WHC_1, T	0.84	0.71

* Fat-O-Meater reflectance value

The pH_1 value may be regarded as a biochemical parameter. The WHC_1 value (Grau and Hamm 1959) according to Hofmann et al (1982) represents a structure parameter for which until now there was no mathematically fixed limit between PSE and non PSE material. A further structure parameter was consulted to get a value independent of pH_1. The relationship between WHC_1 and D_1V can be adequately described by a linear function. Using a third degree polynom in the comparison an inflection was obtained. This inflection in the function of the two structure-related parameters WHC_1 and D_1V marks the limit between PSE and non PSE material and is fixed independently of the pH_1 value.

In series A (n = 980) the WHC_1 limiting value for PSE was 0.64
In series B (n = 1117) the WHC_1 limiting value for PSE was 0.65
In series C (n = 458) the WHC_1 limiting value for PSE was 0.64

For quality control purposes these limits may be entered in the curvilinear function of WHC_1 against pH_1 (Fig 8). The limit 0.64 falls on pH_1 5.55 thereby confirming indirectly the pH_1 limiting value for PSE.

For differentiating between PSE and non PSE material on the slaughterline the pH_1 and WHC_1 values are not practical. However, alternative methods must correlate as closely as possible with these two values.

Traits recorded at 24 h post mortem were less well differentiated than those collected in the first hour post mortem. As the distributions in the frequency polygons demonstrate measurements on cold carcasses did not clearly discriminate between PSE and non PSE material.

Geometric delimitation using statistical data

The geometric delimitation of PSE from non PSE material is only possible on the basis of the pH_1 value. A transition field $5.6 < pH_1 \leqslant 5.8$ (5.9) separated the PSE and non PSE subsamples, defining them more clearly and reducing dispersion. The overlap area of the double standard deviations of the two subsamples, which runs horizontally through the bar charts (Fig 13 and 14), is another uncertainty area for classification in addition to the pH_1 region from 5.6 to 5.8 (5.9). The superiority of D_1V compared with FOM values can be seen in the bar charts by comparing the overlap areas. The limits indicated by a polynom approximation on the respective frequency polygons are also confirmed here (Fig 13 and 14).

An attempt was made to find graphic limits for PSE and non PSE material from the LF_1 value (series B). It is evident in Figure 15 that the LF_1 values for non PSE carcasses fall completely within the double standard deviation of the PSE subsample. As a consequence no definite LF_1 limits for non PSE carcasses can be determined. In series B, no LF_1 values below 0.98 (non PSE area) were found.

Discriminant analysis

For comparative reasons data from series B (n = 651) were used for this analysis. The number of carcasses measured for LF_1 was the limiting factor. A standardisation of the discriminant coefficients (D_1V and FOM values for the PSE or for the non PSE groups) demonstrated that the result of the discriminant analysis was determined by the D_1V value (96.2%) more than by the FOM reflectance value (3.8%).

The complex conductivity (MS Tester) was exchanged for the simple conductivity (LF instrument, type LF DIG 550) and the latter combined with the FOM reflectance value. A standardisation of the discriminant coefficients demonstrated that allocation by means of discriminant analysis was determined by the LF_1 value (81.35%) more than by the FOM reflectance value (18.65%).

Standardisation of discriminant coefficients for D_1V and LF_1 values measured in PSE or in non PSE material ($pH_1 \leqslant 5.6$ and $pH_1 > 5.8$ respectively) demonstrated that 95.5% of the results were obtained by the D_1V value and only 4.5% by the LF_1 value.

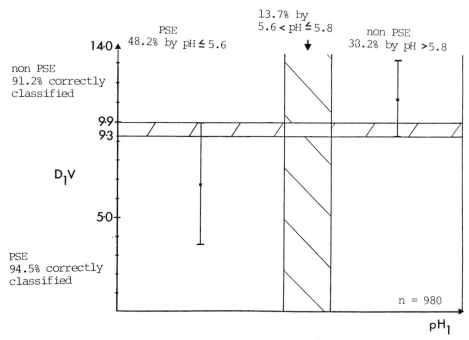

Fig. 13   Bar chart showing D₁V values (means ± 2 SD) for samples classified as PSE or non PSE on the basis of pH₁ values. The overlap region (horizontal shaded zone) indicates uncertain quality.

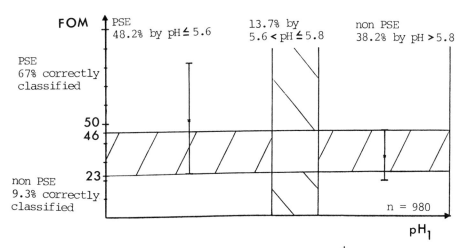

Fig. 14   Bar chart of FOM reflectance values (means ± 2 SD) for samples classified on the basis of pH₁ values. Values in the overlap region (horizontal shaded zone) indicate uncertain meat quality.

188

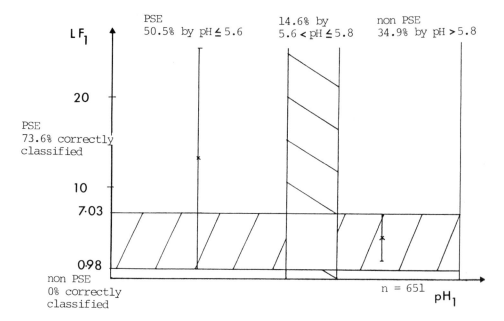

Fig. 15   Bar chart for LF_1 values for samples classified by pH_1 measurements. Values in the overlap region (horizontal shaded zone) represent uncertain meat quality.

CONCLUSIONS

1. Statistical evaluations of measurements for differentiating between PSE and non PSE material should be made on a sufficiently high number of data.

2. The distribution of PSE and non PSE material in the population should be balanced.

3. Parameters chosen for classifying PSE and non PSE material should demonstrate a bimodal curve in the frequency polygon.

4. The pH_1 limits for identifying PSE and non PSE material were confirmed by the statistical results.

5. Independently of pH_1, a limiting value between PSE and non PSE material for water holding capacity was fixed using a mathematical function.

6. There was good agreement between pH_1 and WHC limit values.

7. A curvilinear function described the relationships between most of the traits measured more correctly than a linear one.

8. Traits measured in cold carcasses were inadequate for detecting PSE because they did not meet the demand for bimodal distribution in the frequency polygon.

9. Taking pH_1 values on the slaughterline is not practical. Three alternatives were tested. The FOM reflectance value measuring colour brightness and the LF_1 value measuring electrical conductivity were found to be inadequate. The complex conductivity measured by the MS Tester (Lemgo modification) at 45 minutes post mortem in pretreated M. longissimus dorsi gave the closest fit to pH_1 and WHC_1 data.

REFERENCES

Barton-Gade, P. and Olsen, E.V. 1984. The relationship between water holding capacity and measurements carried out with the automatic Danish meat quality probe. Proc. Scient. Meeting Biophysical PSE-Muscle Analysis, (Ed. H. Pfützner) (Technical University, Vienna) pp 192-202.

Blendl, H.M. and Puff, H. 1978. Kennwerte für eine objektive Erfassung von Fleischqualitätsmängeln beim Schwein. Fleischwirtsch., 10, 1702-1704.

v. Gils, J.H.J. and v. Logtestijn, J.G. 1967. Die pH-Messung des Fleisches. Hilfsmethoden zur Beurteilung des Fleisches von Schlachttieren. Arch. Lebensmittelhyg., 18, 49-56.

Grau, R. and Hamm, R. 1953. Eine einfache Methode zur Bestimmung der Wasserbindung im Muskel. Naturwissenschaften, 40, 29-30.

Hofmann, K., Hamm, R. and Blüchel, E. 1982. Neues über die Bestimmung der Wasserbindung des Fleisches mit Hilfe der Filterpreßmethode. Fleischwirtsch., 62, 87-94.

Honikel, K.O. 1984. Einfluß der Temperatur post mortem auf das Safthaltevermögen von Schweinefleisch. Proc. Scient. Meeting Biophysical PSE-Muscle Analysis (Ed H. Pfützner) (Technical University, Vienna) pp 38-39.

Kallweit, E. 1984. Korrelation zwischen Ergebnissen verschiedener Messungen der Fleischbeschaffenheit. Proc. Scient. Meeting Biophysical Pse-Muscle Analysis, (Ed H. Pfützner) (Technical University, Vienna) pp 270-283.

Kleibel, A. Pfützner, H. and Krause, E. 1983. Messung des dielektrischen Verlustfaktors. Eine im Routinebetrieb anwendbare Methode zur Erkennung von PSE-Muskeln. Fleischwirtsch., 63, 322-328.

Pfützner, H., Fialik, E., Krause, E., Kleibel, A. and Hopferwieser, W. 1981. Routine detection of PSE- Muscle by dielectric measurements. 27th European Meeting of Meat Research Workers, Vienna, p 50.

Pfützner, H., Kleibel, A. and Fialik, E. 1985. PSE-Muskeln. Interpretation des dielektrischen Verlustfaktors. Fleischwirtsch., 65, 46-49.

Ristic, M. 1984. Methoden zur objektiven Beurteilung der Fleischbeschaffenheit. Fleischwirtsch., 64, 1340-1350.

Sack, E., Fischer, K. and Scheper, J. 1983. Vorläufige Untersuchungen zur Aussage des FOM-Reflektionswertes über die Fleischbeschaffenheit bei Schweinehälften. Mitt. bl. der Bundesanstalt für Fleischforschung, No 82, 5687-5689.

Sack, E., Fischer, K., von Canstein, B. and Scheper, J. 1984.

190

Beziehungen des FOM-Reflexionswertes zu Merkmalen der Fleischbeschaffenheit. Proc. Scient. Meeting Biophysical PSE-Muscle Analysis,
(Ed H. Pfützner) (Technical University, Vienna) pp 304-310.

Schmitten, F., Schepers, K.H., Jüngst, H. et al. 1984. Fleischqualität
beim Schwein, Untersuchungen zu deren Erfassung. Fleischwirtsch.,
64, 1238-1242.

Schwörer, D. Blum, J.K. and Rebsamen, A. 1984. Beziehungen zwischen MS-und
FOP-Messwerten sowie diversen anderen Fleischbeschaffenheitsparametern.
Proc. Scient. Meeting Biophysical PSE-Muscle Analysis,
(Ed H. Pfützner) (Technical University, Vienna) pp284-295.

Seidler, D., Bartnick, E., Nowak, B., Gelhausen, A. and Kuhfuß, R. 1984.
PSE-Erfassung mit modifiziertem 'MS-Tester' und FOM-Gerät in der
Schlachtlinie. Proc. Scient. Meeting Biophysical PSE-Muscle Analysis,
(Ed H. Pfützner) (Technical University, Vienna) pp 203-220.

Seidler, D., Nowak, B., Bartnick, E. and Weisemann, M. 1984. PSE-
Diagnostik am Schalchtband. Teil 1: Untersuchungen über ausgewählte
Parameter zur PSE-Diagnostik in der 1. Stunde post mortem.
Fleischwirtsch., 64, 1379-1387.

Seidler, D., Nowak, B., Bartnick, E. and Huesmann, M. 1984. PSE-
Diagnostik am Schlachtband. Teil 2: Gegenüberstellung der
diagnostischen Möglichkeiten mit einem modifizierten MS-Tester und
dem FOM-Gerät auf der Basis biochemisch und strukturell orientierter
Fleischbeschaffenheitsparameter. Fleischwirtsch., 64, 1499-1505.

EVALUATION OF MEAT QUALITY

BY MEASUREMENT OF ELECTRICAL CONDUCTIVITY

F. Schmitten, K.-H. Schepers and A. Festerling

Institut fur Tierzuchtwissenschaft der Universität Bonn
Endenicher Allee 15, D-5300 Bonn, Germany

ABSTRACT

The suitability of electrical conductivity measurements for
determining meat quality in pigs was investigaged. Electrical
conductivity (EC) was measured between 40 and 60 min postmortem and 24 hr
postmortem in different muscles of about 3500 purebred and crossbred
female pigs at the experimental station. In addition this method was
used under practical conditions in four slaughterhouses on 17000 pigs.
The observed accuracy, precision and validity qualifies this method as a
safe and simple measurement for predicting meat quality at 40 min
postmortem and also for quality control in the purchase of chilled
carcasses at 24 hr postmortem. On the basis of the usual methods for
identification of meat quality different grades were derived for the
commercial classification of pig meat quality. The EC method has been
introduced on trial in German pig progeny testing stations.

INTRODUCTION

The marked antagonism between leanness, especially muscle

conformation, and meat quality is the result of economic preference for

extremely meaty pigs with a pronounced shape of the ham. The lack of

consideration of meat quality in the pig carcass grading system has

resulted in an increase in the incidence of meat quality defects. A

suitable method for the identification of meat quality during the grading

of pig carcasses was not available. For this reason the market was

unable to give financial incentives to the pig breeder and producer for

improving meat quality at the desired high level of leanness. Therefore

in 1980 we commenced to develop a practical method for determining meat

quality in the slaughter line by using the physical principles of the

electrical conductivity of muscle, according to the principles described

by Banfield and Callow (1935), Sye (1969) and Swatland (1980). This

paper reviewed the experimental results and their practical application in

the slaughterhouse

MATERIALS AND METHODS

About 3500 pigs of different breeds and crossbreeds at our

experimental station were used. Additionally 17000 pig carcasses were examined in a field test in four commercial slaughterhouses. Breeds of animals tested and sample sizes are presented in Table 1.

TABLE 1   Number and origin of pigs in the experiment

Breed	Sample 1	Sample 2	Sample 3
German Landrace (DL)	261	330	594
Belgian Landrace (LB)	113	203	418
Pietrain (Pi)	108	169	315
Halothane negative			
German Landrace, DL (HN)	103	130	
Pi x DL (HN)		123	541
Other crossbreeds		50	

The pigs were kept under standardized conditions of feeding, housing and slaughter. They were fattened to a final live weight of about 100 kg. The criteria used for measuring meat quality are listed in Table 2.

TABLE 2   Criteria applied for the determination of meat quality

	M. longissimus dorsi	M. semimembranosus
pH_{40} value [1]	x	x
pH_{24} value [2]	x	x
EC_{40} value [3]	x	x
EC_{24} value [4]	x	x
Meat structure tester (MST)	x	x
Göfo value (reflectance)	x	
Elrepho value (reflectance)	x	
Transmission value	x	
Firmness (Rigormeter)	x	
Visual score	x	

[1]pH_{40} = pH value at 40 min postmortem

[2]pH_{24} = pH value at 24 h postmortem

[3]EC_{40} = Electrical conductivity at 40 min postmortem

[4]EC_{24} = Electrical conductivity at 24 h postmortem

The Meat Structure Tester (MST) as originally equipped and operated did not give a precise differentiation of meat quality. By modifying this instrument to use specified steel electrodes and give a digital display of the voltages measured, very high correlations (r = 0.93) were found between the EC values and the MST voltage values. These results confirm the similar physical principles of the two methods concerning the electrical properties of muscle tissue.

The physical basis of electrical conductivity in postmortem muscle is described in detail in a working hypothesis by Swatland (1980) that explains postmortem changes in resistivity and capacitance in muscle. The intact membrane of myofibers operates like an insulator. In postmortem muscle with developing PSE characteristics the rapid decline of pH and diminishing ATP reserves damage the myofiber membrane and its insulating function is lost. The disrupted membranes allow continuity of extracellular and intracellular fluid and cause a generalized reduction of resistivity and a corresponding increase in conductivity.

Electrical conductivity was measured using the Konduktometer LF DIGI 550, operating on a frequency of 4 kHz, and subsequently using a new model LF 191 (1 kHz) (Fa. WIW, Weilheim, Germany). There is a constant difference in the measurements between the two instruments of nearly −0.7 mS for the LF 191 with the lower frequency. The probe consists of two stainless steel needles 6 mm in diameter, 70 mm in length and 15 mm apart. The needles were insulated by sleeves of 48 mm in length to ensure that the surface area of the electrodes in contact with muscle tissue was constant for every measurement.

This insulation prevented contact of the electrodes with intercostal tissue in measurements in M. longissimus dorsi of intact carcasses. For standardized measurements the full length of the electrodes was inserted into the muscle.

The high accuracy and precision of the conductivity measurements was shown by coefficients of repeatability of 1.0 in KCl standard solutions and of 0.97 in carcasses. In contrast to pH measurements, instrument handling is very simple and robust, without need of frequent calibration checks. The validity of the EC method was checked in a comprehensive study by comparison with well known standard reference methods.

RESULTS AND DISCUSSION

1. Influence of time of measurement

Postmortem changes in muscle reflect a dynamic process, therefore the optimum measuring time for differentiation of meat quality was determined.

Figure 1 shows changes over time in the EC values of meat of different quality classes. The threshold EC values for classification of meat quality in three grades (good, uncertain, deficient) were derived by discriminant analysis. In carcasses with superior meat quality (EC $\leqslant$ 5.0 mS) there was only a small increase in conductivity within the time interval of 30 to 60 min postmortem. Also these values showed little variability. The curves of the two other meat quality classes rose markedly within that time interval and showed wide variability in the measurements.

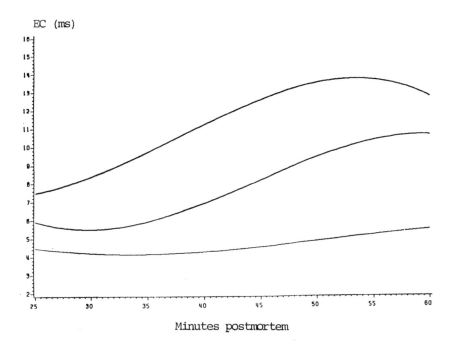

Fig. 1   Changes in electrical conductivity values as a function of time postmortem for muscles of different quality. The threshold EC_{40} values were 5.00 and below = good quality (——);   5.01 to 8.99 = uncertain quality (——);   9.00 and above = deficient quality (——).

The optimum time for differentiation was 40 or 50 minutes postmortem. Practical considerations indicate that measurement at 40 min postmortem would be convenient as that is the time of commercial carcass grading procedures in slaughterhouses. The present results are based on measurements taken at 40 min postmortem. If necessary measurements can be made at 50 or 60 min postmortem, with appropriately redefined threshold values for classification of meat quality.

Threshold values for different times postmortem derived using the present material and conditions are given in Table 3. These values may have to be adapted for individual situations. This is especially true for the EC_{24} results, which can be influenced by several factors occurring after classification, including methods of chilling and transport of carcasses, and others.

TABLE 3   Threshold values for the classification of meat quality by EC measurements at different times postmortem

Meat quality grades	Time of measurement			
	min postmortem*			hr postmortem**
	40	50	60	24
Good	≤ 5.0	≤ 5.5	≤ 6.0	≤ 8.5
Uncertain	5.01-8.99	5.51-10.49	6.01-10.99	8.51-10.49
Deficient (PSE)	≥ 9.0	≥ 10.5	≥ 11	≥ 10.5

*   Temperature of muscle $41 - 39^{o}C$
**   Temperature of muscle $4 - 6^{o}C$

2.   Influence of tissue temperature

Electrical conductivity depends generally on the temperature of the medium. In early measurements made between 30 and 60 minutes postmortem the muscle temperature in carcasses varied by about $2^{o}C$ (from $41 - 39^{o}C$). The calculated correction for the temperature effect results in maximum differences of 0.4 mS which is negligible and does not justify the use of a temperature compensation probe. The measurement of conductivity in carcasses after chilling was carried out at a lower but uniform temperature level ($4 - 6^{o}C$). Consequently in routine measurements at this time technical equipment for compensation of temperature is also not needed, if threshold values for classification of meat quality are applied under standard conditions.

3. Location of measurements

Electrical conductivity was measured in M. semimembranosus, M. gluteus medius and M. longissimus dorsi at various locations. The best results were obtained in M. longissimus dorsi in the middle region (13th/14th rib) of the back. This point of measurement gives the best indication of meat quality especially for detection of PSE in the whole carcass. Measurements in the ham showed differences in meat quality between the two sides of the carcass. This is caused by the triggering effect of unilateral shackling of the leg and elevation immediately after stunning, which leads to higher EC values in this half of the carcass.

2. Validity of methods for determining meat quality based on electrical changes in postmortem muscle

The relative frequencies of meat quality grades obtained using different methods for measuring meat quality in pig carcasses are presented in Table 4.

TABLE 4   Frequencies of meat quality grades obtained using different methods (Measurements were made in M. longissimus dorsi of 169 pig carcasses)

Methods	Meat quality grades		
	Good	Uncertain	Deficient
Meat Structure Tester	81.1	18.3	0.6
EC_{40} value	40.3	18.9	40.8
pH_{40} value	38.5	15.4	46.1
Göfo value	23.7	31.3	45.0
Visual score	25.3	24.7	50.0

Quality grading by EC_{40} was in good agreement with classification by the other reference methods. Quality identification with the original version of the Meat Structure Tester was not satisfactory. As mentioned above, some technical modifications provided results that were compatible with EC measurements.

Correlations between the different meat quality criteria indicate that EC_{40} is a suitable method for characterising meat quality (Table 5). Data for most of the quality criteria were not normally distributed. Before statistical treatment the values were transformed mathematically to eliminate kurtosis and skewness. A suitable transformation for

TABLE 5 Phenotypic correlations between meat quality criteria measured in pig M. longissimus dorsi and M. semimembranosus (n = 2244)

Criteria	M. longissimus dorsi					M. semimembranosus			
	pH_{40}	EC_{24}	Göfo	Transmission	Visual	pH_{40}	pH_{24}	EC_{40}	EC_{24}
M. longissimus dorsi									
EC_{40}	-.56	.60	-.70	.70	-.71	-.43	.04	.45	.40
pH_{40}		-.53	.56	-.52	.54	.56	.02	-.38	-.33
EC_{24}			-.66	.63	-.70	-.42	.00	.34	.54
Göfo				-.81	.84	.44	.11	-.45	-.44
Transmission					-.81	-.37	.01	.38	.44
Visual score						.43	.05	-.41	-.46
M. semimembranosus									
pH_{40}							.15	-.66	-.45
pH_{24}								-.16	-.30
EC_{40}									.45

EC_{40} values was found to be the decadic or natural logarithm.

The relationship between the EC_{40} value and the reference methods for final expression of meat quality at 24 hr is stronger than that between the pH_{40} value and these criteria. A correlation coefficient of 0.45 was found between the EC_{40} values in M. longissimus dorsi and M. semimembranosus. Consequently identification of the PSE condition in the loin muscle does not allow any definite conclusion concerning quality in the ham. The correlation between the EC_{40} value in M. longissimus dorsi and the pH_{24} in M. semimembranosus tends towards zero. This is important with respect to the identification of DFD carcasses.

In the present material and conditions of slaughter no DFD muscles were detected, using the classical limit of pH values at or above 6.2 in the ham. DFD meat occurs mainly in the ham of exhausted stress-susceptible pigs. Such material shows little or no decline in pH between 40 min and 24 hr postmortem in M. semimembranosus. Carcasses were selected with pH values in the ham of 5.9 or above at both times of measurement. It was observed that these pigs had EC_{40} values of more than 5 in M. longissimus dorsi with a tendency towards the deficient quality class.

For selection purposes we have estimated coefficients of heritability for the EC_{40} value. They are in the same range as those of the other reference criteria. The most effective way to breed for better meat quality seems to be the use of independent selection limits for meat quality criteria or with the halothane test.

The meat quality characteristics of pigs graded using EC_{40} threshold values are demonstrated in Tables 6 and 7. A definite discrimination between the different meat quality classes was achieved as shown by the other criteria used to identify meat quality. It was concluded that EC_{40} is an effective measurement for differentiation of meat quality on the slaughter line. Additionally the EC_{24} values is suitable for quality control in cold carcasses and in meat processing.

5. EC measurements under field conditions

The technical performance of the conductometer used was tested in 17000 pigs in four slaughterhouses operating at about 350 pigs per hour. The results are given in Figure 2 as a frequency distribution of meat yield and quality grades.

For the routine application of EC measurements in the slaughterline

TABLE 6   Meat quality criteria in pig carcasses that were graded by the electrical conductivity value at 40 minutes after slaughter (EC_{40}) in M. longissimus dorsi.   Results are presented as mean values and standard deviation (n = 2244)

Criteria	Good ≤5.0 (28.4%)		Uncertain 5.01 - 8.99 (26.0%)		Deficient ≥9.0 (45.6%)	
	x̄	s	x̄	s	x̄	s
pH48	6.02	0.34	5.68	0.26	5.58	0.21
Gö48	62.3	8.94	48.9	8.32	44.5	6.70
Transmission	25.8	19.49	57.1	24.15	73.0	20.92
Visual score	6.9	2.37	3.7	1.94	2.6	1.50

TABLE 7   Meat quality criteria in pig carcasses graded by the electrical conductivity value at 24 hours postmortem (EC_{24}) in M. longissimus dorsi (n = 2244)

Criteria	Good ≤8.5 (22.4%)		Uncertain 8.51 - 10.49 (30.8%)		Deficient ≥10.5 (46.8%)	
	x̄	s	x̄	s	x̄	s
pH48	6.00	0.37	5.72	0.29	5.61	0.24
Gö48	62.3	10.01	50.7	8.85	45.2	7.61
Transmission	27.7	23.97	54.7	25.97	69.3	23.20
Visual score	7.0	2.57	3.7	2.11	2.9	1.75

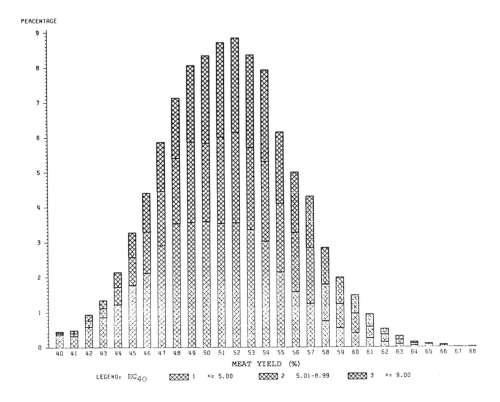

Fig 2.   Frequency distribution of meat yield and meat quality grades of commercially slaughtered pigs (n = 17000).

further technical equipment for data processing combined with carcass grading is desirable. The final aim in pig production is best served by achieving the optimum payment for pigs of the required leanness and good quality meat.

REFERENCES

Banfield, F.H. and Callow, E.H. 1935. The electrical resistance of pork and bacon. 3 parts, J. Soc. Chem. Ind., 13, 411 ff.
Hubbers, B. 1983. Untersuchungen über Zusammenhänge zwischen biochemischen Kriterien des Lipidstoffwechsels und der Schlachtkörperqualität sowie Strebanfälligkeit beim Schwein unter besonderer Berücksichtigung der Leitfähigkeitsmessung zer Erfassung der Fleischbeschaffenheit. Diss. Bonn.
Schmitten, F., Schepers, K.-H., Jüngst, H., Reul, U. und Festerling, A. 1984. Fleischqualität beim Schwein. Untersuchungen zu deren Erfassung. Fleischwirtschaft 64, 1238-1242.
Swatland, H.J. 1980. Postmortem changes in electrical capacitance and resistivity in pork. J. Anim. Sci. 51, 1108-1112.
Sye, Y.-S. 1969. Untersuchungen über die Fleischbeschaffenheitsmangeler-scheinung mittels Leitfähigkeit unter besonderer Berücksichtigung der Fleischqualität beim Schwein. Diss. Wien.

THE MEASUREMENT OF LIGHT SCATTERING PROPERTIES AT 45 min POST MORTEM FOR
PREDICTION OF PORK QUALITY

P.G. van der Wal, H. Nijeboer, G.S.M. Merkus

Research Institute for Animal Production "Schoonoord",
P.O. Box 501, 3700 AM Zeist, The Netherlands

ABSTRACT

Meat quality measurements were carried out on 123 slaughter pigs at
45 min post mortem. The measurements consisted of determinations
of light scattering with carcass grading equipment, i.e. the Danish Fat-
o-Meat'er (FOM) and the New Zealand Hennessy Grading Probe 2 (HGP), and
the Fibre Optic Probe (FOP), in combination with pH measurements in the
lumbar region of M. longissimus and M. semimembranosus. Additionally, meat quality was evaluated visually and by tristimulus colour measurements on M. longissimus lumborum at 24 h post mortem.

The mutual correlations between the various meat quality characteristics were quite low, despite their high levels of significance. The low
correlations could be explained by the small variation in muscle colour
that was present in the material under study. It was concluded that measurements of light scattering have limitations, only a portion of the PSE
carcasses could be detected by using these measurements. Because of the
small variation in muscle colour we found it impossible to clearly distinguish accuracy differences among the three methods studied. A suggestion was made that for predicting pork quality, reflectance measurements
may be improved by a shift of the measuring position from the thoracic to
the lumbar region of M. longissimus.

INTRODUCTION

The Fibre Optic Probe (FOP) can be used as an alternative to pH measurements in the detection of PSE and DFD meat (MacDougall, 1984). This conclusion was based on ultimate FOP measurements 24 h post mortem (p.m.) in
relation to pH values 45 min and 24 h p.m. Somers and Tarrant (1984) already indicated that FOP values 45 min p.m. were significantly correlated
with panel scores for meat quality at two days post slaughter. These findings were in agreement with those on Dutch slaughter-pigs (Van der Wal et
al., 1985).

A recent development is the incorporation of meat quality, determined
using reflectance measurements, in the evaluation of carcass quality.
It has been shown that the Danish grading probe FOM (Fat-o-Meat'er) provides reliable information about the PSE condition of the carcass (Sack et
al., 1984). The New Zealand Hennessy grading probe (HGP2) is currently also
suitable for meat quality measurements. Therefore, it was decided to measure the meat's light scattering with both types of grading equipment and

the FOP at slaughter, and to compare the data with an instrumental colour measurement together with a visual evaluation of pork quality based on wetness, colour and texture, carried out at 24 h post mortem.

MATERIAL AND METHODS

Reflectance measurements with the FOM (Fat-o-Meat'er; SFK, DK 2650 Hvidovre, Denmark), HGP (Hennessy Grading Probe (GP2); Hennessy and Chong Ltd., Auckland, New Zealand) and FOP (Fibre Optic Probe; TBL Fibres Ltd., Leeds LS10 1AU, England) were performed in a commercial slaughter house on   M. longissimus from the left side of 123 pork carcasses at approximately 45 min p.m.

The FOM measurements were carried out in   M. longissimus between the 3rd and 4th lumbar vertebrae (3/4LV) 8 cm from the dorsal midline (FOM-1; FOP-1) and between the 3rd- and 4th-from-last ribs (3/4LR) 6 cm from the dorsal midline (FOM-2; HGP-2; FOP-3). FOM reflectance measurements were performed at position FOM-2. The first HGP measurement was made at the last rib (LR) 8 cm from the dorsal midline (HGP-1; FOP-2), while the second measurement was made at the same location as FOM-2. The HGP measured reflectance at both positions, while an overall reflectance value was also calculated (HGP-O) using both reflectance values. The FOP measurements were made at both FOM and HGP positions. Additionally, pH measurements (Schott Geräte CG 818; electrode: Schott N48A) were carried out in M. semimembranosus (SM) and in   M. longissimus lumborum (LL) at the last lumbar vertebra. All measuring positions are indicated in Fig. 1.

Carcass grading according to the CEC grading system (EEC Regulation no. 2108/70, 20th October 1970) resulted in 23 (18.7 %) carcasses grading EAA, 84 (68.3 %) 1A, 1 (0.8 %) 2A, 1 (0.8 %) 3A and 14 (11.4 %) 1B.

A visual evaluation of pork quality was carried out on the LL of the left half-carcasses at the last lumbar vertebra 24 h p.m. The visual scores, which were based on wetness, colour and texture, ranged from 1 (DFD), 2 (beginning DFD), 3 (normal), 4 (slightly aberrant), 5 (beginning PSE) to 6 (PSE). A comparison of the meat colour with a set of standard models of pork colour (Nakai et al., 1975) was made simultaneously. Furthermore, colour measurements according to the L*, a*, b* scale were performed with a tristimulus colour analyzer (Minolta Chroma II Reflectance; Minolta Camera Co. Ltd., Japan).

Mean values and standard deviations of the various characteristics were calculated per meat quality category and mutual correlation coefficients were determined. The measurements with the HGP and with the FOP have been compared within carcasses using a paired t-test (two-tailed).

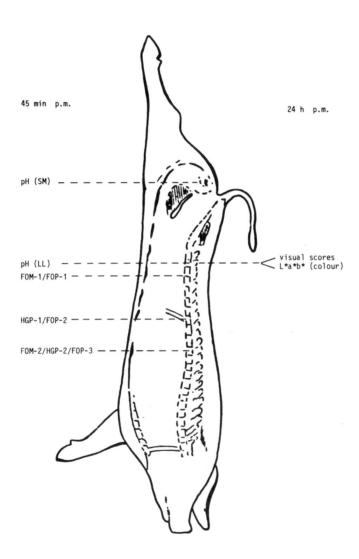

Fig. 1. Measuring positions in the carcasses at 45 min and 24 h post mortem.

RESULTS

Little variation in meat quality at 24 h p.m. was present in the material that was used in the experiment. Mean values and standard deviations of the various characteristics per visually evaluated meat quality category are given in Table 1. No DFD and only a few carcasses with PSE (3.3 %) were observed, while 'beginning DFD' (1.6 %) was scarce. 'Beginning PSE' was present in a higher percentage of the carcasses, viz. 16 %. This category reached 24 % when judged with the colour bar. The majority of the carcasses belonged to the visually evaluated categories 'normal' (24 %) or 'somewhat aberrant' (55 %). Meat quality of most of the latter carcasses showed deteriorations in wetness (94 %) and, at a lower level, in colour (51 %), while texture seldom was affected (3 %).

Based on experience and data from the literature on reflectance values and meat quality, supplemented with the data in Table 1, a division into three classes of light scattering was composed for the FOM, the HGP and the FOP at one, two and three measuring positions, respectively (Table 2). Two of the PSE carcasses, i.e. meat quality category 6, scored FOM reflectance values above 40, while the other two had values below 35. Light scattering determined with the HGP at position 1 (LR) indicated 2 PSE carcasses with a reflectance above 50, but also carcasses in the categories 'normal' (2), 'slightly aberrant' (2) and 'beginning PSE' (1). Comparable results were obtained with HGP measurements at position 2 (3/4LR). Calculated HGP reflectance values (HGP-0) above 20 indicated 1 PSE carcass, while in the reflectance class 15 to 20 two other PSE carcasses occurred together with one carcass each in categories 'normal' and 'slightly aberrant'. The FOP detected only one PSE carcass at position 1 (3/4LV), two at position 2 (LR), but none at position 3 (3/4LR), using the highest reflectance class. At the FOP positions 1 and 2, however, three non PSE carcasses were indicated. The mutual correlations between the various characteristics are given in Table 3. Despite the high levels of significance, the values were quite low.

It was observed that the HGP reflectance values at position 1 (LR) were significantly higher ($P < .05$) than at position 2 (3/4LR), viz. 39.2 ± 6.9 and 37.6 ± 6.5, respectively. The comparisons were made within carcasses. Comparable observations were made for the FOP values at the three positions, which were 107.6 ± 8.8 (3/4LV), 105.2 ± 9.4 (LR) and 102.9 ± 5.1. The differences between the values were highly significant ($P \leq .001$).

TABLE 1   Mean values ($\bar{x}$) and standard deviations (SD) of the various meat
           quality characteristics grouped according to the visual quality
           scores at 24 h p.m. The measuring positions are indicated in
           Fig. 1.

Meat quality category n[1]	2		3		4		5		6	
	2		29		68		20		4	
	$\bar{x}$	SD	$\bar{x}$	SD	$\bar{x}$	SD	$\bar{x}$	SD	$\bar{x}$	SD
45 min p.m.										
Reflectance										
FOM-2	30	1	30	3	30	3	32	5	38	6
HGP-1	40	4	39	7	38	5	42	5	49	24
HGP-2	38	1	36	7	37	6	39	5	49	12
HGP-0	10	1	9	3	9	2	10	1	18	8
FOP-1	108	4	107	11	106	6	111	10	120	11
FOP-2	106	4	104	7	104	6	107	6	133	29
FOP-3	106	2	102	5	102	5	104	4	112	7
pH										
SM	6.6	0.2	6.3	0.3	6.2	0.3	6.0	0.3	5.6	1.4
LL	6.5	0.2	6.3	0.3	6.2	0.3	6.0	0.3	5.8	0.2
24 h p.m.										
Colour bar	4.0	0.0	3.1	0.1	2.6	0.3	1.8	0.4	1.0	0.0
Tristimulus										
L*	45	7	51	3	53	3	58	3	62	4
a*	14	0	14	2	14	1	15	1	15	1
b*	6	2	8	1	8	1	10	1	13	1

[1]Meat quality: 1 (DFD), 2 (beginning DFD), 3 (normal), 4 (slightly
              aberrant), 5 (beginning PSE) and 6 (PSE)

FOP-1: 3/4LV

FOP-2, HGP-1: LR

FOP-3, HGP-2, FOM-2: 3/4LR

pH: M. semimembranosus (SM) and M. longissimus lumborum (LL)

Colour bar: 1 (PSE) to 6 (DFD)

Tristimulus colour values according to the L*a*b* system

HGP-0: calculated overall HGP reflectance value

TABLE 2 A division into light scattering classes of the carcasses belonging to the various meat quality categories.

Meat quality category[1]	2	3	4	5	6
n	2	29	68	20	4
FOM-2 ≥ 40				1	2
≥ 35, < 40		4	3	4	
< 35	2	25	65	15	2
HGP-1 ≥ 50		2	2	1	2
≥ 45, < 50		4	4	4	1
< 45	2	23	62	15	1
HGP-2 ≥ 50		1	2	1	2
≥ 45, < 50		2	3	1	
< 45	2	26	63	18	2
HGP-0 ≥ 20					1
≥ 15, < 20		1	1		2
< 15	2	28	67	20	1
FOP-1 ≥ 130		1		1	1
≥ 125, <130		2		1	
< 125	2	26	68	18	3
FOP-2 ≥ 130			1		2
≥ 120, <130				2	
< 120	2	29	67	18	2
FOP-3 ≥ 125					
≥ 120, < 125			1		
< 120	2	29	67	20	4

[1]Meat quality: 1 (DFD), 2 (beginning DFD), 3 (normal), 4 (slightly aberrant), 5 (beginning PSE) and 6 (PSE)

FOP-1: 3/4LV

FOP-2, HGP-1: LR

FOP-3, HGP-2, FOM-2: 3/4LR

HGP-0: calculated overall HGP reflectance value

TABLE 3   The correlation matrix of the quality characteristics determined on 123 pork carcasses.

	1	2	3	4	5	6	7	8	9	10	11	12	13
1. FOM-2													
2. HGP-1	.27												
3. HGP-2	.47	.33											
4. HGP-0	.57	.44	.87										
5. FOP-1	.34	.24	.32	.43									
6. FOP-2	.41	.26	.45	.71	.60								
7. FOP-3	.47	.49	.50	.64	.49	.70							
8. pH SM	.22	.34	.30	.38	.23	.35	.34						
9. pH LL	.17	.43	.25	.32	.22	.30	.33	.69					
10. quality	.29	.23	.29	.37	.24	.34	.23	.40	.39				
11. c.bar.	.30	.25	.31	.40	.27	.34	.32	.41	.36	.84			
12. L*	.28	.21	.31	.36	.23	.28	.24	.37	.33	.69	.66		
13. a*	.06	.08	.02	.47	.09	.18	.13	.26	.21	.07	.07	.24	
14. b*	.28	.20	.30	.39	.25	.40	.29	.44	.40	.66	.65	.69	.41

1-7 Reflectance values with FOM, HGP and FOP; the measuring positions are indicated in Fig. 1.

8-9 pH values in   M. semimembranosus (SM) and M. longissimus lumborum (LL)

10 Visual score for pork quality

11 Colour bar: 1 (PSE) to 6 (DFD)

12-14 Tristimulus colour values according to the L*a*b* system

$r = .18$: $P \leq .05$

$r = .23$: $P \leq .01$

$r = .29$: $P \leq .001$

The differences between the pH values that were determined in the SM and the LL were significant (P < .05); the SM's pH was lower.

DISCUSSION

A poorer meat quality corresponded with higher reflectance values and an increased lightness expressed by higher tristimulus L* values. The mutual correlations between the FOM, HGP and FOP reflectance values were quite low compared to other experiments with FOM/HGP and FOP equipment (Lundström et al., 1985; Van der Wal and Nijeboer, 1985). Comparable low correlations were found between reflectance values and the visual meat quality scores and tristimulus colour measurements. An explanation for the poor correlations is probably due to the lack of sufficient carcasses with a DFD or PSE appearance. The colour of 8 (6.5 %) carcasses was evaluated as very pale, while a somewhat dark colour was found in only 2 (1.6 %) carcasses. The majority of the carcasses scored normal for colour (50.4 %) or slightly aberrant (41.5 %). Failures in meat quality not only depend on colour; they can also be the result of excessive wetness and loose texture. In our study most of the carcasses in meat quality category 'slightly aberrant' (cat. 4) and all 'beginning PSE' and PSE carcasses, being 71.5 % of the total material, showed deteriorations in wetness. However, wetness is not necessarily detected directly with equipment specificially designed for measuring the scatter of light. The failure to measure wetness, which contributes to the visual quality score is another explanation for the low correlations between the reflectance values and the quality scores. Despite the poor correlations between the various reflectance values and the other characteristics, most of them were highly significant. It can be concluded that measurements of light scattering have limitations, especially when the variation in muscle colour is limited. This also can be deduced from Table 2. Only a portion of the PSE carcasses could be detected by using the higher reflectance classes. The limits for the three reflectance classes, however, were arbitrary. They have to be changed when more information from other experiments conducted under slaughterhouse conditions becomes available. Based on the data presented above it is not possible to clearly distinguish accuracy differences in meat quality determinations by reflectance measurements among the three methods that were studied.

Thw HGP and FOP measurements confirmed that meat quality, indicated by colour in particular, is not constant along M. longissimus. The cranial part of the muscle is less susceptible to an aberrant quality than the lumbar region (Lundström et al., 1984; Van der Wal, 1985). These data suggest that for predicting pork quality, reflectance measurements may be improved by a shift in the measuring positions, i.e. from the thoracic to the lumbar region of M. longissimus.

REFERENCES

Lundström, K., Bjärstorp, G. and Malmfors, G., 1984. Comparison of different methods for evaluation of pig meat quality. Proc. Scient. Meeting 'Biophysical PSE-muscle analysis' (Ed. H. Pfützner) (Technical University, Vienna) pp 311-322.

Lundström, K., Hansson, I. and Bjärstorp, G., 1985. The relationship between slaughter-line and 24-hour measurement of pig meat colour and light scattering by the use of Hennessy grading and fibre optic probe. In: Proc. E C Seminar 'Evaluation and control of meat quality in pigs', Dublin, Nov, 1985.

MacDougall, D.B., 1984. 'Detection of PSE-meat by the MRI Fibre Optic Probe (FOP). Proc. Scient. Meeting 'Biophysical PSE-muscle analysis' (Ed. H. Pfützner) (Technical University, Vienna) pp 162-168.

Nakai, H., Saito, F., Ikeda, T., Ando, S. and Komatsu, A., 1975. Standard models of pork-colour. Bull. Nat. Inst. Animal Industry, 29, 69-74.

Sack, E., Fischer, K., Canstein, B. von and Scheper, J., 1984. Relationship between meat characteristics and FOM-reflection values. Proc. Scient. Meeting 'Biophysical PSE-muscle analysis' (Ed. H. Pfützner) (Technical University, Vienna) pp 304-310.

Somers, C., Tarrant, P.V. and Sherington, J., 1984. Evaluation of some objective methods for measuring pork quality. Proc. Scient. Meeting 'Biophysical PSE- muscle analysis' (Ed. H. Pfützner) (Technical University, Vienna) pp 230-242.

Wal, P.G. van der, 1985. Fibre Optic Probe (reflectie)-waarden in ham en karbonadestreng. Vleesdistr. & Vleestechnol., 20(6), 16-19.

Wal, P.G. van der, Eikelenboom, G. and Custers, J.P., 1985. Detectie van PSE in varkenskarkassen met behulp van de Fibre Optic Probe. Vleesdistr. & Vleestechnol., 20 (3), 18-19.

Wal, P.G. van der en Nijeboer, H., 1985. Metingen met classificatie-apparatuur in combinatie met lichtreflectie-metingen met de Fibre Optic Probe. Res. Inst. Anim. Prod. 'Schoonoord', Zeist. Report B-261.

EVALUATION OF MEAT QUALITY CHARACTERISTICS IN THE ITALIAN HEAVY PIG

V. Russo, P. Bosi, L. Nanni Costa

Istituto Allevamenti Zootecnici, Universita di Bologna
Villa Levi, Coviolo, 42100 Reggio Emilia, Italy

ABSTRACT

Two surveys were made on 1008 and 205 carcasses to study several meat quality traits and to assess qualitative aspects of meat from Italian heavy pigs. Measurements were made of pH_1, pH_u, water holding capacity (WHC) by Filter Paper Press (FPP) and Kapillar Volumeter (KV), drip loss and colour (Hunter $\underline{L}, \underline{a}, \underline{b}$). Values for pH_1, pH_u and FPP were taken in several muscles and muscle sites. The carcasses were subjectively classified as normal, PSE or DFD.

The incidence of PSE was 4.1 percent and very little DFD was found. Almost all of the meat quality traits were correlated with each other. Abattoir effects strongly influenced the parameters tested. Measurements of pH_1, FPP and KV only provided a rough index for PSE estimation. An advantage of pH_1 is that it can be measured more easily, quickly and cheaply than the other parameters.

Drip loss averaged 3.9 percent with a range from 1.2 to 10.2 percent and a non normal distribution. Spearman's rank correlation coefficients showed a link between drip loss and other parameters. The high correlation with the $\underline{L}$ value of the $\underline{L}, \underline{a}, \underline{b}$ system is of special interest for estimating drip loss in the early post-slaughter period using portable instruments.

INTRODUCTION

Italian pig breeding differs from that of other European countries, producing chiefly a heavy pig, all of which, except for the loin, is used for processing. The main aim of heavy pig production is to obtain typical seasoned products, in particular hams and quality salami. In Italy, then, meat quality is extremely important, not only from the point of view of nutrition and taste, but for processing technology (Russo, 1984).

Our goal was firstly to assess meat quality in a typical area of heavy pig production, the Po Valley, and secondly to determine the effect of different sources of variation including sex, genetic type, abattoir, carcass weight and slaughtering season. A third aim was to evaluate the incidence of PSE and DFD and to find out whether limit values for these conditions can be established for the normal population. To achieve these aims, a first survey assessed meat quality, pH_1, pH_u and water holding capacity (WHC) in a sample of 1008 carcasses. In a second trial, drip loss and its relationship with pH value, WHC and colour was evaluated. The effect of different sources of variation has been discussed in a previous paper (Bosi et al., 1985).

MATERIALS AND METHODS

In the first survey, 1008 heavy pig carcasses were randomly chosen, over a two-year period, from five abattoirs located in the Po Valley. Knowledge of genetic type was the only restriction. In the second survey 205 heavy carcasses from different farms in the Po Valley and slaughtered in the same abattoir, were used.

About 25 carcasses per day were examined in the first survey and about 6 per day in the second survey. Maximum transportation time from farm to abattoir was 2h. Almost all the pigs were slaughtered after a night's rest. Sex, genetic type and carcass weight were recorded. The average carcass weight was 132 ± 23 kg in the first test and 135 ± 14 kg in the second. The predominant genetic type was Large White x Landrace, but there were also Large White, Landrace and several crosses with Duroc and Spotted breeds.

In the first trial, meat quality was measured on the right side of the carcasses. The pH_1 and pH_u were taken on M. semimembranosus (SM), M. biceps femoris (BF), and M longissimus dorsi (LD). In the LD pH values were taken at the 6th rib (6r), where the loin cut ('lombata') separates from the neck cut ('coppa') and at the last rib (Lr). Two portable Digipok Radiometer pH meters fitted with Ingold probe-type electrodes were used. At 1 h post mortem a 0.5 kg sample of LD was taken from the 6th rib and four hours later, water holding capacity (WHC) was tested with a Kapillar Volumeter (KV) and by the filter paper press (FPP) method. WHC was also measured on the BF by FPP in the same area where pH values had been taken, 5 and 30 h post mortem. The KV was described by Hofmann (1975). Readings were taken at 30, 60, 90 and 120 seconds but only the 60 second data are reported here. From an operational point of view, this seems to be the best time as it was more highly correlated with the readings taken at 90 and 120 seconds than was the 30 second reading and was less time consuming than the former (Russo et al, 1982). The FPP area was determined according to Grau and Hamm (1957) with a meat sample of 0.3 g and a 1 kg weight applied for 5 minutes. The values obtained correspond to the difference between the whole wet area and the meat area. The carcasses were classified as PSE or DFD if they clearly showed traits of these conditions in the loin or ham when subjectively assessed.

In the second trial, which also aimed to study the meat quality/ quantity ratio (Russo et al, 1985), pH values of the left side were taken in the same way as those taken in the first trial, except that the cranial measurement on the LD was taken at the 7th rib (7r). WHC was measured using the FPP method on the LD at 5 h post mortem. Colour reflectance and drip loss were measured on an LD sample. Colour was measured by the L,a,b method, using a Hunterlab Colorimeter (Santoro et al., 1978). Drip losses were calculated according to the method of Honikel (personal communication) which consisted of suspending the LD sample (about 2.5 cm thick) inside a plastic bag, at 0 to 4°C for 48 hours. Drip losses were expressed as the percentage decrease in the weight of the sample compared to initial weight.

Statistical method

All variables were tested for degree of fit   to a normal distribution. Only the FPP area on the BF at 30 h post mortem and pH_1 at the 6r of the LD were normal. All the others differed significantly from normal distribution (P < 0.05). Suitable transformations were used to normalise the other variables except the FPP at 5 h post mortem and drip loss.

In the first survey an analysis of variance for the data relative to the three meat quality groups was made. Where variables were not normally distributed, transformed values were used. For FFP areas at 5 h post mortem this was not possible and consequently an analysis of variance was not done.

Overall correlations between meat quality variables were also determined. The correlations between other variables and FPP at 5 h post mortem and drip loss were made with the non-parametric test of Spearman (Snedecor and Cockran, 1980).

RESULTS

Meat quality traits for the whole sample in the first survey are reported in Table 1. The mean pH_1 values were different in the various muscles (P < 0.01) and between the two LD sites (P < 0.01). The value for the BF was the highest. The pH_u was less variable than the pH_1 in the different muscles and measurement sites. All WHC methods showed a high variability and included values equal to zero. For the FPP method at 5 h post mortem a high percentage of zero values, amounting to 23.2% for the LD and 39.2% for the BF was found. The distributions of these measurements were significantly different from normal and it was not possible to normalise them with the usual transformations.

Meat quality traits for carcasses subjectively assessed to be normal, PSE or DFD are reported in Table 2. In the whole sample there were 4.1% and 0.5% PSE and DFD carcasses respectively. The average values for pH and WHC in the three quality groups showed statistically significant differences. Similarly, significant differences were found within each quality group for the different muscles and measurement sites. In the PSE group zero values for FPP areas did not arise. Of the five DFD samples only two showed zero values for FPP areas at 5 h post mortem.

TABLE 1   Meat quality characteristics in the carcasses of Italian heavy pigs from the first survey. Measurements of pH and water holding capacity (WHC) using the Kapillar Volumeter (KV) and filter paper press (FPP) methods were carried out on M. longissimus dorsi (LD), M. semimembranosus (SM) and M. biceps femoris (BF)

	Number of carcasses	Mean	S D[*]	C V.	Range
pH_1					
LD, 6th rib	999	6.09	0.35	5.7	5.21 – 7.04
LD, last rib	915	6.27	0.32	5.1	5.23 – 7.00
SM	992	6.33	0.33	5.3	5.23 – 7.04
BF	940	6.38	0.29	4.5	5.41 – 7.09
pH_u					
LD, 6th rib	952	5.65	0.17	3.0	5.10 – 6.51
LD, last rib	917	5.68	0.19	3.3	5.10 – 6.90
SM	979	5.75	0.21	3.7	5.18 – 6.99
BF	942	5.74	0.20	3.5	5.22 – 6.70
WHC with KV					
LD, 6th rib (μl)	983	40.1	16.9	42.1	0 – 115
WHC with FPP					
LD, 6th rib (cm^2)	1000	2.46	1.89	76.8	0 – 8.89
BF, 5 h (cm^2)	426	1.22	1.49	122.1	0 – 5.59
BF, 30 h (cm^2)	963	3.17	1.15	36.3	0 – 7.42

*   Standard deviations. Values are included for variables that were not normally distributed.

Meat quality traits for the three quality groups at the five abattoirs are reported in Table 3. Three of the five DFD carcasses were found in a single abattoir. The percentage of PSE was also found to be different in the five abattoirs. Indeed, a high variability was found between abattoirs for all the parameters examined.

Threshold pH_1 and WHC values that include different percentages of PSE carcasses, and the percentage of normal carcasses included within these limits, are given in Table 4. For all the parameters considered, the percentage of normal carcasses within the threshold values that included all of the PSE carcasses was extremely high. This percentage decreased

TABLE 2   Meat quality characteristics in carcasses of Italian heavy pigs that were subjectively assessed as normal, PSE or DFD. Mean values and standard deviations are shown. The abbreviations used are explained in Table 1.

	Normal		PSE		DFD	
	Mean	S D	Mean	S D	Mean	S D
Number of carcasses	962		41		5	
Percentage	95.4		4.1		0.5	
pH_1						
LD, 6th rib	6.11	0.35	5.63	0.20	6.24	0.29
LD, last rib	6.29	0.30	5.86	0.31	6.43	0.29
SM	6.35	0.28	6.00	0.36	6.44	0.21
BF	6.39	0.28	6.06	0.32	6.44	0.24
pH_u						
LD, 6th rib	5.66	0.17	5.55	0.15	5.85	0.07
LD, last rib	5.69	0.18	5.58	0.20	6.22	0.41
SM	5.75	0.20	5.67	0.19	6.40	0.26
BF	5.74	0.20	5.67	0.20	6.24	0.23
WHC with KV						
LD, 6th rib (μl)	39.4	16.5	57.2	16.7	23.2	9.1
WHC with FPP						
- LD, 6th rib (cm^2)	2.37	1.84	4.72	1.41	2.59	1.96
- BF, 30 h (cm^2)	3.16	1.14	3.50	1.03	2.24	1.67

- All means were significantly different at $P < 0.001$ except for FPP in BF where the level of significance was 0.05 and FPP in LD where an analysis of variance was not performed.

TABLE 3 - Meat quality characteristics in the carcasses of Italian heavy pigs that were slaughtered at five different abattoirs. The abbreviations used are explained in Table 1.

	1			2			3			4			5		
	Normal	PSE	DFD	Normal	PSE	DFD	Normal	PSE	DFD	Normal	PSE	DFD	Normal	PSE	DFD
Number of carcasses	146	3		143	7	3	485	17	1	93	11		92	3	1
pH_1															
LD,6r	6.13	5.71		5.92	5.51	6.13	6.14	5.64	6.04	5.80	5.62		6.57	5.86	6.68
LD,Lr	6.24	5.80		6.33	5.67	6.51	6.24	5.85	5.98	6.26	6.11		6.53	5.73	6.66
SM	6.25	5.81		6.35	5.74	6.56	6.36	6.02	6.16	6.30	6.18		6.51	6.07	6.35
BF	6.18	5.90		6.40	5.83	6.50	6.39	6.05	6.04	6.37	6.25		6.61	6.17	6.67
pH_u															
LD,6r	5.57	5.45		5.66	5.48	5.85	5.66	5.53	5.80	5.70	5.64		5.68	5.58	5.88
LD,Lr	5.65	5.46		5.77	5.57	6.37	5.65	5.51	6.00	5.78	5.71		5.69	5.69	5.98
SM	5.73	5.58		5.77	5.56	6.49	5.71	5.59	6.22	5.89	5.85		5.79	5.74	6.40
BF	5.65	5.51		5.76	5.55	6.18	5.71	5.59	6.06	5.90	5.89		5.77	5.78	6.59
WHC with KV															
LD,6r	56.9	79.0		43.3	42.9	24.7	34.0	55.5	16.0	41.2	57.9		31.5	50.7	26.0
WHC with FPP															
LD,6r	1.53	3.26		3.47	5.85	2.99	2.24	4.27	3.96	3.89	4.96		1.07	4.49	0
BF,30 h	3.94	3.89		3.56	4.58	2.90	2.82	2.98	C.12	3.14	3.41		2.07	3.55	2.37

with a less rigorous threshold, but the possibility of classifying PSE carcasses as normal increased.

The correlation coefficients between meat quality measurements are given in Table 5. Correlations between pH_1 values were significant but not very high, pH_1 also showed significant but lower correlations with pH_u. There was a negative correlation between pH_1 and WHC at 5 h post mortem, whereas with WHC at 30 h post mortem, the correlation was lower or non-existent. The coefficients of correlation between pH_1 and WHC taken on the same muscle were higher than those taken on different muscles. The correlations between pH_1 and WHC were higher for the FPP method than for the KV method. Correlations between different pH_u measurements were very high. There was only a few, very low, significant correlations between pH_u and WHC. WHC measurements with the FPP method carried out at various times post mortem on the same muscle, were moderately correlated, whereas those taken at the same time on different muscles showed quite a good correlation. A low correlation was found between WHC values determined by the FPP and KV methods.

TABLE 4   Percentages of normal carcasses that would be misclassified as PSE, using threshold values that include different percentages of real (subjectively confirmed) PSE carcasses. The abbreviations used are explained in Table 1.

	Percentages of real PSE carcasses					
	100%		90%		80%	
	Threshold values	Percent normal carcasses	Threshold values	Percent normal carcasses	Threshold values	Percent normal carcasses
pH_1						
mean	6.35	57.3	6.20	30.3	6.10	26.5
LD,6r	6.10	49.1	5.95	33.6	5.90	29.5
LD,Lr	6.30	49.1	6.15	32.1	6.10	25.0
SM	6.70	91.4	6.50	66.6	6.35	46.2
BF	6.80	96.8	6.50	63.4	6.30	35.7
WHC with KV						
LD,6r	30	68.0	39	48.0	43	39.1
WHC with FPP						
LD,6r	2.0	57.8	3.3	35.5	3.9	24.0

TABLE 5 - Overall correlation coefficients between meat quality traits in the carcasses of Italian heavy pigs. The abbreviations used are given in Table 1.

	pH_1				pH_u				KV LD,	FPP	
	LD,6r	LD,Lr	SM	BF	LD,6r	LD,Lr	SM	BF	6r	LD,6r	BF,5 h
FPP											
BF, 30 h	-.05NS	-.04NS	-.16**	-.15**	-.12**	-.03NS	-.09**	-.11**	+.23**	+.19**	+.17**
BF, 5 h[1]	-.37**	-.44**	-.43**	-.55**	+.01NS	-.01NS	+.03NS	-.03NS	+.29**	+.46**	-
LD,6r[1]	-.63**	-.44**	-.34**	-.25**	-.21**	-.07NS	-.05NS	-.05NS	+.29**	-	
KV											
LD,6r	-.32*	-.23**	-.25**	-.25**	-.25**	-.05NS	-.05NS	-.11**	-		
pH_u											
BF	+.09**	+.16**	+.15**	+.18**	+.60**	+.66**	+.81**	-			
SM	+.09**	+.16**	+.15**	+.12**	+.60**	+.73**	-				
LD,Lr	+.15**	+.24**	+.18**	+.17**	+.79**	-					
LD,6r	+.34**	+.26**	+.26**	+.25**	-						
pH_1											
BF	+.37**	+.56**	+.68**	-							
SM	+.50**	+.68**	-								
LD,Lr	+.58**	-									

[1] Spearman's rank correlation coefficients.

*: P<0.05; **: P<0.01.

TABLE 6 - Meat quality characteristics in the carcasses of Italian heavy pigs in the second survey. The abbreviations used are explained in Table 1.

		Number of carcasses	Mean	S D [*]	C V	Range
Drip loss	(%)	205	3.91	2.14	54.7	1.07 – 10.18
WHC	(cm^2)	"	3.74	2.02	54.0	0 – 10.10
Colour	L	"	45.21	4.81	10.6	34.05 – 62.03
	a	"	8.36	1.20	14.3	4.63 – 12.82
	b	"	7.34	1.41	19.2	3.30 – 11.25
pH_1	LD,6th rib	201	6.08	0.36	5.9	5.21 – 6.95
	LD,last rib	205	6.14	0.34	5.5	5.30 – 6.88
	SM	"	6.38	0.31	4.9	5.44 – 7.42
	BF	201	6.23	0.30	4.8	5.17 – 6.89
pH_u	LD,6th rib	205	5.81	0.14	2.4	5.51 – 6.51
	LD,last rib	197	5.83	0.17	2.9	5.45 – 6.65
	SM	205	5.86	0.17	2.9	5.52 – 6.51
	BF	"	5.86	0.16	2.7	5.53 – 6.47

[*] Standard deviations are also reported for variables that had not got a normal distribution.

TABLE 7   Spearman's rank correlation coefficients between drip loss
in a slice of M. longissimus dorsi and   other measurements of meat
quality.   The abbreviations used are explained in Table 1.

			r
WHC[1]			0.783**
Colour[2]	L		0.715**
	a		0.132
	b		0.549**
pH_1	LD	6th rib	−0.571**
	LD	last rib	−0.541**
	SM		−0.328**
	BF		−0.359**
pH_u	LD	6th rib	−0.296**
	LD	last rib	−0.286**
	SM		−0.096
	BF		−0.135

[1] Water holding capacity was measured in the LD using the filter paper
press method of Grau and Hamm (1957).

[2] Colour was measured in the LD using the Hunterlab colorimeter.

** $P \leq 0.01$

Mean values and coefficients of variation for meat quality
measurements in the second survey are given in Table 6. In this trial
also, pH_1 values were different between muscles, although differences
between the two LD measurement sites were more moderate. Variability for
drip loss and FPP area was very high. Drip losses averaged 3.9% with
values ranging from 1.1% to 10.2%; the distribution tailed considerably
towards the highest values and was, as mentioned, significantly different
from normal (P < 0.01).

The coefficients of the correlation of Spearman between drip loss
and other meat quality traits are reported in Table 7. With the
exception of pH_u measurements on the ham muscles and a colour value, drip
losses were significantly correlated with the other traits (P < 0.01).
WHC measured with FPP and L colour value showed the highest correlations.

DISCUSSION

The meat quality of the heavy pig can be judged in terms of the percentage of the carcasses identified as PSE or DFD, as shown in the first survey. While DFD was very limited, the percentage of carcasses with PSE could not be ignored, although it did not reach as high an incidence as in other countries.

The incidence of PSE was in agreement with the percentage of halothane reactors found in Italian genetic control centres for boars of the Landrace (11%) and Large White (0.4%) breeds (Russo 1984). These two breeds either purebred or crossbred provide the majority of commercial heavy pigs.

Mean values of all parameters were clearly different for normal, PSE and DFD carcasses. No parameter, however, was able to discriminate with certainty between a PSE and a normal carcass. Indeed, threshold values which included all PSE samples were unhelpful. Even in the best case, that of pH_1 on the LD, they wrongly classified 49% of the normal carcasses as PSE (Table 4). With lower threshold values the number of carcasses wrongly classified as PSE would be reduced but the number of PSE carcasses considered normal would be increased. The considerable discrepancy between the number of carcasses with pH_1 values lower than the commonly accepted PSE threshold values (potentially PSE) and the number of carcasses which really present this condition may be explained by the difference in immediate post-slaughter treatment of Italian heavy pig carcasses compared to that of the light pig produced in other European countries. Within about half an hour of slaughter, the heavy pig carcass is divided into commercial cuts and the dorsal lard of the loin is removed. Consequently the muscles, in particular M longissimus dorsi, undergo more rapid cooling than occurs when the side remains whole. Even when there is rapid glycolysis, which lowers muscle pH to below normal values, more rapid cooling could prevent PSE symptoms in carcasses with a predisposition to the condition.

On the other hand, the effect of various factors acting immediately before, during and after slaughter, which are partly responsible for the variability found between abattoirs for all the parameters reported, represents a major obstacle to the definition of valid threshold values for identifying PSE. Bosi et al. (1985) found that abattoir and season of slaughter have a statistically significant effect on pH_1 in the heavy pig. This confirms previous observations in Italy (Bosi et al., 1983; Russo et al. 1983) and in the U.K. (Kempster and Cuthbertson, 1975, Smith and Wilson, 1978; Kempster et al. 1982) on the importance of the effect of these two factors and slaughtering day on meat quality attributes.

pH_1 and FPP on the LD taken at the 6th rib, as shown by the large range of variation from abattoir to abattoir, seem most influenced by slaughter conditions and particularly by the way the carcass was transported from stunning to splitting. Indeed in abattoirs 2 and 4 where the pigs were hung up from the start, mean pH_1 values were below 6, whereas in the other three abattoirs where transportation from stunning to dehairing was carried out on a conveyor belt, mean pH_1 values were above 6.

These observations show that pH_1 does not allow a precise estimate of the final meat characteristics at the slaughterline. It provides a rough estimate and is easy, fast and cheap. The best use of pH_1 is to assess batches rather than to estimate meat quality in single carcasses. These results agree with those of Evans et al. (1978) and Barton-Gade (1981). Readings of pH_1 taken on the SM and BF muscles showed less discriminatory capacity between normal and PSE carcasses than readings taken on the LD. Discriminatory capacity did not improve if the mean value of the four pH_1 values was used. Recognising the real limitations, the most suitable site for estimation of pH_1 was the LD at the level of the last rib. WHC, regardless of the method used, was correlated with pH_1. This correlation was higher when the two parameters were measured on the same muscle and at the same site, whereas the correlation was lowered when the same parameters were measured on different muscles. The correlations between pH_1 and FPP area were generally higher than those between pH_1 and KV value. This, together with the low correlation found between the two methods of WHC analysis, shows the need to better understand their mechanisms.

All things considered, WHC (however it is determined) does not improve the precision of estimating PSE in carcasses and furthermore, especially in the case of FPP, is a more difficult and lengthy procedure than pH. Finally it should be emphasised that WHC measurements are not normally distributed and often cannot be normalised with suitable transformations for statistical analysis, due to the high frequency of zero values. Consequently, non-parametric methods were used to calculate the coefficients of correlation.

The results of the second trial show that weight loss due to drip can be considerable, even in the heavy pig. On average, drip was about 4% of the muscle sample weight and varied considerably from carcass to carcass, regardless of the presence or absence of PSE. Since drip losses are directly translated into economic losses for the slaughter and processing industry, their importance in meat quality assessment is obvious. However, they can not be determined very quickly and this is clearly a disadvantage. Some significant correlations were found, particularly between drip and WHC, and drip and L colour value measured by L,a,b system. As far as WHC is concerned, the correlation found was higher than that obtained by Lundström et al. (1979) and by Somers et al. (1984), whereas the results for colour coincide with those found by Somers et al (1984) and Lundström et al (1984). The coefficients of correlation are not directly comparable, since in the present case, having ascertained that the distribution of values was statistically different from the normal one, the non-parametric method of Spearman was used. The correlation with colour appears particularly interesting as a means of estimating drip loss in the first few hours after slaughter. Meat lightness can be measured on the slaughterline using portable instruments, although the problems posed by the abnormal distribution of the values must be resolved.

REFERENCES

Barton-Gade, P.A. 1981. The measurement of meat quality in pigs post mortem. In 'Porcine stress and Meat Quality' (Ed. T. Froystein, E. Slinde, N. Standal) (Agric. Food Res. Society, Ås) pp 205-218.

Bosi, P., Casini, L. and Russo, V. 1983. Effetti della stagione sulla capacità di ritenzione idrica del suino pesante, Selez. Vet., 24, 1323-1330.

Bosi, P., Russo, V., Casini, L., Nanni Costa, L. 1985. Indagine sul pH_1 di alcuni muscoli di suino pesante. Zoot. Nutr. Anim. (in press).

Evans, D.G., Kempster, A.J. and Steane, D.E. 1978. Meat quality of British crossbred pigs. Livest. Prod. Sci., 5, 265-275.

Grau, R. and Hamm, R. 1957. Über das Wasserbindungsvermögen des Säugetiermuskels. II. Mitt. Über die Bestimmung der Wasserbindung des Muskels. Z. Lebensmittel-Untersuch. u. Forsch., 105, 446-460.

Hofmann, K. 1975. Ein neues Gerät zur Bestimmung der Wasserbindung des Fleisches: Das 'Kapillar-Volumeter'. Fleischwirtschaft, 55, 25-30.

Kempster, A.J. and Cuthbertson, A. 1975. A national survey of muscle pH values in commercial pig carcasses. J. Fd. Technol., 10, 73-80.

Kempster, A.J., Chadwick, J.P. and Evans, D.G. 1982. Trends in the incidence of pale soft exudative muscle in the national pig population. Winter Meeting of Brit. Soc. Anim. Prod., 29-31 March, Harrogate, 7.

Lundström, K., Nilsson, H. and Malmfors, B. 1979. Interrelations between meat quality characteristics in pigs. Acta Agric. Scand. Suppl. 21, 71-80.

Lundström, K., Bjärstorp, G. and Malmfors, B. 1984. Comparison of different methods for evaluation of pig meat quality. Proc. Scient. Meeting 'Biophysical PSE-Muscle Analysis', (Technical University, Wien) p. 311, (Ed. H. Pfützner).

Russo, V. 1984. Relazione tra quantità e qualità della carne nella specie suina. Implicazione per il miglioramento genetico. Atti XIX° Simp. Int. Zoot. Soc. I. Prog. Zoot., Milano, 15 Apr.

Russo, V., Casini, L., Bosi, P., Nanni Costa L. and Del Monte P. 1982. Qualità della carne del suino pesante. Impiego del Kapillar Volumeter per la determinazione del potere di ritenzione idrica. Suinicoltura 23 (10), 41-46.

Russo, V., Bosi, P. and Casini, L. 1983. Variazione stagionale del pH in alcuni muscoli del suino pesante Suinicoltura 24 (5), 45-50.

Russo, V., Nanni Costa, L., Lo Fiego, D.P., Santoro, P. 1985. Ricerche sulle perdite di sgocciolamento della carne del suino pesante. Atti VI° Conv. Naz. Assoc. Scient. Prod. Anim. Perugia, 28 May - 1 June.

Santoro, P., Rizzi, L. and Mantovani, C. 1978. Colore e contenuto di mioglobina in carni di tacchini macellati a diverse età. Zoot. Nutr. Anim., 4, 147-154.

Smith, W.C. and Wilson, L.A. 1978. A note on some factors influencing muscle pH_1 values in commercial pig carcasses. Anim. Production 26, 229-232.

Snedecor, G.W. and Cochran, W.G. 1980. Statistical Methods. 7th edition (Iowa State University Press, Ames).

Somers, C., Tarrant, P.V. 1984. Evaluation of some objective methods for measuring pork quality. Proc. Scient. Meeting 'Biophysical PSE - Muscle Analysis'. (Technical University, Vienna), (Ed. H. Pfützner), pp 230-242.

MEAT QUALITY IN PROGENY TEST PIGS IN NORTHERN IRELAND

B W Moss

Agricultural and Food Chemistry Research Division
Department of Agriculture for Northern Ireland
and
Department of Agricultural and Food Chemistry
The Queen's University of Belfast

ABSTRACT
The routine method of progeny testing pigs in N. Ireland is described and the results of additional measurements on meat quality reported. There is little evidence to suggest that in the last 10 years the selection for leanness has resulted in poorer meat quality. The results show that classification of carcases as PSE on the basis of $pH_1 < 5.9$ is likely to overestimate the problem. This overestimation is much greater in the Landrace then Large White breed. Drip loss measurements showed a poor correlation with pH_1 measurements and EEL (reflectance) values. Some pigs classified as 'non PSE' i.e. $pH_1 > 5.9$ had drip loss as great as 'PSE' i.e. $pH_1 < 5.9$ pigs. The results reported suggest that classification of carcases as PSE should be on the basis of pH_1 and colour and not pH_1 alone.

INTRODUCTION

It has been known for some time that leaner breeds of pigs tend to be more stress sensitive with a higher incidence of pale soft exudative muscle (PSE) (Jensen et al. 1967, MacDougall & Disney 1967). It is important in any breeding programme, particularly when selecting for decreased backfat and/or increased lean content, that the desired carcase characteristics are obtained without detrimentally affecting meat quality (Lister 1979). There is some evidence that in less stress sensitive strains of pigs selection for lean tissue growth or less backfat has no detrimental effect on meat quality (Standal 1979, Wood et al. 1981).

The economic loss due to PSE has been assessed in some studies (Smith & Lesser 1979, Taylor, Dant & French 1973) however a reliable figure for economic weighting in progeny test schemes is not available. The relatively small amount of information on the heritability coefficients of meat quality parameters tends to indicate these are in the low range (Tarrant et al, 1979).

Most workers classify carcases as PSE if the pH_1 is less than 6.0, however this may not be a reliable method of classification (Moss 1980, Barton-Gade 1979, Swatland 1982). Alternative methods of measuring PSE i.e. colour, water holding capacity, transmission values (Hart 1962) and protein solubility cannot be carried out until 24 h post mortem and may not be suitable for commercial use. The use of such methods in progeny test schemes may also be limited by the equipment and scientific skills required.

Background to progeny testing in Northern Ireland

Progeny testing of pigs in Northern Ireland started in 1958 with an Accredited Pig Herds Scheme, since then there have been several changes in the testing system. The present system combines progeny and performance testing and started with the revision of the scheme in 1966. Performance testing requires two littermate boars which are tested for rate of liveweight gain, food conversion and ultrasonic backfat together with 2 gilt sibs with associated carcase data (Table 1). The progeny test result for a sire combines the performance testing of 5 litter groups, each comprising 2 boars and 2 gilts, with detailed carcase assessment available for the gilts. The test is carried out from 34 to 91 kg liveweight, the two gilts being penned together and the boars penned singly. Pigs are currently fed a pelleted ration once a day an amount adjusted to liveweight. Although pigs are not fed on the morning of slaughter most pigs have a small amount of unconsumed food in their trough from the previous day's feed. Prior to 1980 pigs were meal fed twice a day and from 1980 to 1983 were fed meal once a day an amount of feed each could consume in approximately 30 minutes at each feed (the implications for meat quality are discussed later). The gilts are taken for slaughter when they reach 91 kg. The carcase measurements taken are given in Table 1.

The combined testing system is designed for selection of bacon type pigs and conditions are standardised as much as possible to ensure that any ensuing differences between pigs reflects their genetic merit. The method of preslaughter and post slaughter practices requires further considerations in relation to meat quality assessment. Gilts selected for slaughter are mixed on the lorry and transported to the abattoir as a mixed group. The numbers in a load varies from 6 to 30 pigs with space allocations during transport varying from approximately 2 to 0.45 m^2/pig. The transport journey is approximately 1 h and on arrival the pigs are penned as one group in the lairage. The pigs are slaughtered after holding in lairage for 1 to 1.5 h, using an automated restrainer shute and electrical stunning. The pigs are not fed on the morning prior to slaughter, their last feed being at 08.00 h the previous day, a lapse of 27 to 28 h from last feed to slaughter.

TABLE 1 List of carcase measurements taken routinely on progeny test pigs in N. Ireland

Parameter	Comments
Carcase length	
Backfat[1]	shoulder, middle, loin, C, K
Eye muscle area	area of cross section of M longissimus dorsi at the level of the last rib
Percentage lean in rump back	
Percentage prime cuts	percentage of weight of side in rump back, rib back and ham joints
Subjective scores	Neatness of shoulder, quality of ham, back rasher score, streak quality, general bacon type, colour score of LD, firmness of fat

[1] These measurements are taken on boars only using ultrasonic equipment.

Routine meat quality assessment of the pigs consists of initial pH (pH_1) measurements of M longissimus dorsi (LD) and M adductor (ham) taken at 35 min post mortem using a combined electrode probe (Russell CMSWL/TB). After a rapid initial chilling one side of each carcase is transported to the pig testing station in an insulated (but not refrigerated) van, these sides are then held at 1° C until the following day when ultimate pH (pH_u) measurements are taken in the LD and the Ham. EEL reflectance readings are taken on the cut surface of the LD at the mid point along the back where the carcase is cut for dissection of the rump back joint. The eye muscle area (EMA) is also taken at the mid point along the back.

In addition to the routine measurements of meat quality other methods of assessment including transmission values, expressible juice and bag drip have been undertaken for consideration into the routine scheme. In the following discussion three aspects of meat quality will be considered (1) trends in selection (2) classification of PSE carcases and (3) use of bag drip in routine testing.

RESULTS AND DISCUSSION

(1) Trends in Selection

The results in Table 2 show that in both 1975 and 1984 the difference in leanness between the two breeds is accounted for mainly by the difference in shoulder fat depth. The leaner Landrace pigs (LR) had greater EMA in 1975 but no difference in dissectable lean in the rump back joint. On the basis of Table 2 it would be difficult to account for differences in meat quality between the two breeds by differences in leanness. Examination of the trends from 1970 to 1984 (see Figure 1) shows that there has been considerable improvement in lean meat content as measured by EMA and percentage dissectable lean in rump back joint. There is very little evidence that this increased selection for leanness has resulted in increased PSE as assessed by mean pH_1 values in the LD or EEL values and is in agreement with reports of other workers (Wood, et al. 1981, Standal 1979).

Fig 1.   Trends in lean meat content and meat quality in Landrace and Large White pigs

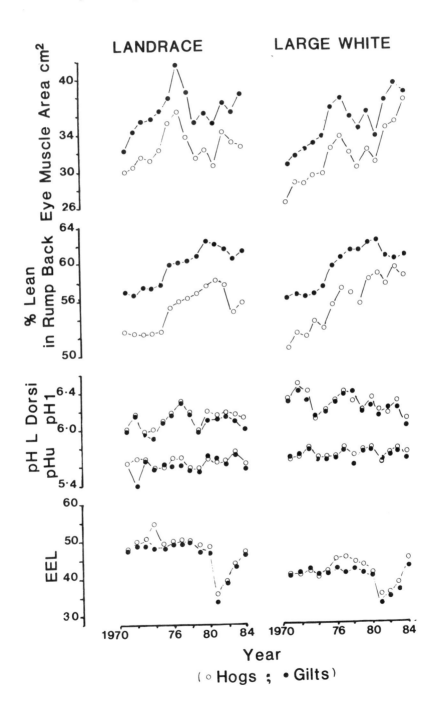

Year

(○ Hogs ; ● Gilts)

TABLE 2   A comparison of parameters associated with leanness in progeny
test pigs in 1975 and 1984

	Landrace		Large White	
	1975	1984	1975	1981
Daily gain (kg day^{-1})	0.75	0.90	0.76	0.88
Food conversion (kg feed kg gain^{-1})	3.04	2.70	2.98	2.66
Carcase length (mm)	812	813	811	793
Backfat thickness (mm)				
Shoulder	37	29	40	32
Middle	19	14	18	15
Loin	18	12	20	14
C	14	10	14	10
K	17	11	18	12
Eye muscle area (cm^2)	33.7	38.1	31.4	38.2
Percentage lean in rump back	55.2	61.05	55.3	60.7
Percentage lean meat in side	59.2	62.7	59.2	62.7
Percentage prime cuts	51.3	50.4	50.7	50.6
Percentage carccase yield	63.0	65.6	62.6	66.4

The annual trends presented in Figure 1 require more detailed
consideration since factors other than genetic pressures may contribute
to differences in meat quality. A particular example of this is the
low mean pH_1 values in 1974 in both breeds, where over 44% of LR and
20% of LW pigs had pH_1 values of 5.9 or less. This high incidence of
'PSE' may be accounted for by changes from a CO_2 stunning system to the
current electrical stunning system in the abattoir used for slaughter
of these pigs. The marked drop in mean EEL values from 1980 to 1981
may be more readily accounted for by the use of a different EEL
reflectance meter with a corresponding change in calibration procedure.

The decrease in EEL values from 1980 to 1981 cannot be satisfactorily accounted for by changes in pH_1 or pH_u values. Examining the data from 1981 to 1984 inclusive there does appear to be an increase in paleness (from 35 to 45 EEL units associated with an increase in EMA (from 33 to 38 cm^2). There does not appear to be any decrease in pH_1 values in the LD or ham over this period.

The pH_u values in LR and LW pigs are approximately 0.2 units higher than reported by other workers (Tarrant et al. 1979) and may be due to the 27 h time lapse between last feed and slaughter. Campbell (1984) showed that the change in routine from twice a day feeding with feed offerred on the morning of slaughter to once a day feeding at the pig testing station not only affected the incidence of DFD but may also increase pH_1 values due to substrate limitation on the rate of post mortem glycolysis. The current incidence of DFD (i.e. $pH_u > 6.2$ in the LD) in progeny test pigs is less than 3%. The time from last feed to slaughter may increase the aggressiveness of pigs and this may also influence the incidence of PSE.

(2) Classification of PSE carcases

Most workers classify carcases as PSE if the pH_1 values are 6.0 or less. Problems inherent in the methodology of pH measurements have been discussed by Bendall (1973), however, problems may also arise in that in commercial situations the convenience of position and conditions prevailing in the abattoir on the line may dictate the time post mortem for pH_1 measurements. The pH_1 values on progeny test pigs given in Figure 1 were taken 35 min post mortem and a cut off point of pH_1 of <5.9 was chosen. Classification of carcases on the basis of pH_1 measurements of the LD with further sub-classification on the basis of EEL values highlights the problems inherent in this method of classification (see Table 3). In 1975 the percentage of carcases classified as 'PSE' on the basis of $pH_1 < 5.9$ for LR and LW pigs was 20.2 and 11.7% respectively. The general concensus of the carcase dissectors was that only carcases with EEL values of 50 or greater actually appeared pale and watery and unacceptable. On this basis a true estimate of PSE condition in 1975 might be 15.1 and 3.1% in LR and LW respectively. The important point to note is that of these carcases classified as PSE using pH_1 values 75% of the LR were considered pale and watery (i.e. EEL>50) but only 26% of LW carcases were considered pale and watery. In 1984 the percentage of carcases with pH_1<5.9 was similar in the two breeds (22.2% in LR and 19.2% in LW). It may not be valid to make direct comparisons of 1975 and 1984 data due to differences in calibration of EEL meter mentioned previously. It is to be noted however that as in 1975 a greater proportion of LR pigs classified as PSE on the basis of pH_1 values had higher EEL values than LW carcases classified similarly. These results are in agreement with those of Martin et al. (1975) and Yang et al. (1984) who show that pH_1 only accounts for a small proportion in the variation in meat quality.

(3) Drip loss as a measure of meat quality

The drip loss was measured on a sample joint of between 400 to 600 g taken from the mid back region. The sample joint was weighed, placed in a net bag and suspended inside a polythene bag from which most of the air had been excluded. The bag containing the sample joint was then hung in a cold room at 1° C for 2 days, after which time the amount of fluid collected in the bag was determined and the joint dissected into lean, fat and bone. Drip loss was calculated as a percentage of the initial sample joint weight and also as a percentage of the dissected lean.

The mean drip loss expressed both as a percentage of chop weight and dissectable lean was significantly higher (P<0.01) in the Landrace than Large White pigs as might be expected (see Table 4). The mean drip loss of PSE carcases (pH_1≤5.9) LR pigs and 'non PSE' (pH_1>5.9) was greater than the corresponding means for LW pigs. In both breeds some carcases classified as 'non PSE' (pH_1>5.9) had drip loss as great a 'PSE' (pH_1≤5.9) (see Figure 2) and correlations of drip loss with pH_1 and EEL values were very low in both breeds. The high proportion of pale wet carcases in the 'non PSE' classification could be accounted for by two main factors 1) problems associated with pH_1 measurements on the factory line and 2) the rate of chilling of the carcases. It has been suggested that the incidence of pale wet muscle may arise in pigs which have a rapid post mortem glycolysis and those which have a

TABLE 3:   Classification of pig carcasses on the basis of pH_1, pH_u and colour (EEL reflectance values) of M. longissimus dorsi

| | Percentage of pigs in meat quality class | | | |
| | Landrace | | Large White | |
	1975	1984	1975	1984
Acceptable carcasses pH_1, >5.9 and pH_u<6.2	78.1	75.8	85.3	79.3
DFD carcasses pH_u≥6.2 with EEL[1]				
less than 30	1.1	0	1.5	0
greater than 30	0.6	2.0	1.5	1.5
PSE carcasses pH_1≤5.9 with EEL[1]	(20.2)	(22.2)	(11.7)	(19.2)
30 to 39	0.4	2.0	2.0	2.9
40 to 49	4.7	13.8	6.6	12.6
50 to 59	8.9	5.3	3.1	3.7
60 and greater	6.2	1.1	0	0
Number of pigs in sample	538	531	184	137

[1] Higher EEL values indicate paler muscles

high glycolytic potential (Monin and Sellier 1985). A proportion of pigs classified as 'PSE' in this report may have low glycolytic potential at the time of slaughter either due to the time from last feed or glycogen depletion due to their excitable nature and aggressive

TABLE 4:   Drip loss and colour (EEL reflectance values) measurements of progeny test pigs

	Landrace		Significance	Large White	
Drip (percent chop weight)					
All Pigs	1.28	0.56	***	0.94	0.57
	(321)			(83)	
Carcases with LD pH$_1$<5.9	1.42	0.55	NS	1.13	0.62
	(51)			(12)	
Carcases with LD pH$_1$>5.9	1.12	0.56	**	0.91	0.56
	(270)			(71)	
Drip as % lean weight					
All pigs	2.03	0.83	***	1.65	0.94
	(319)			(83)	
Carcases with LD pH$_1$<5.9	2.42	0.91	NS	2.01	1.06
	(51)			(12)	
Carcases with LD pH$_1$>5.9	1.96	0.91	**	1.60	0.92
	(268)			(71)	
EEL Value					
All Pigs	45.46	6.63	**	43.47	5.24
	(544)			(128)	
Carcases with LD pH$_1$<5.9	46.9	6.43	NS	44.16	5.73
	(121)			(25)	
Carcases with LD pH$_1$>5.9	45.06	6.69	*	43.30	5.12
	(423)			(103)	

Values given are mean and standard deviation, number of observations is given in parentheses

* P<0.05, ** P<0.01, *** P<0.001, NS not statistially significant

Fig.2. Frequency Distributions of Drip Loss and EEL Reflectance Values in Landrace and Large White pigs.

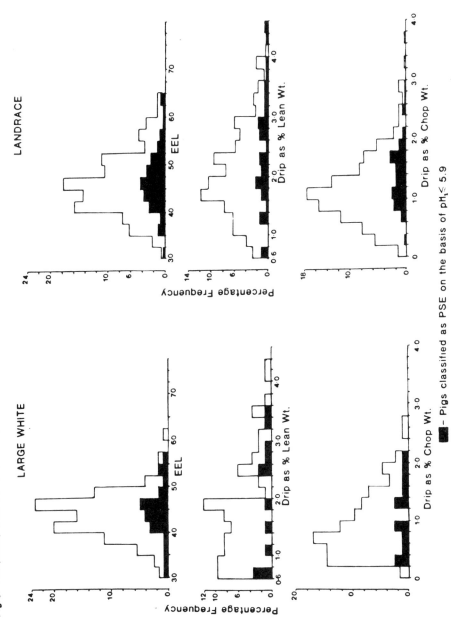

LARGE WHITE

LANDRACE

■ - Pigs classified as PSE on the basis of $pH_1 \leqslant 5.9$

activity preslaughter. A number of pigs have been identified with pH_1 values of 5.9 and pH_u values of 5.8 which would support this hypothesis.

In summary it appears that pH_1 values are not a sufficient discriminator of the PSE condition in progeny test pigs of the Landrace and Large White breeds. An accurate assessment of PSE requires a combination of both pH_1 and colour measurements. The selection pressures applied to pigs in Northern Ireland do not appear to have markedly increased the incidence of PSE in the Landrace breed but may have increased PSE in the Large White breed. A number of parameters need to be assessed as to their suitability in discrimination of the PSE condition, particularly in pig populations of breeds less stress sensitive than the Pietrain.

ACKNOWLEDGEMENTS

The author thanks Mr I Heaney and Mr I Stockdale and the staff of the Pig Testing Station, Antrim, for their cooperation in the work reported.

REFERENCES

Barton-Gade, P.A. 1979 Some experience on measuring the meat quality of pig carcases. Acta Agric. Scand. Suppl. 21, 61-70.

Bendall, J.R. 1973 Post mortem changes in muscle in 'The Structure and Function of Muscle' (Ed. G.H. Bourne) (Academic Press, New York and London) 2, 243-309.

Campbell, W.S. 1984 Ph D thesis, The Queen's University of Belfast.

Hart, P.C. 1962 Physico-chemical characteristics of degenerated meat in pigs II. Tijdschr. Diergeneesh 87, 156-167.

Jensen, P., Craig, H.B. & Robinson, O.W. 1967 Phenotypic and genetic associations among carcass traits of swine. J. Anim. Sci. 26, 1252-1260.

Lister, D. 1979 Some physiological aspects of the stress cycle in relation to muscle function and meat quality. Acta Agric Scand. 21, 281-289.

Martin, A.H., Freedeen, H.T. & L'Hirondelle, P.J.R. 1975 Muscle temperature, pH and rate of rigor development in relation to quality and quantity characteristics of pig carcases. Can. J. Anim. Sci. 55, 527-532.

Monin, G. & Sellier, P. 1985 Pork of low technological quality with a normal rate of muscle pH fall in the immediate post-mortem period: The case of the Hampshire breed. Meat Science, 13, 49-63.

Moss, B.W. 1980 The effects of mixing, transport and duration of lairage on carcass characteristics in commercial bacon weight pigs. J. Sci. Fd. Agric. 31, 308-315.

McDougall, D.B. & Disney, J.G. 1967   Quality characteristics of pork with special reference to Pietrain, Pietrain Landrace and Landrace pigs at different weights. J. Fd. Technol. 2 285-297.

Smith, W.C. & Lesser, D. 1979 An economic assessment of pale soft exudative muscle in the pig carcass. Anim. Prod. 28 p 442.

Standal, N. 1979. Selection for low backfat and high growth rate and vice versa for 9 generations: Effect of quality and quantity of lean meat. Acta. Agric. Scand. Suppl. 21, 117-121.

Swatland, H.J.1982. Meat colour of pork chops in relation to pH and adductor capacitance of intact carcases.   J. Anim. Sci. 54 262-267.

Tarrant, P.V., Gallwey, W.J. & McGloughlin, P. 1979 Carcass pH values in Irish Landrace and Large White Pigs.   Ir. J. Agric. Res.   18 167-172.

Taylor, A.McM, Dant, S.J. & French, J.W.L. 1973 Processing Wiltshire bacon from PSE prone pigs. J. Fd. Technol. 8 167-174.

Wood, J.D., Mottram, D.S. & Brown, A.J. 1981 A note on the eating quality of pork from lean pigs. Anim. Prod. 32 117-120.

Yang, T.S., Hawrysh, Z.J., Price, M.A. & Aherne, F.X. 1984 Identification of PSE in the Longissimus muscle of pigs stunned by captive bolt. Meat Science 10 243-251.

DISCUSSION - Session II

<u>P V Tarrant</u>   It is clear from the papers in this session that there are two aspects of paramount importance deserving of our attention. The first concerns standardisation of methods for measuring meat quality and agreement on appropriate standards and objective values. One of our tasks is to try to achieve as much uniformity as possible throughout the Community. This will allow closer collaboration between research teams and better use of manpower available to support the meat industry. It is an entirely reasonable objective to seek to harmonise methods and set quality standards where meat research and breeding programmes are concerned. However, in relation to commercial quality control there may always be differences because there the criteria that we are measuring are dictated by the consumer and by the buyer and that will always vary between different countries and different sides of the industry.

My second point concerns the very active situation we have in the pigmeat industry at present. Pigmeat is unique among fresh meats for the large number of electrical and optical probes designed to measure pork quality. Prof Kallweit found that meat quality criteria are not used in pig carcass grading systems in the EC at present. Nonetheless, since April 1984 Sweden has used the Hennessy grading probe for all pig carcasses and Dr Lundström says that leading cutting plants now want to classify for meat quality at carcass grading. So we have a situation where the industry may be outstripping the scientists, whose opinions' differ regarding what method is best for use on the slaughterline. The German work suggests that electrical properties may be useful for this purpose. Dr Swatland sees advantages in carrying out both electrical and light measurements to improve predictive power on the slaughterline. Meanwhile, the industry are going ahead with optical systems, such as the Hennessy grading probe from New Zealand and the Danish Fat O Meater, which have the decisive advantage of combining carcass grading and meat quality data in a single operation. Dr Lundström's work shows very clearly that this method misses the late developing PSE carcasses.

These are two points deserving of our attention: the harmonisation of methods for quality evaluation, and what are the appropriate methods and procedures for industrial application.

<u>D E Hood</u>   I would like to refer to Dr Kallweit's survey of meat quality evaluation methods currently used in European laboratories. Some years ago the Carcass and Meat Quality working group, within the EEC Beef Programme, established some standardisation in beef meat quality measurement. This was intended specifically to help research workers in production orientated experiments but the agreed standard methods had more general application. Perhaps it is time to do something similar now in relation to pigmeat and I suggest this as one of the areas for future consideration when it comes to identifying priorities for research or coordinated activities. It is necessary in any such discussion to point out that the post mortem carcass handling must be specified. In addition to pre and peri-slaughter stress the condition of post slaughter handling and chilling is of such fundamental importance that it cannot be omitted. I am surprised that this point was not a part of Dr Kallweit's original survey nor was it mentioned by Dr Swatland, for example, as a contributory factor in the development of PSE when he spoke of major differences in the incidence of

PSE in different commercial abattoirs. The temperature profile of the muscle must also be taken into account.

H J Swatland   The data that I presented were to illustrate a practical application of fibre optic spectrophotometry. Obviously, temperature is a major factor in the development of PSE. For accuracy, all electronic measurements of meat quality must be corrected for sample temperature. This has not yet been found to be necessary for fibre optic measurements.

A J Kempster   I agree with Dr Hood that we should attempt to standardise techniques. But, it is important to distinguish between base-line methods for measuring colour and water holding capacity and the equipment used to measure them. There are many commercial developers of equipment and fair competition should be encouraged so that we get the piece of equipment which is cost-effective. At this stage it would be difficult to identify the best techniques for general application.

R G Kauffman   Which mechanism is more important in determining the water holding capacity of muscle, sarcomere length and its associated swelling phenomenon, or chemical binding of water molecules to the charged positions on the myofibrillar proteins?

K O Honikel   Firstly, concerning drip loss rather than water holding capacity in general, lower pH values, down to pH 5.3, result in higher drip loss. As the isoelectric point of the myofibrillar proteins is approached, filament swelling is reduced, thus releasing free water into the sarcoplasm which is then apparently available to move out of the cell into the extracellular space and to appear as drip at the cut surface. Furthermore, muscle contraction decreases the interfibrillar space, leading to a similar sequence of water movements if muscle goes into rigor in the contracted state, irrespective of the cause of contracture. Both pH and contraction influence drip loss. Other methods of determining water holding capacity, such as cooking loss and the Grau Hamm filter paper press method depend on pH, both pH_1 and ultimate pH, but shortening does not influence the results obtained with these methods.

G Eikelenboom   Concerning the method for measuring drip loss proposed by Prof Honikel, rather than hang the sample in a sealed bag, I wonder if we could pre-pack simply as it is done in butcher shops and supermarkets and record the weight loss. Then we could look at colour and colour stability in the same sample.

P Allen   PSE is characterised not only by a low pH_u and high reflectance or light scatter value, but also by a rapid post mortem fall in pH and a rapid rise in reflectance. It seems likely that measurements of these rates, i.e. the difference between two measurements at a fixed time interval, might be a better estimate of meat quality than measurements on a single occasion. Has Dr Lundström or anyone else examined this?

K Lundström   It is a very good approach but we havn't tried it. I can however see some practical problems with the timing of the measurements if the speed of the slaughter is not quite constant.

P A Barton-Gade   We have actually measured pH values in a small number of pigs at various times post mortem, i.e. just before the scalding tank, after the dehairing machine and 50 minutes after slaughter. We found that pigs with a pH of 5.8 at 50 minutes could vary enormously with respect to pH measured earlier, some would be as low as 5.8 before the scalding tank, others 6.4. There is no doubt that there are quite different patterns of pH fall in the immediate post slaughter period. It might therefore be of interest at least from the research point of view to look at rate of fall in pH or increase in light scatter.

P Allen   It seems likely that WHC will be related to intramuscular fat content since this is not only inversely related to water content but may also have a binding effect. Are you aware of any work where this has been taken into account since the bias will always be towards increased drip in leaner carcasses.

K O Honikel   I am not aware of any literature reference, but it is certainly true that more fat means less water and therefore less drip in relation to the muscle weight. Additionally fat may prevent the release of exudate from meat in the uncooked and in the cooked state. This problem is under investigation at the moment, but it is difficult to study as the distribution of fat in the tissue is very variable and the results depend on the degree of marbling.

K Lundström   I would like to ask Dr Barton-Gade about the relationship between halothane sensitivity and DFD status, particularly the difference in this relationship between the Danish Landrace and Large White breeds. Also, the Danish work showed a lot of DFD in the halothane positive Landrace pigs whereas Dr Monin found no relationship between the halothane status of pigs and DFD in France.

P Barton-Gade   It's quite difficult to answer that question. We have always found that Landrace pigs have a higher frequency of higher than normal ultimate pH values (DFD meat) compared to other breeds, such as Large White, Duroc and Hampshire, and that Landrace show large increases in DFD meat with increasing time in transport and lairage. Moreover, the pre-slaughter treatment of our pigs is standardised and under these conditions there is no doubt that halothane positive Landrace pigs do have a higher DFD frequency than halothane positive Large Whites.

Dr Monin   Our results were obtained from more than 1,000 pigs of four breeds: Large White, French Landrace, Belgian Landrace and Pietrain with more than 100 for each breed. The incidence of Halothane susceptibility was about 70% in the two Belgian breeds, 15 to 20% in French Landrace and 0% in Large White. However, the incidence of DFD meat was higher in Belgian Landrace and lower in Pietrains than in Large White. So we concluded that the relationship between halothane susceptibility and the DFD condition is really not clear.

B W Moss   If I understand the results of Dr Barton-Gade correctly, halothane positive pigs exhibit a high proportion of PSE and DFD in the 'white' muscles. Examination of progeny test pigs in Northern Ireland showed that a large proportion of pigs with pH_1 of 5.9 had pH_u values of around 5.8. One explanation for this might be that stress suscepible pigs are excitable and this excitability may lead to glycogen depletion preslaughter. Thus stress susceptible pigs may exhibit both PSE and DFD.

P A Barton-Gade   Yes, I think I would agree with that - at least that the same environmental influences lead to a faster depletion of glycogen in stress-susceptible animals. We must remember that the pigs have received a short standardised pre-slaughter treatment, which in fact should preclude the development of higher than normal ultimate pH values in the pigs. But you have not really answered the question, unless you postulate that halothane sensitive Large White pigs are not excitable, because they just do not show DFD to the same extent as Landrace pigs.

A J Webb   Could I make a quick general comment. We heard a whole series of techniques for measuring meat quality and we did not have much discussion on the application of those measurements. There are three distinct applications of meat quality measurements : in breeding programmes, in payment, and in research. In breeding programmes in Britain, already the breeding companies do not slaughter any sibs at all as part of the routine selection process and it will be extremely difficult to justify slaughter in the future just to measure meat quality. So there is clearly one piece of technology which is missing and which we are going to need in the future and this is how to measure meat quality in the live animal.

SESSION III

CONTROL OF MEAT QUALITY IN TRANSPORT, LAIRAGE,
SLAUGHTER AND CHILLING

Chairman: P. Barton-Gade

THE EFFECT OF TIME AND CONDITIONS OF TRANSPORT
AND LAIRAGE ON PIG MEAT QUALITY.

P D Warriss

Agricultural and Food Research Council, Food Research Institute-Bristol
Langford, Bristol BS18 7DY, UK

ABSTRACT

The meat quality attributes of colour and the amount of exudate lost during storage are affected by preslaughter handling through the latter's influence on the rate and extent of acidification of the muscles after death. As a generalization, the two extremes of quality, PSE and DFD, are produced, respectively, by acute and chronic stress around the time of slaughter. In practice, interactions between different handling procedures and between handling and genotype affect how an individual pig will respond to stress. Depletion of glycogen by chronic stress will prevent the formation of PSE meat in animals potentially susceptible to its production. Very stress-susceptible pigs may not respond favourably to improved or even "ideal" preslaughter handling because the trauma of slaughter itself is sufficient to initiate development of the PSE condition. These interactions possibly explain much of the variation between results from different studies which have examined how transport and lairage affect meat quality. Some variation may also be due to the influences of factors such as season and stocking density. Most work indicates that extended transport will lead to an increased likelihood of DFD meat, while in non-fatigued, very stress-susceptible pigs, extension of even very short transport times may promote the development of PSE. Lairage for up to 6h may allow some recovery from the stress of transport in stress-susceptible or highly stressed animals. Otherwise there is little evidence that long lairage is desirable; it may simply prolong a period of fatigue and inanition and can promote DFD meat. This is particularly so when unfamiliar groups of pigs have been mixed, leading to fighting.

INTRODUCTION

Several of the components of meat quality can be influenced by how the live pig is handled before slaughter. In the extreme case animals may die during transport or lairage. If fighting occurs between individuals the skin surface may be blemished by lacerations and bruising can be produced by rough handling. Lastly, the quality of the lean meat itself may be affected. It is mainly the latter effect that will be discussed here. The major attributes of pig lean meat quality are colour and the amount of exudate lost during storage. Colour, specifically paleness or darkness, affects saleability (Topel et al., 1976; Wachholz et al., 1978) while exudate reduces yield (Kauffman et al., 1978; Smith and Lesser, 1982) and, if excessive, detracts from appearance at the retail level (Wachholz et al., 1978). This is particularly so when the meat is presented in pre-packs. There is also some evidence that pork which is pale and produces large amounts of exudate (PSE meat) tastes less juicy after cooking (Bennett et al., 1973; Jeremiah, 1984). Dark, firm, dry (DFD) meat is also discriminated against by consumers, but more importantly, is prone to spoilage (Key et al., 1976; Newton and Gill, 1981).

THE INFLUENCE OF PRESLAUGHTER HANDLING ON LEAN MEAT QUALITY

Meat quality is affected by preslaughter handling through its influence on the rate and extent of acidification of the muscles after death (Lister et al., 1981). If the rate is rapid, so that the pH reaches a low value when the temperature of the carcass is still high, PSE meat can result. The extent of the pH fall is determined by the amount of glycogen present in the muscles at death. If this is depleted before slaughter the extent is limited and DFD meat may result. The initial rate of acidification is largely determined by the degree of muscle stimulation produced immediately before and at slaughter.

As a generalization it may be said that chronic stress preslaughter leads to DFD problems and acute stress leads to PSE problems. However, this is an over-simplification for several reasons. To some degree different preslaughter handling factors interact, the genotype of the pig will influence the response and the effect of a particular factor may be determined by previous factors.

The interaction between genotype and handling seems particularly important. Animals of a very stress-resistant nature may hardly ever produce PSE meat irrespective of poor handling. Conversely, genotypes which are extremely sensitive to stress may produce PSE meat no matter how careful the preslaughter handling is, because the trauma of slaughter alone is sufficient to initiate its development (Briskey and Lister, 1968). Genotypes between these extremes will respond more or less favourably to improved handling. In any particular type of pig therefore the contribution made to meat quality by genotype and handling stress will vary. In the UK, recent surveys have shown that about 12% of carcasses are likely to have been PSE (Chadwick and Kempster, 1983; Kempster et al., 1984). The incidence of carcasses with low muscle pH has apparently doubled over the ten years from 1973 to 1983 (Chadwick and Kempster, 1983) and a quarter of the increase seems attributable to change in genotype within the British pig population. The remaining three-quarters must relate to environmental factors of which poorer preslaughter handling seems

most likely. Evidence for the importance of handling comes also from the
marked differences in the frequency of PSE carcasses between different
slaughterhouses (Chadwick and Kempster, 1983). The relative contribution
of genotype to meat quality is much greater in countries where the pig herd
has a high incidence of stress-susceptibility (Augustini, 1983). Whether
the incidence of DFD meat is related to genotype is less clear. There is
evidence of breed differences in the ultimate pH of meat (Gallwey and
Tarrant, 1979; Warriss and Akers, 1980) and Heinze et al. (1984) recorded a
lower incidence of DFD meat in fatter carcasses. However, handling is
likely to be much more important than genotype in determining the extent of
muscle glycogen depletion preslaughter.

There are other factors which complicate interpretation of much of the
information available on the effects of preslaughter handling on meat
quality. Some information has been collected from small-scale experiments
where factors can be carefully controlled but numbers of animals are
restricted; some has been derived from very large surveys of commercial
handling where, while confounding factors are present, the results relate
directly to the normal handling of pigs and are based on large sample
sizes. Many results come from experiments where the primary aim was not
the investigation of preslaughter handling effects. These are also often
confounded by other factors which makes interpretation difficult. There
are no universally agreed criteria for assessing meat quality, particularly
PSE and DFD defects. Some investigations have used only pH measurements.
These correlate well with subjective assessment over numbers of animals but
are imprecise for individual carcasses. Different investigators have used
different muscles on which to base their assessment and these may differ in
their propensity towards PSE and DFD; the difference may also be compounded
by genotype. The limiting pH values which have been used to define poor
meat quality vary. Other parts of the slaughter process such as stunning
may influence meat quality. Stunning with carbon dioxide may increase the
incidence of PSE to an extent which might overshadow any previous handling
effects.

The preslaughter handling of pigs can be divided into two major time
periods. The first is that spent in transit from the production unit to
the slaughter plant; the second is that spent in lairage prior to killing.
Potentially important factors in determining meat quality are therefore
transport and lairage times and the environmental conditions during these
periods. Related ancillary factors are the stocking density and group

sizes in the lorry and holding pens, whether unfamiliar pigs are mixed together and how animals are moved. The latter encompasses loading and unloading procedures associated with transport and the movement of pigs within the lairage and immediately before stunning.

TRANSPORT AND LAIRAGE TIMES

Very little published information is available about the distances travelled to slaughter and the time spent in lairage by pigs. Warriss and Bevis (1985) have recorded data for British slaughter pigs killed at five plants. Some pigs could, exceptionally, spend up to 11 h in transit on journeys of nearly 400 miles, but 99% travelled only up to 40 miles with a journey time of not more than 2 h. Nearly half of the pigs were killed after no more than 2 h in lairage but about 30% were held overnight and slaughtered the next day (ie a lairage time of > 14 h). Just over half of all animals were killed within 5 h of leaving the farm and three-quarters within 10 h. More than a fifth spent more than 18 h in transit and lairage before slaughter. In some British slaughter plants the average handling times may be longer than this. Clark (1973) recorded that only 20% of the total intake of one bacon factory was derived from within a 50 mile radius of the plant and that many consignments of pigs travelled in excess of 300 miles, the average per pig being about 120 miles. This plant however seems atypical of the UK.

Moss and Robb (1978) mentioned that the commercially-slaughtered Irish pigs in their samples had been transported from 8 to 30 miles, more than three quarters travelling from 10 to 12 miles. Normal lairage time ranged from 2 to 4h. for Danish pigs slaughtered in 23 centres, Barton (1971) records transport times ranging from 4 minutes to over 2.5 h and holding times from 5 minutes to more than 15 h.

In contrast to these within-country marketing procedures, some international journeys on mainland Europe can be very long. For example, pigs are exported from the Netherlands to Italy, southern France or Spain for slaughter, distances of about 1500 km which may take up to 40 h (van Putten and Elshof, 1978; Lambooy, 1984).

The overall range of current handling practices is therefore very large. No individual experiments have examined the whole range, most being confined to distances and times which reflect the local handling methods.

INFLUENCES OF TRANSPORT TIME ON MEAT QUALITY

During transportation, animals are exposed to various stressors, particularly noise, unfamiliar odours, deprivation of food and water, vibration and changes in acceleration, and possibly extremes of temperature. Often there will have been a breakdown of social groupings and a degree of confinement and overcrowding greater than during rearing. Using the technique of operant conditioning and a transport simulator which mimicked the noise and vibration of a moving vehicle, Ingram et al. (1983) found pigs would learn to switch off the machine and preferred to keep it immobile for about three-quarters of the time. The animals' aversion to transport was increased if they had eaten a large meal before the test. Later work showed that exposure to the simulated transport for 0.5 hours, or enforced exercise for 5 minutes, raised circulating vasopressin levels (Forsling et al., 1984). As well as the apprehension occasioned by the treatment and the exercise itself, the dehydration seen in transported pigs (Warriss, Dudley and Brown, 1983) could be associated with this hormonal change. Simulated loading procedures (Spencer, et al., 1984) and the combined effects of transport and lairage (Moss, 1984) have been shown to provoke rapid and sometimes prolonged changes in the circulating levels of metabolites and hormones such as cortisol and thyroxine. Based on elevation of cortisol levels, Spencer et al. (1984) suggested that transport itself is an additional stress to that caused by loading procedures. Loading, and to a lesser degree unloading, certainly often appear to be the most stressful part of the transport process and the observation is consistent with results from experiments in which heart rate has been monitored (van Putten and Elshof, 1978; Augustini and Fischer, 1982). Associated with this period of exertion is a rise in body temperature (Augustini and Fischer, 1982) and this is greater at higher stocking densities (von Mickwitz, 1982). The increased temperature will be reflected in higher muscle temperatures immediately post-mortem, leading to an increased likelihood of PSE (van der Wal and Eikelenboom, 1984). The apparent association between the physiological responses of the pig to stress and ultimate meat quality have led Augustini, Fischer and Schon (1977) to try to define the maximum heart rate (<85 beats per minute) and rectal temperature (<38.8°C) at slaughter commensurate with desirable meat. PSE meat was associated with heart rates >136 beats per minute and temperatures >40.1°C suggesting that treatments which promote values greater than these should be avoided.

Pigs subjected to simulated transport also show an increase in plasma corticosteroids after 1 h but the levels decrease to near-basal after 6 h due to habituation of the pituitary-adrenal axis (Dantzer, 1982). Various investigators have also suggested that pigs transported over short distances appear more stressed than those in transit for several hours (Cuthbertson and Pomeroy, 1970; Anon, 1972; Buchter, 1974).

It might be expected that the time and conditions of transport, by having such obvious effects on the physiology of pigs, would influence meat quality, and generally act to its detriment. This does not necessarily appear so, probably because of the interaction with genotype. The exact nature of the interaction may be complex. Schworer et al. (1981) found that a short transport to the slaughterhouse (18km) combined with the other stresses associated with commercial slaughter, had little effect on the meat quality of stress-resistant pigs but reduced meat quality in stress-susceptible animals. In this work stress-susceptibility was assessed using plasma creatine phosphokinase levels. In recent Danish work (Barton-Gade, 1984) characterisation of stress-susceptibility was based on the halothane test. Stress-susceptible (nn) pigs gave poor meat quality irrespective of preslaughter handling. In contrast, stress-resistant (NN) pigs showed improved meat quality with less stressful handling and this was also true, but to a lesser extent, for the heterozygote.

Generally in the UK, where the majority of pigs tend to be relatively stress-resistant, experiments in which transport has been examined have not shown commercially-important effects on meat quality. Cuthbertson and Pomeroy (1970) could find no material differences between the quality of pork from pigs transported for either 0.5 or 8h preslaughter. Similar lack of influence of transport was noted by Moss (1980), Warriss et al. (1983) and by Elliott and Patton (1968) who concluded that journeys up to 100 miles were not particularly stress-inducing although there was some evidence of slightly increased ultimate pH values and darker meat from pigs given a long transport. Lengerken et al. (1977) expressed the view that, with good conditions during transport, distance had little effect on meat quality.

In contrast, larger influences have been shown in Dutch and Danish pigs. In Dutch pigs long (200 - 300km) transport led to a higher incidence of PSE carcasses than short (20 - 80 km) journeys (Lendfers, 1968). In Danish Landrace pigs however, longer transport, from 0.25h up to 3h, increased very slightly the incidence of meat with high ultimate pH but led to a substantial reduction in the incidence of PSE meat. This was 40% after 0.25h but only 10% after 3h (Buchter, 1974). A similar effect in these pigs was later shown by Barton-Gade et al. (1982) in an experiment to define acceptable transport conditions for progeny-test pigs. Overall, the incidence of PSE was highest with 0.5h transport and lowest with 1.3h and moreover there was a pronounced genotype interaction, with progeny from three different boars reacting differently. The longer transport also raised the ultimate pH of several red muscles.

The possibility of transport fatiguing pigs to the extent of leading to glycogen depletion is illustrated by the work of Malmfors (1982). Lengthy transport combined with a long (24h) lairage resulted in a high incidence of DFD meat. With transport distances >90km the incidence averaged about double that after short distances (<35km). As the transport time was reduced there was a corresponding increase in the incidence of PSE, particularly when pigs were killed immediately on arrival at the factory; after a journey of <35km 22% of carcasses were judged PSE. The influence of longer transport on the incidence of meat with high ultimate pH is also seen from the results of Heinze et al. (1984). These authors showed that as transport distance increased from <50km to >100km the percentage of carcasses with an ultimate pH >6 in the LD increased from 23 to >30%. In German Landrace pigs transported between 5 and 500km there was a gradual increase in the incidence of low initial pH loins (indicative of PSE) and a smaller but similar effect in the ham. At the same time the incidence of meat with high ultimate pH generally increased in both the loin and, more especially, in the ham (Scheper, 1971).

In international transport animals may travel very considerable distances and ambient conditions may vary a lot throughout the journey. Some of these conditions may be extreme. The effects of very long transport times on Dutch pigs have been investigated by Lambooy (1984). A journey of up to 1500km took 24h with the animals being unloaded and rested for 1h after eight and 16h transport. During these rest periods the pigs were offered food and/or water but their consumption of both was much reduced compared with control groups and it was suggested that this might have been because they were fatigued. The pigs lost up to 5% of their live weight during transport. However, despite these effects, meat quality was not materially or consistently different between groups transported for 8, 16 or 24h. Overall, quality was good with group mean pH values in the normal range and no evidence of PSE or DFD.

The actual conditions under which pigs are transported may be more important than journey time and improvement of conditions may improve meat quality. Nielsen (1979) found that use of a lorry equipped with non-skid floors, partitions and mechanical ventilation reduced the incidence of DFD (by reducing the need for physical activity and the opportunity for fighting) but had no effect on the percentage of PSE carcasses, when compared with an unimproved vehicle. Mechanical ventilation of lorries also reduced the total deaths during transport and lairage by nearly 40% from the level in passively-ventilated vehicles. Lendfers (1968) recorded poorer meat quality in pigs transported at higher stocking densities (0.35 m^2/pig) in poorly ventilated lorries compared with those given more room (0.66 m^2/pig) with good ventilation. The incidence of PSE was reduced from 24% to 6%. Much of the value of good ventilation rests with its effect on the temperature of the pigs' immediate environment. Transport at an ambient temperature of 19°C was less stressful (based on elevation of the heart rate) than at 2°C or 29°C (Augustini and Fischer, 1982) although the results were confounded by relative humidity and loading density. Higher ambient temperatures (25°C) have been shown to be associated with poorer meat quality than normal temperatures (15°C) in Dutch slaughter pigs (Lendfers, 1968) and a significant effect of season was seen in Danish work (Barton, 1971). In the latter, the poorer meat quality in summer and autumn was largely due, not to more pigs having meat quality deviating from normal, but to an increase in the extent of the deviation. Conversely, Scheper (1971) observed higher incidences of PSE and DFD in autumn and winter, than in spring and summer in Germany. Perhaps this is a reflection of differences in climate and the stressful nature of both high and low temperatures, as indicated by Augustini and Fischer's (1982) work.

While use of tail-gate lifts seems to be less stressful to pigs than being walked up a ramp (Augustini and Fischer, 1982), it may (Lendfers, 1968) or may not (Nielsen, 1979) improve meat quality. Both too-low and too-high loading densities may be undesirable. With low densities pigs can be thrown about with changes in velocity but pigs often lie down during transport and therefore this is less of a problem than overstocking. An area of 0.5 m^2/100kg will enable all pigs to lie down (CEC, 1985) and this agrees with Meat and Livestock Commission (MLC, 1980) recommendations of 0.4 to 0.5 m^2 per animal for UK pigs which are generally slaughtered at slightly lower weights. However, to some degree optimal stocking density will obviously depend on ambient conditions. Higher stocking densities are associated with a higher incidence of PSE meat (Blomquist and Jorgensen, 1963).

Nielsen (1982) emphasized the importance of the personal qualities of the vehicle driver and there is direct evidence of the deleterious influence of a poor driver on meat quality (Lendfers, 1968). Associated with this is the type of road on which the vehicle is driven. Scharner et al. (1979) found that the heart rate of pigs was higher when driven along normal roads, with consequent frequent changes in velocity, compared with transport along major trunk roads. Rail transport of pigs does not appear to be common in Europe nowadays although it is recorded in Poland (Czyrek, 1972). In a comparison of rail with road transport in South Africa, the carcasses of pigs transported by road had a higher ultimate pH than those transported by rail for similar distances and this was reflected in a higher incidence of DFD (33.6% against 24.2%; Heinze et al., 1984).

Pigs should be unloaded immediately on arrival and prefer walking up, rather than down, slopes. This can make unloading an even more stressful experience than loading, particularly for pigs carried on the upper decks of multiple-tier vehicles, and there is some evidence that these lorries induce more fatigue and higher ultimate pH values in the muscles (Elliott and Patton, 1968).

In general there is agreement that transport, particularly loading and unloading, is a potentially very stressful experience for pigs. How they react to it may depend greatly on their stress-susceptibility. Very stress-susceptible pigs may produce increased incidences of PSE meat after only short transport distances. More stress-resistant animals may show little or no reaction over moderate distances under good conditions, especially of ventilation. Most pigs will show some evidence of fatigue and muscle glycogen depletion after extended transport and particularly under poor conditions, leading to increased incidence of DFD meat. A problem with interpreting much information is the interaction of handling with genotype and the fact that most investigators have considered only a limited range of transport times or distances under various environmental conditions and have not usually included control groups of un-transported animals.

Improvement of the conditions of transport of pigs will often lead to improvements of meat quality as well as being desirable from the point of view of decreasing the incidence of deaths in transport (Hails, 1978) and in the interests of better animal welfare.

INFLUENCES OF LAIRAGE TIME ON MEAT QUALITY

The prime purpose of a lairage is to act as a holding area where a reservoir of animals can be maintained. This enables the slaughterline to operate at a more-or-less constant speed irrespective of variations in the delivery schedule from producers or markets. Its secondary function is to allow animals which are stressed or fatigued by their previous transport to recover. Tradition has it that this will lead to the production of better meat, as well as being desirable from the humanitarian point of view. Under modern conditions this may or may not be correct. Holding in lairage may have negligible or no influence on meat quality, particularly in pigs which have not been subjected to very stressful previous handling or are resistant to stress. Prolonged holding may even lead to a reduction in meat quality.

Bendall et al. (1966) noted an increased average ultimate pH in British pigs killed after longer holding time when this ranged from 0.5 to 1.5h. They also found that extended times increased the variation in initial pH values and reduced the overall mean, increasing the incidence of PSE meat. However, the size of these effects was very small, possibly because of the stress-resistant nature of the pigs. Similarly, Taylor (1966) and Cuthbertson and Pomeroy (1970) recorded negligible influences on meat quality of lairage for periods up to overnight (about 20h). From the results of several trials the latter authors inferred that the behaviour and handling of the animals during transport could influence the response to rest in lairage.

After prolonged rail transport (54h), rest for 24h preslaughter increased the incidence of watery meat as well as reducing carcass yield (Czyrek, 1972), and slaughter as soon as possible was recommended. For pigs travelling by road, similar recommendations were made by Ivanov et al. (1983) for journeys > 60km and < 2h and for pigs in the UK by Elliott and Patton (1968) who concluded that if pigs were not stressed on arrival at the factory, then resting in lairage would not improve meat quality. However, if pigs were stressed then rest could alleviate the effects and a period of at least 2h was suggested for groups of mixed animals. Vrchlabsky (197) also suggested that if pigs were transported correctly then they did not need to be held for more than 3 to 6h preslaughter to recover from the effects of travel. In a large survey of the influence of lairage time (0, 4 or 24h) on Polish pigs, which appear fairly stress-resistant, Wajda and Denaburski (1982) found no significant effect on initial pH (and hence the incidence of PSE meat) but the ultimate pH increased significantly after 24h and there was a general improvement in water holding capacity with longer lairage. Lairage for 0 or 4h was recommended in preference to 24h.

With more stress-susceptible pigs, such as many continental breeds, the value of a short period of lairage in reducing the incidence of PSE is more apparent. Stein (1978) compared lairage for 0, 8 or 24h. The lowest amount of PSE was recorded after 8h rest and the highest after no lairage. It was recommended that pigs be held for >6h for the best meat quality. A rest of 8h was claimed as optimal by Lengerken (1977); periods shorter than this were not sufficient for recovery from the effects of transport, longer periods tended to lead to further stress. Augustini and Fischer (1981), in a large field experiment, found that increasing lairage times up to 6h decreased the incidence of carcasses with low initial pH values. With longer holding times there was a slight increase in the incidence of DFD carcasses. Pigs with an ultimate pH >6 increased from about 3% in the group held for < 0.5h to over 10% in pigs laired 6h. An overnight lairage, as well as increasing the incidence of DFD, also caused a slight increase in PSE. This latter effect seems difficult to explain. The complication of different effects being observed on different metabolic types of muscle is illustrated by Danish work on Landrace pigs (Barton, 1971). Short periods of stress promoted the development of PSE, particularly in the M. longissimus dorsi (LD) but also to a lesser extent in the M. biceps femoris (BF). In the pig the LD is a white muscle which has a metabolism tending to be glycolytic whereas the BF contains more myoglobin and has an oxidative metabolism (Beecher et al., 1965; Bendall, 1975). Longer periods of stress reduced the incidence of PSE in both muscles but this was associated with a corresponding increase in the incidence of DFD. The author's conclusion was that holding Danish Landrace pigs in lairage for 1 to 2h preslaughter did not necessarily improve meat quality. A comprehensive investigation of preslaughter handling on Danish pigs was later carried out by Nielsen (1979). Holding times of 0, 2, 6 and 22h were compared for pigs which had been transported for about 2h. The number of PSE carcasses fell as the holding period increased while the number of DFD carcasses increased. The incidence of PSE did not begin to fall appreciably until pigs had been rested for 6h. However, the amount of DFD had already begun to increase after only 2h in lairage. These results agree closely with findings from later work on Swedish pigs by Malmfors (1982) who concluded that, irrespective of transport time, lairage for 4h (compared with 0 or 24h) was beneficial for meat quality in substantially reducing PSE while not increasing DFD materially. Moss and Robb (1978) found similar effects in Irish Landrace pigs. The incidence of PSE was higher after short lairage (1 to 1.5h) than after an overnight lairage and the incidence of DFD was lower.   DFD was more prevalent in boars, particularly after overnight lairage. This was attributed to the more excitable and aggressive nature of entire males leading to increased fighting with consequent lowering of muscle glycogen stores. This is confirmed by recent work by Warriss and Brown (1985). The incidence of PSE did not appear to be influenced to the same degree in crossbred pigs which were possibly less stress-susceptible. However, the effect of overnight lairage on these animals was similar to the purebred Landrace.

Generally high incidences of DFD pig meat have been recorded in both Northern Ireland (Moss, 1980) and Eire (Gallwey and Tarrant, 1978). The latter authors recorded 18% of carcases having high ultimate pH in the loin. The origin of the pigs (markets, dealers, large or small fattening units) had a large influence on this. Up to 30% of pigs from markets and those purchased through dealers had loins with ultimate pH >6. However, lairage design was also important. Bacon factories with long narrow lairage pens with intervening steel bars had the highest incidence of carcases with high ultimate pH, factories with small lairage pens with solid walls had least. These differences were attributed to the greater possibility of animals in the large pens having come from mixed groups, leading to fighting, and the extra possibility of inter-animal interactions with non-solid pen walls. Poor ventilation in the lairage may also be deleterious. A 3.5h lairage under good conditions decreased the incidence of PSE from 36 to 2% in Dutch pigs but long lairage in a badly ventilated pen increased it (Lendfers, 1968).

It is evident that generally, rather than allowing recovery from stress occasioned by previous handling, long lairage tends to prolong or promote stress and the depletion of body glycogen stores. Nevertheless, this appears not always the case. Long lairage times reduced the incidence of DFD in South African pigs from 33% with an ultimate pH >6 in the LD in pigs held for less than 19h to 12% in pigs kept for >48h, most effect occurring after 48h (Heinze et al., 1984). It is often not clear from reports whether animals held for long periods (> 20h say) have received

food in lairage, however, where reduced DFD has resulted, as above, this
seems likely. Even when offered food, if this is not in the form to which
the pigs have been accustomed then intake may be restricted. Under these
conditions depletion of glycogen stores is likely.

To conclude, lairage may or may not provide a suitable environment for
recovery from the stress of transport and this is reflected in the
influences it can have on meat quality. In some pigs; particularly those
of a fairly stress-resistant nature, lairage seems to have little material
influence on meat quality unless prolonged when an increase in DFD meat may
occur. This is also true if pigs are not stressed on arrival at the
slaughterhouse. In this case, immediate slaughter may produce the best
meat quality. Nevertheless, in highly stressed, or stress-susceptible
animals, a short period in lairage may allow some recovery from the
previous handling and reduce the incidence of PSE carcasses. The length of
this short lairage is ill-defined for different types of pig but appears to
be of the order of up to 6h. If this time is extended, muscle glycogen
depletion may lead to problems with DFD meat, particularly in pigs already
fatigued by transport, fasting or the fighting which frequently ensues on
mixing unfamiliar animals. By reducing muscle glycogen stores the
potential for formation of PSE meat is reduced and it is unclear whether
this or the actual resting in lairage is more important in reducing the
incidence of meat with a low initial pH. Longer lairage times are also
associated with an increase in the incidence of deaths associated with the
stress of previous transport (Allen and Smith, 1974).

THE CONSEQUENCES OF MIXING UNFAMILIAR ANIMALS

An important consequence of mixing pigs from different rearing groups
during transport and in lairage is that they often fight. This leads to
unsightly lacerations on the skin. In bacon carcasses in the UK this may
lead to downgrading as they are unsuitable for production of rind-on bacon.
The overall incidence of downgrading may reach 7-8% in some plants
(Warriss, 1984; MLC, 1985) and tends to be higher in boars than in gilts
or castrates. Even when the severity of fighting damage is insufficient to
cause downgrading, other consequences may be economically important. The
physical exertion may deplete muscle glycogen stores and lead to the
production of DFD meat (Gallwey and Tarrant, 1979; Warriss and Brown,
1985) and, if the stress occurs immediately preslaughter, it may also
increase the incidence of PSE meat (Wismer-Pedersen and Riemann, 1960).

Boars seem more prone than castrates to the DFD condition, probably because they are more excitable and aggressive (Moss and Robb, 1978; Tarrant et al., 1978).

METHODS TO COUNTERACT UNDESIRABLE CONSEQUENCES OF TRANSPORT AND LAIRAGE:

To eliminate much of the stress of loading and unloading, the use of special containers for transport has been suggested (Ring and Blendl, 1984). An added advantage is that carcass damage caused by bruising and biting is reduced since pigs from different rearing groups are not mixed (Tatulov et al., 1982).

Many agents to mask or disguise pig body odours, and so prevent fighting, have been suggested but with little or no evidence for their effectiveness. Recently however, Braathen and Johansen (1984) have documented the use of liquid smoke aroma sprayed onto pigs before transport to reduce fighting. The procedure also appeared to reduce the incidence of PSE meat. In the late fifties the Danish Meat Research Institute developed a halter which when fitted individually to pigs prevented them from biting one another and reduced aggression. As well as resulting in fewer lacerations on the carcass, use of the halter gave meat with higher initial pH values and improved colour (Blomquist and Jorgensen, 1963). Sprinkling or spraying pigs with a fine mist of water in lairage has also been claimed to reduce aggression and generally quieten the animals. By lowering body temperature it reduces the frequency of PSE carcasses under warm ambient conditions (15-20°C) but the effect is reduced or even counterproductive on cold (9-11°C) days (van Logtestijn et al., 1977; Smulders et al., 1982).

The use of sugar-feeding in lairage to counteract the effects of fatigue and fasting, and reduce the incidence of DFD meat, is well documented (Ingram, 1964; Clark, 1973; Gallwey et al., 1977 Gallwey and Tarrant, 1979; Gardner and Cooper, 1979; Fernandes et al., 1979). With pigs which would potentially give PSE carcasses but for the preslaughter glycogen depletion occasioned by fatigue, feeding sugar also tends to increase the frequency of PSE meat (Nielsen, 1979).

Lister and Ratcliff (1971) described the use of injected magnesium salts to retard the rate of postmortem glycolysis and improve meat quality. More recently magnesium aspartate hydrochloride, which can be added as a supplement to feed, has been claimed to reduce the incidence of both PSE and DFD meat, at least in stress resistant pigs (Schmitten et al., 1984), and reduce mortality in transit (Schneider, 1984). The tranquillizer, azaperone, will improve meat quality (Oldigs and Unshelm, 1971; Devloo et al., 1971) and reduce fighting (Knightly and Connolly, 1971) in pigs. The beta-blocking agent carazolol has also been shown to markedly improve meat quality in stress-susceptible pigs (Warriss and Lister, 1982). Problems with pharmacological treatments are their administration and the possibility of harmful residues remaining in the meat. Because of the latter, their use immediately preslaughter is prohibited in some countries.

Malmfors (1982) found that pigs arriving at the slaughterhouse could be sorted visually into different groups, ranging from those that were not stressed to those that were severely stressed. Stressed pigs shivered and showed reddening of the skin, bulging eyes and limp tails. The groups classified as more stressed had poorer meat quality. Malmfors suggested that animals recognised as severely stressed after transport could be kept in lairage for longer to recover, with a concomitant improvement in meat quality.

CONCLUSIONS

There appears to be concern at the quality of some pigmeat. Periodic surveys of the incidence of PSE meat show that it has tended to increase even in countries where the pig herd is relatively free of stress-susceptibility. A proportion of the increase is thought to relate to generally poorer handling of the live animal preslaughter. Certainly, the importance of environmental influences at the time of slaughter is apparent from variation between slaughterhouses and time periods in surveys of meat quality. Although there is no information on exactly how conditions have changed, a contributing factor in countries such as the UK might be the trend towards fewer, larger slaughterhouses. As well as this lack of information about the changes in how we treat pigs preslaughter there is not complete agreement on the effects of different handling procedures on meat quality. A major reason for this seems to be the interaction between genotype and handling. All but the most recent investigations on handling effects have not included this interaction or have failed to define sufficiently precisely the inherent meat quality characteristics of the animals used. With the increasing costs of production of meat, its quality becomes even more important. Yet in many pigs we are still some way from understanding exactly how to ensure this quality or even what factors are most important in influencing it, particularly in the preslaughter period.

REFERENCES

Allen, W.M. and Smith, L.P. 1974. Deaths of pigs in Great Britain during and after their transportation. Proc. European Meeting Meat Research Workers 20, 49-51.

Anon. 1972. ARC Meat Research Institute Annual Report 1972-1973, p29.

Augustini, C. 1983. Ursachen unerwunschter Fleischbeschaffenheit beim Schwein. Fleischwirtschaft 63, 297-307.

Augustini, C. and Fischer, K. 1981. Behandlung der Schlachtschweine und Fleischbeschaffenheiteine Felduntersuchung. Fleischwirtschaft 61, 775-785.

Augustini, C. and Fischer, K. 1982. Physiological reaction of slaughter animals during transport. In: Transport of animals intended for breeding, production and slaughter (Ed R. Moss) (The Hague, Netherlands; Martinus Nijhoff Publishers) pp.125-135.

Augustini, C., Fischer, K. and Schon, L. 1977. Welche Information konnen unmittelbar vor der Schlachtung erhobene physiologische Messwerte uber die zu erwartende Fleischbeschaffenheit geben? Fleischwirtschaft 57, 1028-1033.

Barton, P.A. 1971. Some experience on the effect of preslaughter treatment on the meat quality of pigs with low stress resistance. Proc. 2nd Int. Symp. Cond. meat quality of pigs, Zeist. Pudoc, Wageningen, Netherlands, pp. 180-190.

Barton-Gade, P. 1984. Influence of halothane genotype on meat quality in pigs subjected to various preslaughter treatments. Proc. European Meeting of Meat Research Workers 30, 8-9.

Barton-Gade, P.A., Busk, H. and Pedersen, O.K. 1982. Proc. European Meeting of Meat Research Workers 28, paper 1.07.

Beecher, G.R., Cassens, R.G., Hoekstra, W.G. and Briskey, E.J. 1965. Red and white fibre content and associated post mortem properties of seven porcine muscles. J. Fd. Sci. 30, 969-976.

Bendall, J.R. 1975. Cold contracture and ATP-turnover in the red and white musculature of the pig, post mortem. J. Sci. Food Agric. 26, 55-71.

Bendall, J.R., Cuthbertson, A. and Gatherum, D.P. 1966. A survey of pH, and ultimate pH values of British progeny-test pigs. J. Fd. Technol. 1, 201-214.

Bennett, M.E., Bramblett, V.D., Aberle, E.D. and Harrington, R.B. 1973. Muscle quality, cooking method and aging vs. palatability of pork loin chops. J. Fd. Sci. 38, 536-538.

Blomquist, S.M. and Jorgensen, T.W. 1963. Transport of live pigs from farm to bacon factory. Proc. 3rd Symp. World Ass. Vet. Fd. Hygienists, Nice, France 1962. pp.108-120.

Braathen, O.S. and Johansen, J. 1984. Liquid smoke aroma for smell masking gives reduced fighting and stress among slaughter pigs. Proc. European Meeting Meat Research Workers 30, 12.

Briskey, E.J. and Lister, D. 1968. Influence of stress syndrome on chemical and physical characteristics of muscle post mortem. In: The Pork Industry: Problems and Progress (Ed. D.G. Topel) (Iowa State University Pres, Ames) pp. 177-186.

Buchter, L. 1974. Slaughter of meat animals. In: Meat (Eds. D.J.A. Cole and R.A. Lawrie). London, Butterworths, pp.133-148.

CEC 1985. Commission of the European Communities Communication COM (95) 70 of 5 March 1985.

Chadwick, J.P. and Kempster, A.J. 1983. A repeat national survey (ten years on) of muscle pH values in commercial bacon carcasses. Meat Sci. 9, 101-111.

Clark, J.B.K. 1973. The effect of preslaughter influences on quality. Proc. Inst. Fd. Sci. Technol. (UK) 6, 136-144.

Cuthbertson, A. and Pomeroy, R.W. 1970. The effect of length of journey by road to abattoir, resting and feeding before slaughter on carcass characteristics in bacon weight pigs. Anim. Prod. 12, 37-46.

Czyrek, B. 1972. Further about pre-slaughter resting of pigs and its costs. Medycyna Weterynaryjna 28, 560-563.

Dantzer, R. 1982. Research on farm animal transport in France: a survey. In: Transport of animals intended for breeding, production and slaughter (Ed. R. Moss). The Hague, Netherlands., Martinus Nijhoff Publishers. pp. 218-230.

Devloo, S., Geerts, H. and Symoens, J. 1971. Effect of azaperone on mortality and meat quality after transport of pigs for slaughter. Proc. 2nd Int. Symp. Cond. meat quality of pigs, Zeist, Pudoc, Wageningen, Netherlands. pp. 215-224.

Elliott, R.J. and Patton, J. 1968. The effects of road transportation and lairage treatment on pig muscle. Proc. European Meeting of Meat Research Workers 14, 397-405.

Fernandes, T.H., Smith, W.C., Ellis, M., Clark, J.B.K. and Armstrong, D.G. The administration of sugar solutions to pigs immediately prior to slaughter. 2. Effect on carcass yield, liver weight and muscle quality in commercial pigs. Anim. Prod. 29, 223-230.

Forsling, M.L., Sharman, D.F. and Stephens, D.B. Vasopressin in the blood plasma of pigs and calves exposed to noise and vibration comparable with that experienced during transport. J. Physiol. 357, 96p.

Gallwey, W.J. and Tarrant, P.V. 1978. Pre-slaughter management affects pigmeat quality. Farm and Food Research 9, 30-32.

Gallwey, W.J. and Tarrant, P.V. 1979. Influence of environmental and genetic factors on ultimate pH in commercial and purebred pigs. Acta agric. Scand. Suppl. 21, 32-38.

Gallwey, W.J., Tarrant, P.V. and McMahon, P. 1977. Pigmeat quality and yield in relation to preslaughter sugar feeding. Ir. J. Fd. Sci. Technol. 1, 71-77.

Gardner, G.A. and Cooper, T.J.R. 1979. An evaluation of feeding liquid sugar to pigs lairaged overnight before slaughter. Proc. European Meeting of Meat Research Workers 25, 5-10.

Hails, M.R. 1978. Transport stress in animals: a review. Anim. Reg. Stud. 1, 289-343.

Heinze, P.H., Gouws, P.J. and Naudé, R.T. 1984. The influence of various factors on the occurrence of high ultimate pH values as an indication of dark, firm and dry (DFD) pork at a South African bacon factory. S. Afr. J. Anim. Sci. 14, 97-104.

Ingram, D.L., Sharman, D.F. and Stephens, D.B. 1983. Apparatus for investigating the effects of vibration and noise on pigs using operant conditioning. J. Physiol. 343, 2-3.

Ingram, M. 1964. Feeding meat animals before slaughter. Vet. Rec. 76, 1305-1310.

Ivanov, L. Kaloyanov, I., Katsarov, D., Grozdanov, A., Monov, G., Dilova, N., Radeva, M., Rizvanov, S., Nenov, K. and Draganova, L. 1983. On the quality of the meat of pigs slaughtered without lairage. Mesopromishlenost, 16 (4), 75-77.

Jeremiah, L.E. 1984. A note on the influence of inherent muscle quality on cooking losses and palatability attributes of pork loin chops. Can. J. Anim. Sci. 64, 773-775.

Kauffman, R.G., Wachholz, D., Henderson, D. and Lochner, J.V. 1978. Shrinkage of PSE, normal and DFD hams during transit and processing. J. Anim. Sci. 46, 1236-1240.

Kempster, A.J., Evans, D.G. and Chadwick, J.P. 1984. The effects of source population, feeding regimen, sex and day of slaughter on the muscle quality characteristics of British crossbred pigs. Anim. Prod. 39, 455-464.

Knightly, M. and Connolly, P. 1971. Evaluation of azaperone for the prevention of damage, due to fighting, in slaughter pigs. Irish. Vet. J. 25, 65-66.

Lambooy, E. 1984. Watering and feeding pigs during road transport for 24 hours. Proc. European Meeting of Meat Research Workers 30, 6-7.

Lendfers, L.H.H.M. 1968. Differences in meat quality by varying pre-slaughter conditions. Proc. European Meeting Meat Research Workers 14, 493-502.

Lengerken, G. von, Stein, H.J. and Pfeiffer, H. 1977. Einfluss der Ausruhzeit vor der Schlachtung auf die Fleischbeschaffenheit. MH. Vet. Med. 32, 376-380.

Lister, D. and Ratcliffe, P.W. 1971. The effect of preslaughter injection of magnesium sulphate on glycolysis and meat quality in the pig. Proc. 2nd Int. Symp. Cond. meat quality of pigs, Zeist, Pudoc, Wageningen, Netherlands, pp.139-144.

Lister, D., Gregory, N.G. and Warriss, P.D. 1981. Stress in meat animals. In: Developments in Meat Science-2 (Ed. R.A. Lawrie) (London, Applied Science Publishers pp.61-92.

Logtestijn, J.G. van, Corstiaensen, G.P. and Kruift, J.M. de 1977. Showering of slaughter pigs. Proc. European Meeting of Meat Research Workers 23, 1-15.

Malmfors, G. 1982. Studies on some factors affecting pig meat quality. Proc. European Meeting of Meat Research Workers 28, 21-23.

Mickwitz, G. von 1982. Various transport conditions and their influence on physiological reactions. In: Transport of animals intended for breeding, production and slaughter (Ed. R. Moss) (The Hague, Netherlands, Martinus Nijhof Publishers) pp. 45-54.

MLC 1980. Handling pigs from farm to slaughterhouse. Technical Bulletin No.14. Meat and Livestock Commission, Bletchley, UK.

MLC 1985. Concern at rindside damage in pigs. Meat and Livestock Commission, Meat and Marketing Technical Notes No.4, March 1985, Bletchley, UK.

Moss, B.W. 1980. The effects of mixing, transport and duration of lairage on carcass characteristics in commercial bacon weight pigs. J. Sci. Food Agric. 31, 308-315.

Moss, B.W. 1984. The effect of preslaughter stressors on the blood profiles of pigs. Proc. European Meeting of Meat Research Workers 30, 20-21.

Moss, B.W. and Robb, J.D. 1978. The effect of preslaughter lairage on serum thyroxine and cortisol levels at slaughter, and meat quality of boars, hogs and gilts. J. Sci. Food Agric. 29, 689-696.

Newton, K.G. and Gill, C.O. 1981. The microbiology of DFD fresh meats: a review. Meat Sci. 5, 223-232.

Nielsen, N.J. 1979. The influence of preslaughter treatment on meat quality in pigs. Acta. agric. Scand. Suppl. 21, 91-102.

Nielsen, N.J. 1982. Recent results from investigations of transportation of pigs for slaughter. In: Transport of animals intended for breeding, production and slaughter (Ed. R. Moss) (The Hague, Netherlands; Martinus Nijhoff Publishers) pp.115-124.

Oldigs, B. and Unshelm, J. 1971. Influence of a stress-reducing medical treatment before transport on meat quality of pigs. Proc. 2nd Int. Symp. Cond. meat quality of pigs, Zeist, Pudoc, Wageningen, Netherland, pp.205-207.

Putten, van, G. and Elshof, W.J. 1978. Observations on the effect of
    transport on the well-being and lean quality of slaughter pigs. Anim.
    Reg. Stud. 1, 247-271.
Rey, C.R., Kraft, A.A., Topel, D.G. Parrish, F.C. and Hotchkiss, D.K. 1976.
    Microbiology of pale, dark and normal pork. J. Fd. Sci. 41, 111-116.
Ring, C. and Blendl, H.-M. 1984. Containertransport von Schlachtschweinen.
    Fleischwirtschaft 64, 1058-1062.
Scharner, E., Schiefer, G., Prell, M. and Luhn, E. 1979.
    Elekrokardiographische Untersuchungen an Schlachtschweinen Während des
    LKW-Transports. Proc. European Meeting of Meat Research Workers 25,
    paper 1.3.
Scheper, J. 1971. Research to determine the limits of normal and aberrant
    meat quality (PSE and DFD) in pork. Proc. 2nd. Int Symp. Cond. Meat
    Quality pigs, Zeist, 1971. Pudoc, Wageningen
Schmitten, F., Jüngst, H., Schepers, K.-H. and Festerling, A. 1984.
    Untersuchungen uber die Wirkung des magnesiumhaltigen
    Erganzungsfuttermittels "Cytran" auf die Fleischbeschaffenheit von
    stressresistenten und stressanfälligen Schweinen. Deutsche
    Tierärztliche W. 91, 149-151.
Schneider, W. 1984. Magnesium for pigs? Pig News and Info. 5, 83-84.
Schworer, D., Blum, J.K. and Rebsamen, A. 1981. Einfluss der
    Belastungsverhaltnisse kurz vor der Schlachtung auf die
    Fleischbeschaffenheit. Proc. European Meeting of Meat Research Workers
    27, 46-49.
Smith, W.C. and Lesser, D. 1982. An economic assessment of pale, soft,
    exudative musculature in the fresh and cured pig carcass. Anim. Prod.
    34, 291-299.
Smulders, F.J.M., Romme, A.M.C.S., Woolthuis, C.H.J., Kruijf, J.M. de,
    Eikelenboom, G. and Corstiaensen, G.P. Prestunning treatment during
    lairage and pork quality. In: Stunning of Animals for Slaughter (Ed.
    G. Eikelenboom) The Hague, The Netherlands, Martinus Nijhoff
    Publishers, pp.90-95.
Spencer, G.S.G., Wilkins, L.J. and Hallett, K.G. 1984. Hormonal and
    metabolite changes in the blood of pigs following loading and during
    transport and their possible relationship with subsequent meat
    quality. Proc. European Meeting of Meat Research Workers 30, 15-16.
Stein, H.J. 1978. Möglichkeiten zur Verbesserung der Fleischqualität von
    Schlachtschweinen. Fleisch 32, 33-34.
Tarrant, P.V., Gallwey, W.J. and McGloughlin, P. 1979. Carcass pH values in
    Irish Landrace and Large White pigs. Ir.J. Agric. Res. 18, 167-172.
Tatulov, Y.V., Nemtchinova, I.P. and Mironov, N.G. 1982. Proc. European
    Meeting of Meat Research Workers 28, paper 1.06.
Taylor, A. McM. 1966. The incidence of watery muscle in commercial British
    pigs. J. Fd. Technol. 1, 193-199.
Topel, D.G., Miller, J.A., Berger, P.J., Rust, R.E., Parrish, F.C. and Ono,
    K. 1976. Palatability and visual acceptance of dark, normal and pale
    colored porcine m. longissimus. J. Fd. Sci. 41, 628-630.
Vrchlabsky, J. 1979. Importance of transportation and pre-slaughter
    treatment of slaughter animals. Zpravodaj Masneho Prumyslu No.3,
    17-24.
Wachholz, D., Kauffman, R.G., Henderson, D. and Lochner, J.V. 1978.
    Consumer discrimination of pork color at the market place. J. Fd. Sci.
    43, 1150-1152.
Wajda, S. and Denaburski, J. 1982. Schlachtwert von Mastschweinen aus
    industriemassigen Grossmastanlagen in Abhangigkeit zu
    unterschiedlichen Standzeiten am Schlachthof. Fleischwirtschaft 62,
    1168-1172.

Wal, van der, P.G. and Eikelenboom, G. 1984. Effect of muscle temperature soon after slaughter on pork quality: a pilot study. Neth. J. Agric. Sci., 32, 245-247.

Warriss, P.D. 1984. Incidence of carcass damage to slaughter pigs. Proc. European Meeting of Meat Research Workers 30, 17-18.

Warriss, P.D. and Akers, J.M. 1980. The effect of sex, breed and initial carcass pH on the quality of cure in bacon J. Fd. Technol. 15, 629-636.

Warriss, P.D. and Bevis, E.A. 1985. Transport and lairage times in British slaughter pigs. Br. Vet. J. 141, (in press).

Warriss, P.D. and Brown, S.N. 1985. The physiological responses to fighting in pigs and the consequences for meat quality. J. Sci. Food Agric. 36, 87-92.

Warriss, P.D. and Lister, D. 1982. Improvement of meat quality in pigs by beta-adrenergic blockade. Meat Sci. 7, 183-187.

Warriss, P.D., Dudley, C.P. and Brown, S.N. 1983. Reduction of carcass yield in transported pigs. J. Sci. Food Agric. 34, 351-356.

Wismer-Pedersen, J. and Riemann, H. 1960. Preslaughter treatment of pigs as it influences meat quality and stability. Proc. 12th Res. Conf. Amer. Meat Institute Foundation, Chicago. pp.89-106.

EFFECT OF STUNNING ON CARCASS AND MEAT QUALITY

N.G. Gregory

AFRC, Food Research Institute Bristol, Langford, Bristol
BS18 7DY, UK

ABSTRACT

Meat traders sometimes tell us that it is the traditional methods of preparation that provide the best meat. This review examines whether this is true in the case of stunning methods and pigmeat.

The physical spasms that occur during and following stunning can play a part in the production of broken bones, blood splash, bruising and PSE meat. Particular stunning methods are apt to produce particular defects, and the worst methods seem to be those that cause the greatest convulsions.

HISTORICAL ASPECTS

Pigs are put through two distinct procedures when they are killed for meat consumption. First of all they are stunned, and then they are slaughtered. Electricity and carbon dioxide are the principal stunning agents used today, but there are occasions when the captive bolt, percussion bolt and free bullet are used. Exsanguination (or sticking) is the commonest slaughtering method, but inducing a cardiac arrest has been introduced recently as an alternative.

At the turn of the 20th century, stunning was only performed in the bigger abattoirs and in killing large boars and sows. A hammer or mallet was used for small pigs and a pole axe for adults, whilst the smaller abattoirs and butchers simply stuck their pigs without any preslaughter stunning. During the 1910s and '20s various makes of captive bolt pistol became available, and it was found they required fewer repeat attempts when stunning the animals (Howarth et al, 1925). It was widely held in the meat trade though, that the pistols carried distinct disadvantages such as: insufficient bleeding of the carcass to allow it to be used in bacon production, increased incidence of blood splash in the meat, unduly violent convulsions after shooting, and, the whole procedure slowed down the slaughtering operation. Butchers and slaughtermen considered that a carcass had to be well-bled to avoid premature spoilage of the meat, and this belief no doubt arose during the days when carcass chilling facilities were non-existent or at best rudimentary. Nevertheless, the humanitarian considerations took precedence and in Britain the captive bolt method was condoned by the Ministry of Health in its model by-law of 1920. Thirteen years later the pole axe, hammer and mallet were banned by the Slaughter of Animals Act (1933).

During the 1930s, electrical stunning took the place of the captive bolt. Electricity was preferred because it was easier to administer, and in the case of pigs the stunned animal was also easier to handle (Muller, 1932). It was developed as a low voltage system varying between 40 and 120 volts depending on the country where it was used and the frequency of the current. In Britain the early manufacturers of the equipment recommended that the current should first be applied behind the animal's ears and, as soon as the pig fell to the ground, the tongs should be replaced with one electrode on the pig's forehead and the other behind an ear. The total duration of current flow for the two applications had to last 10 to 15 seconds when using 40 to 70 volt AC. From experience it was found that a total application time of 5 to 7 seconds seemed to be effective in stunning pigs. Some problems were encountered with broken shoulderblades but this was largely overcome by using only one current application behind the animal's ears (Anthony, 1932). Bacon produced from pigs which were stunned in this way, was said to be indistinguishable from bacon produced from pigs bled without prior stunning. Thus, the low voltage system developed as a subjective and pragmatic solution to the humanitarian and commercial problems previously encountered with other stunning methods.

Low voltage electrical stunning has been the preferred method for stunning pigs in Britain since the 1930s, and it has usually been performed with the animals free-standing on the floor of a stunning pen. The practical difficulties of stunning an animal whilst it is unrestrained has led to the development of restraining conveyors. Another noteable development in recent years has been the high voltage system. For a number of years there has been concern that pigs are not adequately stunned by the low voltage system and this led to research into the minimum current required to induce the stunned state as determined by the electroplectic fit in the pig's brain (Hoenderken, 1978). This led to the introduction of systems employing not less than 240V. Meat traders sometimes claim, however, that such voltages have their drawbacks, and these include blood splash, broken shoulderblades and greater carcass kicking.

Head to back electrical stunning has been used in a few pig abattoirs in the UK, and it is based on a method that was originally tested in pigs during the late 1920s (Ducksbury and Anthony, 1929). In the early version, a 200 volt current was applied between the forehead and midback regions for 6 to 10 seconds whilst the pig was held in a restrainer. The idea was dropped shortly after these preliminary trials because of problems with blood splash and haemorrhaging near the backbone and in the offal. It was found that haemorrhaging at the backbone could be avoided by placing the back electrode in the shoulder region, but this then gave rise to problems with broken scapulae.

Carbon dioxide stunning methods were developed in the USA during the 1940s, and this is now the principal method used in Denmark.

THE SCIENTIFIC EVIDENCE

The carcass and meat quality defects which are influenced by stunning and slaughter have been reviewed by Warrington (1974), Marple (1977) and Warriss (1977; 1984). They include bone fractures, blood splash, bruising, inadequate bleeding and PSE meat.

Although there are no exact figures on the incidence of bone fractures which are caused at stunning, it is felt that this problem has increased during recent years. It is associated with electrical stunning, and depends to some extent on the way the animal is restrained during the stunning procedure. Stunning pigs with high voltages can lead to broken shoulderblades if the animal is not supported off the floor. This is because the force of the initial body contraction sends a shock wave up the forelimb when the foot makes impact with the ground (van der Wal, 1976). Characteristic star-like fractures are found in the glenoid process of the scapula, and there is often a complete fracture at the neck of the bone. The incidence of the condition varies considerably between abattoirs and it is usually noticed in plants which bone-out the shoulder joint. The associated haemorrhaging requires trimming, which takes time, but the problem can be avoided by either lifting the animal off the floor prior to stunning using a restraining conveyor, or restricting the duration of current application to a short period (2 sec when using 320 volt, Braathen and Johansen 1984).

High stunning currents in restrained pigs are said to cause bone fractures in other parts of the carcass and these include the thoracic vertebrae and pelvis. In rats it is known that the larger, faster growing animal is more likely to develop fractures during electroconvulsive seizures (Stern et al, 1957), but this relationship has not been examined in the pig. Head to back stunning can produce fractures in the sacrum and slipped discs in the thoracic and lumbar vertebrae, which may or may not be associated with haemorrhaging. The incidence of the fractures probably depends on the voltage used.

Blood splash is a more nebulous problem, and the phrase "here today, gone tomorrow" is sometimes applied to this condition by the meat trade. It appears as small bleps of blood in the meat which vary in incidence according to the muscle in the carcass. It can be greatest in the gracilis and quadriceps muscles (van der Wal, 1978) and may only be noticed when the carcass is jointed. At various times it has been claimed that all stunning methods contribute towards blood splash formation. It should be pointed out, however, that religious slaughter in ruminants (which involves no prior stunning) is also associated with this defect especially if the animal is restrained for too long whilst awaiting exsanguination. There is very little documented evidence on the physical and physiological factors which cause the condition, but it is likely that any muscular contractions which cause aneurysms in the blood vessels whilst systolic pressure is high enough to force blood through the perforations, are of importance. In pigs high voltage stunning in a restraining conveyor, and captive bolt stunning, lead to intense muscular contractions and a high incidence of blood splash (Larsen, 1982; Burson et al, 1983). Cooper et al (1980), when using blood splash in lungs as an index of meat blood splash, concluded that CO_2 stunning produced a lower incidence of these conditions than electrical stunning. This was subsequently confirmed by Larsen (1982). High voltage head to back stunning (240V, 50 Hz) gave a reduction in the severity of blood splash in comparison with either low voltage high frequency (97V, 1700 Hz) or low voltage low frequency (80-125V, 50 Hz) head only stunning. The reason for this effect probably rests in the fact that head to back stunning would have induced a cardiac arrest in some of the pigs. Presumably a beating heart was required to force blood through the ruptured blood vessels. Larsen (1982) recorded less severe blood splashing when pigs were stunned at 700V in comparison with 300V, and here again the reason for the improvement probably rests in the fact that the higher voltage would have been associated with a greater incidence of cardiac arrests. Shortening the time between stunning and sticking also reduces the expression of blood splash, by minimising the period during which blood can leak through the perforated blood vessels (Burson et al, 1983). High frequency electrical stunning is said to reduce the incidence of blood splash, but the reason for this is less clear (Hatton and Ratcliffe, 1973).

The handling procedures used at slaughtering can lead to bruising in the carcass. In the case of sheep it has been found that the heart has to be beating normally for significant bruising to develop between stunning and sticking, and doubtless this also applies to pigs (Gregory and Wilkins, 1984). Thus inducing a cardiac arrest at stunning could help to reduce the incidence of bruising in pigs, which in recent years has risen by about 40% in England and Wales (Blamire et al, 1980). The electrical stunning-cardiac arrest technique does not seem to affect the rate of bleeding out from the sticking wound in pigs, nor does it influence the amount of blood apparently retained in the carcass (Warriss and Wotton, 1981). With CO_2 stunning, the pigs have a profound bradycardia by the time they leave the stunning unit, and heart rate only gathers pace on exsanguination. Bleeding efficiency may be slightly impaired in this situation (Leest et al, 1970). Problems with bleeding, however, are more serious when the sticking knife enters the shoulder joint or penetrates the pleura (Sheard et al, 1981). Shoulder sticking causes blood to flow into a pocket within the shoulder joint giving it a bruised appearance, whilst penetrating the pleura allows blood to enter the thoracic cavity where it adheres to the lining membranes.

Excessive stimulation of the animal prior to sticking can exacerbate the PSE condition. This has been demonstrated with most stunning methods, and so the underlying requirement is to minimise stress and muscular activity in every stunning and slaughtering method. Stress exerts its effects through the sympathoadrenomedullary and somatomotor nervous systems. Adrenergic agonists released from the sympathoadrenomedullary system activate muscle glycogenolysis and thereby increase the availability of substrates for acid formation. Cholinergic stimuli bring about acid formation in a more direct way by initiating muscle contractions and mitochondrial metabolism, and the two systems interact when catecholamines enhance the strength and duration of somatomotor contractions (Gregory 1981).

Carbon dioxide stunned pigs were thought to be particularly prone to producing PSE meat because of the stress associated with this gas, but this view has changed in recent years with the introduction of the Compact CO_2 stunning system. Three types of CO_2 stunning unit have been used: Oval tunnel, Dip Lift and Compact stunner (Grandin, 1980). The early versions based on the Oval tunnel were often associated with PSE meat (Bendall et al., 1966; Leest et al., 1970; Smith and Wilson, 1978). However, Sybesma and Groen (1970) observed that putting pigs through an Oval tunnel filled with air followed by electrical stunning resulted in the same incidence (9%) of PSE meat as CO_2 stunning alone. Electrical stunning without passage through the CO_2 unit produced only 1% PSE meat, as determined by the $pH_{45\ min}$ values in the semimembranosus. The conclusion, therefore, was that the stress inherent in the Oval tunnel unit was sufficient to induce the higher incidence of PSE meat. The Compact stunner operates along the lines of a Ferris Wheel and by contrast produces less PSE meat than electrical stunning (Larsen, 1982; Barton-Gade, 1984). It is presumed that this is because it is less stressful to the pigs than the Oval tunnel method.

Captive bolt stunning is rarely used in pigs because of the associated convulsions. These can be particularly intense during the first minute following stunning in pork weight pigs, and the increased activity results in a rapid initial postmortem muscle glycolysis, sometimes producing a muscle pH of less than 6.1 within 5 min. The net effect is that a high proportion of the carcasses have meat that is PSE (Klingbiel and Naude, 1977; Schoberlein et al., 1979).

There have been a number of reports on the effect of different electrical stunning procedures on meat quality (Braathen and Johansen, 1984; Hatton and Ratcliffe, 1973; Leest et al., 1970; Nielsen, 1977; Overstreet et al., 1975; Rothfuss et al., 1984; Schutt Abraham et al., 1983; van der Wal, 1978). Leest et al (1970) and van der Wal (1978) reported that there was no difference in $pH_{45\ min}$ value between 300V stunning for 1.5 sec and 70V applied for 15 sec. Schutt Abraham et al (1983) found that there was no difference in semimembranosus pH 1h portmortem between 300V and 90V head only stunning, and Overstreet et al (1975) obtained a similar finding for longissimus dorsi muscle. Nielsen (1977) and Hatton and Ratcliffe (1973) observed that high frequency electrical stunning had no effect on the incidence of PSE meat in comparison with 50 Hz alternating currents when both were applied at low voltages (up to 100V). Larsen (1982) compared 700V with 300V electrical stunning and found a 15 and 18% incidence of PSE respectively. Rothfuss et al (1984) found that stunning with 180V or more for 19 sec was effective in producing PSE meat in the majority of pigs, regardless of whether they were positive or negative halothane reactors. Braathen and Johansen (1984) reported that 320V for 2 sec gave a lower rate of postmortem muscle glycolysis than 12 sec stunning at the same voltage. The overall conclusion from these studies is that excessive stimulation by electric currents can lead to the PSE condition, but it is unlikely that the high voltage systems as they are practised today produce any more or less PSE meat than the traditional low voltage methods.

REFERENCES

Anthony, D.J. 1932. Electricity for the slaughter of animals. Vet. Rec. $\underline{12}$, 380-386.

Barton-Gade, P. 1984. Influence of halothane genotype on meat quality in pigs subjected to various preslaughter treatments. Proc. 30th Europ. Meet. Meat Res. Wkrs. 8-9.

Bendall, J.R., Cuthbertson, A. and Gatherum, D.P. 1966. A survey of pH_1 and ultimate pH values of British progeny-test pigs. J. Fd. Technol. $\underline{1}$, 201-214.

Blamire, R.V., Goodhand, R.H. and Taylor, K.C. 1980. A review of some animal diseases encountered at meat inspections in England and Wales, 1969 to 1978. Vet. Rec. $\underline{106}$, 195-199.

Braathen, O.S. and Johansen, J. 1984. The effect of short or long electrical stunning times upon pork quality. Proc. 30th Europ. Meet. Meat Res. Wkrs. 22-23.

Burson, D.E., Hunt, M/C., Schafer, D.E., Beckwith, D. and Garrison, J.R. 1983. Effects of stunning method and time interval from stunning to exsanguination on blood splashing in pork. J. Anim. Sci., $\underline{57}$, 918-921.

Cooper, T.J,.R., Gardner, G.A. and Patton, J. 1980. An evaluation of high voltage stunning of pigs in a restrainer. Proc. 26th Europ. Meet. Meat Res. Wkrs. 102-104.

Ducksbury, C.H. and Anthony, D.J. 1929. Stunning of the pig by electricity before slaughter. Vet. Rec. $\underline{9}$, 433-434.

Grandin, T. 1980. Mechanical, electrical and anesthetic stunning methods for livestock. Int. J. Stud. Anim. Prob. $\underline{1}$, 242-263.

Gregory, N.G. 1981. Neurological control of muscle metabolism and growth in stress sensitive pigs. In: Porcine Stress and Meat Quality. Ed. T. Froystein, E. Slinde and N. Standal. Agric. Fd. Res. Soc., As, Norway, pp. 11-20.

Gregory, N.G. 1983. Recent development in slaughtering methods. Meat Hygienist $\underline{40}$, 22-23.

Gregory, N.G. and Wilkins, L.J. 1984. Effect of cardiac arrest on susceptibility to carcass bruising in sheep. J. Sci. Food Agric. $\underline{35}$, 671-676.

Hatton, M. and Ratcliffe, P.M. 1973. Some observations on electrical stunning techniques in relation to biochemical and quality factors in pork. Proc. 19th Europ. Meet. Meat Res. Wkrs. 57-61.

Hoenderken, R. 1978. Electrical stunning of pigs. In: Hearing on pre-slaughter stunning. Meat Res. Cen., Kavlinge, Sweden. pp,. 29-38.

Howarth, W.J., Hayhurst, R.J. and Young, T.D. 1925. Report on the humane slaughtering of animals. Corp. London Public Heath Department, 16 pp.

Klingbiel, J.F.G. and Naude, R.T. 1977. Die invloed van proteienvoeding en bedwelminstegnieke op die pH en temperatuur van varkspiere. Agroanimalia 9, 31-35.

Larsen, H.K. 1982. Comparison of 300 volt manual stunning, 700 volt automatic stunning and CO_2 Compact stunning, with respect to quality parameters, blood splashing, fractures and meat quality. In: Stunning of Animals for Slaughter. Ed. G. Eikelenboom. Martinus Nijhoff. pp. 73-81.

Leest, J.A.; van Roon, P.S. and Brower, H.A. 1970. The influence of stunning methods on the properties and quality of pig meat. Proc. 16th Europ. Meet. Meat Res. Wkrs. 240-247.

Marple, D.N. 1977. The effect of slaughter and stunning methods on meat quality. Proc. Meat Ind. Res. Conf. 141-146.

Muller, M. 1932. The electric stunning of pigs from the standpoint of the meat industry and animal protection. Vet. J. 88, 452-454.

Nielsen, N.J. 1977. The influence of pre-slaughter treatment on meat quality in pigs. Slagteriernes Forskningsinstitut, Svin-Kodvalitet, MS 561E, 14pp.

Overstreet, J.W., Marple, D.N., Huffman, D.L. and Nachreiner, R.F. 1975. Effect of stunning methods on porcine muscle glycolysis. J. Anim. Sci. 41, 1014-1020.

Rothfuss, U., Muller, E. and Grashorn, M. 1984. Einfluss unterschiedlicher betaubungsspannungen auf parameter des stoffwechsels und der fleischbeschaffenheit beim schwein. Fleischwirtsch. 64, 833-837.

Schoberlein, L., Renatus, K., Lengerken, G.V., Hennebach, H. and Albrecht, V., 1979. Untersunchungen zum einfluss der betaubung auf die muskelfeischqualitat des schweines. Arch. exper. Vet. Med. 33, 337-345.

Schutt-Abraham, I., Levetzow, R., Wormuth, H-J. and Weise, E. 1983. Aspekta der Hochvoltbetaubung von schlachtschweinen. Fleischwirtsch. 63, 387-397.

Sheard, B., Scott, P. and White, M. 1981. In: An illustrated guide to meat inspection. Northwood Books, London. pp. 103-108.

Smith, W.C. and Wilson, A. 1978. A note on some factors influencing muscle pH values in commercial pig carcasses. Anim. Prod. 26, 229-232.

Stern, J.A., McDonald, D.G. and Werboff, J. 1957. Relationship between development of fractures during ECS, type of convulsion and weight of animals. Am. J. Physiol. 189, 381-383.

Sybesma, W. and Groen, W. 1970. Stunning procedures and meat quality. Proc. 16th Europ. Meet. Meat Res. Wkrs. 341-350.

van der Wal, P.G. 1976. Bone fractures in pigs as a consequence of electrical stunning. Proc. 22nd Europ. Meet Meat Res. Wkrs. C3: 1-4.

van der Wal, P.G. 1978. Meat quality aspects of stunning methods. In: Hearing on pre-slaughter stunning. Ed. S. Fabiansson; A. Rutegard. The National Food Administration, Uppsala. pp.39-49.

Warrington, R. 1974. Electrical stunning: A review of the literature. Vet. Bul. 44, 617-635.

Warriss, P.D. 1977. The residual blood content of meat - A review. J. Sci. Fd. Agric. 28, 457-462.

Warriss, P.D. 1984. Exsanguination of animals at slaughter and the residual blood content of meat. Vet. Rec. 115, 292-295.

Warriss, P.D. and Wotton, S.B. 1981. Effect of cardiac arrest on exsanguination in pigs. Res. Vet. Sci. 31, 82-86.

INFLUENCE OF CHILLING ON MEAT QUALITY ATTRIBUTES OF FAST GLYCOLYSING PORK MUSCLES

K.O. Honikel

Federal Centre for Meat Research, Kulmbach
Federal Republic of Germany

ABSTRACT

Pork muscles with a fast post mortem glycolysis ($pH_1 < 5.8$) lead at prevailing high temperatures ($> 35°C$) to meat with PSE characteristics. In these muscles no sarcomere contracture occurs as is observed in normal glycolysing muscle under the same temperature/pH conditions. Contracture increases drip loss. The exudative and pale appearance of PSE muscles is due to protein denaturation and membrane leakage which occur slowly at low pH and temperatures above 35°C. Rapid chilling of PSE-prone muscles shortens the time of coinciding low pH and high temperatures. In this paper we show that rapid chilling of fast glycolysing muscles reduces the   rapidly   occuring drip loss and leads to meat with less pale colour. The quality attributes of normal glycolysing muscles, however, cannot be obtained.

INTRODUCTION

In muscles of freshly slaughtered animals the pH falls. At the same time the temperature of the muscles decreases. In conventionally chilled pork carcasses with a normal post mortem glycolysis the pH falls in M. longissimus dorsi within 3 hours post mortem to pH 5.9 - 6.3. In the same period under conventional chilling conditions the temperature in these in-carcass muscles decreases to 20 - 25°C. Rigor mortis occurs then at pH 6.0 - 6.1 and at about 20° - 23°C. This temperature/pH relationship reduces myofibrillar contracture to a minimum (Honikel et al., 1986). Faster or slower chilling in relation to pH fall induces cold or rigor contracture. Contracture of sarcomeres increases drip loss in beef and pork muscles (Honikel et al., 1986).

The knowledge of these relationships is especially necessary for performing   hot boning as small hot boned cuts are easily and usually more rapidly chilled than muscles remaining in the carcass. But in pig carcasses a considerable part of the primal cuts, especially M. long. dorsi, show PSE characteristics with their rapid and increased release of drip and their pale appearance. PSE characteristics are initiated by a rapid glycolysis post mortem with a pH at 45 min post mortem below 5.8 and temperatures well above 35°C. This temperature/pH relationship, especially if the temperature decrease is slow in the following hours as happens in muscles within the carcass, induces membrane leakage and protein denaturation, the direct causes of the exudative and pale appearance of PSE meat. In PSE-prone muscles the hot boning and rapid chilling of cuts may provide the possibility of diminishing the qualitative and economic disadvantages of PSE meat. In our most recent studies we evaluated the optimum conditions for the chilling of normal and fast glycolysing muscles with the intention of looking for the best way of combining hot boning of normal and PSE-prone muscles which appear often in one carcass. This paper concerns the temperature/pH/time relationships and their influence on meat quality in PSE-prone muscles.

MATERIAL

The following muscles were dissected from the carcasses of German Landrace pigs (slaughterweight 80 - 100 kg): M. longissimus dorsi, M. psoas major, M. supraspinam, M. cleidooccipitalis, M. semimembranosus, M. semitendinosus, M. gluteus medius and M. adductor. Most of the experiments were carried out with M. long. dorsi which was obtained either within 35 to 45 min post mortem or deboned after 24 hours post mortem. After deboning, the meat was stored under various temperature/time conditions as specified in the text.

METHODS

1. pH measurements

The pH was measured in hot boned muscles by homogenizing about 3 g of meat in 10 ml of distilled water for 10 sec followed by immediate pH measurement. In muscles within the carcass the pH electrode was inserted into the meat. All pH measurements were made using combined glass electrodes (Fa. Ingold, Frankfurt/Main, Germany).

2. Determination of drip loss

Drip loss was determined in slices of M. long. dorsi of about 70 - 90 g which were weighed and sealed under atmospheric pressure in plastic pouches and stored under different temperature/time conditions. At certain times of storage the meat was taken from the pouches, dried gently with absorbent tissue and reweighed. The difference in weight in relation to initial weight was expressed as percent drip loss. The method is described by Honikel (1986) in detail.

3. Measurement of colour brightness

Colour brightness (L) was measured with Zeiss Elrephomat (Fa. Zeiss, Oberkochen, Germany) at 5 days post mortem in the slices used for drip loss measurements.

4. Determination of sarcomere length

Sarcomere length was determined with a Helium-Neon laser (wave length 632.8 nm) according to Voyle (1971).

RESULTS and DISCUSSION

Changes in muscles during post mortem glycolysis

Normal and PSE-prone muscles contain in the living animal energy-rich compounds like creatine-phosphate, ATP and glycogen. These energy-rich compounds are used up post mortem for maintaining a quasi-living state for some time. Glycogen is broken down anaerobicly to lactic acid. The velocity of lactic acid production depends on the speed of ATP turnover (Scopes, 1974). In normal glycolysing pork muscles the biochemical changes post mortem end at a pH of 5.6 - 5.4 within 5 - 9 hours after death. During this time the temperature of the muscle falls more or less rapidly according to the chilling conditions. Rapid chilling to temperatures below 10°C at pH values above 6.1 causes cold shortening, slow chilling to temperatures above 25°C and pH around 6.0 induces rigor shortening (Honikel et al., 1986). Sarcomere shortening in either case increases drip loss (Honikel, 1985). Also tenderness is affected by these temperature/pH conditions (Honikel and Reagan, 1986). In normal glycolysing muscles rigor shortening is observed when temperature/pH conditions are changed to those of PSE muscles. The pH fall in PSE muscles, however, is much faster during the first hour post mortem. Also, contrary to normal muscles, in PSE muscles rigor shortening does not occur as shown in Table 1 for 8 different muscles. PSE muscles with a pH_1 below 5.8 had in all cases longer (unshortened) sarcomeres in comparison to muscles with a pH_1 above 5.8. Therefore, the high release of exudate in PSE muscles cannot be induced by shortening of sarcomeres. Furthermore the increased drip loss of shortened muscles is released with a lag phase after the second day post mortem as shown by Honikel et al. (1986) and Fig. 1, whereas PSE muscles released their drip within the first day post mortem, levelling off afterwards (Fig. 1). We explain this fast release of drip in PSE muscles by membrane leakage and protein denaturation due to the temperature/pH/time conditions in these muscles (Honikel and Kim, 1985).

TABLE 1 Sarcomere lengths of pork muscles with different glycolytic rates, measured at 3 days post mortem (muscles were stored at 20 - 24°C from 1 to 24 hours post mortem; then at 0°C)

muscle	pH_1	number	sarcomere length (µm)		
			$\bar{x}$	s	rel.length
long. dorsi	< 5.8	43	1.72	0.22	1
	5.81-6.10	19	1.65	0.08	0.96
	> 6.10	6	1.46	0.26	0.85
supraspinam	< 5.8	1	1.82	-	1
	5.81-6.10	3	1.54	0.23	0.85
	> 6.10	8	1.49	0.18	0.82
cleidooccipitalis	< 5.8	3	1.85	0.27	1
	5.81-6.10	2	1.70	-	0.92
	> 6.10	7	1.57	0.13	0.85
semimembranosus	< 5.8	5	1.72	0.13	1
	5.81-6.10	5	1.64	0.13	0.95
	> 6.10	2	1.56	-	0.91
semitendinosus	< 5.8	9	1.77	0.14	1
	5.81-6.10	8	1.58	0.23	0.89
	> 6.10	7	1.60	0.06	0.90
adductor	< 5.8	5	1.77	0.16	1
	5.81-6.10	7	1.62	0.29	0.91
	> 6.10	9	1.45	0.21	0.82
gluteus medius	< 5.8	8	1.56	0.33	1
	5.81-6.10	6	1.32	0.22	0.85
	> 6.10	11	1.36	0.17	0.87
psoas major	< 5.8	16	2.71	0.13	1
	5.81-6.10	9	2.35	0.51	0.87
	> 6.10	1	2.35	-	0.87

A pH of 5.8 and lower is reached in PSE muscles within 45 min post mortem at temperatures well above 35°C. The temperature of rapidly glycolysing muscles within a carcass falls slowly reaching 30°C in M.long.dorsi after 2 - 3 hours. This means that high temperatures at low pH values last for some time. Under these conditions the proteins of the myofibre denature, leading to exudative meat (Fischer et al., 1979). Also the pale appearance is explained by protein denaturation (Bendall and Wismer-Pedersen, 1962). Protein denaturation in PSE muscles was confirmed by Stabursvik et al. (1984)

and Honikel and Kim (1985) using differential scanning calorimetry (DSC), protein solubility and myofibrillar ATPase activity measurements.

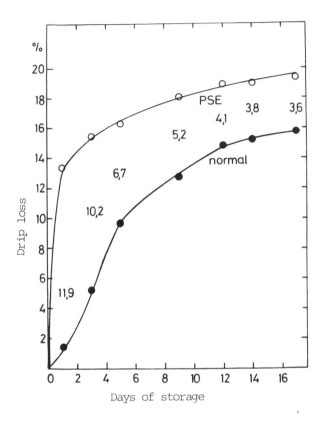

Fig. 1 Drip loss during storage of slices of "normal" and "PSE" M. longissimus dorsi with a pH_1 value of 6.0.

Slices of M. long. dorsi were obtained within 45 min post mortem with a pH of 6.0. Two slices were stored at 38°C between 45 min and 2 hours, at 35°C between 2 and 3 hours, and at 33°C between 3 and 4 hours p.m. From 4 hours up to 17 days post mortem the temperature of storage was 0°C. These conditions are called "PSE". Two further slices of the same muscles were stored at 20°C between 45 min and 4 hours post mortem. Then the temperature was 0°C up to day 17 post mortem. These conditions are called "normal". Drip loss was measured at the days indicated. The figures between the curves are the differences in drip loss at the days of measurement in percent. Data of drip loss are the mean of the two slices used.

Proteins denature rather slowly. Rapid chilling of PSE-prone muscles by diminishing the time of low pH at high temperature should at least partially prevent protein denaturation, consequently diminishing drip loss and meat paleness.

Influence of chilling conditions on drip loss of PSE-prone muscles

To determine the influence of prevailing high temperatures at low
pH values on drip loss, M. long. dorsi muscles of pork were cut from
the carcass within 35 - 45 min post mortem. In the first set of experiments
we used M. long. dorsi with a pH_1 of 6.0 and incubated adjacent slices under
different temperature conditions. The first two slices were incubated
between 45 min and 2 hours at 38°C, then the temperature was lowered
to 35°C at 2 to 3 hours, from 3 to 4 hours post mortem the slices were
kept at 33°C. In Fig. 1 these are the "PSE" conditions. Two further slices
were stored at 20°C between 45 min and 4 hours post mortem, representing
"normal" conditions. As expected the "PSE" muscle slices had a high drip
loss of 13 % in the first 24 hours while the "normal" muscles lost only
1.5 % as drip. Storage at 0°C for 17 days showed that the "PSE" muscle
slices lost a total of 19.5 % drip, the "normal" slices lost 15.9 % of
exudate. This 23 % increase of drip in the "PSE" vs. "normal" meat is
attributed to protein denaturation as we found the same percentage of
protein denaturation with DSC measurements and in the reduced protein
solubility of PSE muscles (Honikel and Kim, 1985). The rapid release at
day 1 may be additionally due to membrane leakage, as we observed in some
cases denatured myosin in the exudate (Honikel and Kim, 1985). It is
interesting to note, that the difference in drip loss of the differently
incubated slices of the same muscle is largest at day 1 and still very
high at day 5 at the time when pork is usually sold and consumed.

A similar experimental set up, but with 3 different fast glycolysing
M. long. dorsi of pH_1 of 5.4, 5.5 and 5.7 is shown in Fig. 2. In these
experiments again we took adjacent slices of the muscles and lowered
the temperature with different velocities increasing the chilling rates
from experiment 1 to 6. The conditions used in experiments 1 and 2
simulated the first 4 hours post mortem in muscles within the carcass
in a conventional chilling slaughterhouse. Experiments 3 to 6 investigated
conditions relevant to hot boning. It can be seen that the faster the
muscle slices were chilled the lower was the drip loss.

Again under "PSE-in-carcass" conditions (number 1 and 2) the drip on
day 1 was very high in relation to "hot boning" conditions (number 3 - 6).
The results in Fig. 2 obtained with fast glycolysing muscles confirmed
the results of Fig. 1. Pork is usually sold and consumed within 5 - 6 days
post mortem.

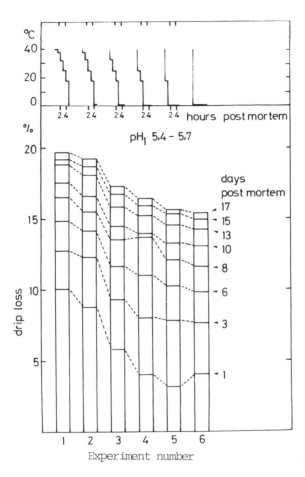

Fig. 2 Influence of chilling conditions on drip loss in slices of M.long.
dorsi with pH_1 values between 5.4 and 5.7. Six different chilling regimes
were used between 45 min and 5 hours post mortem and these are given below.
The drip loss values are the means of the measurements on three M.long.
dorsi of different carcasses which were stored under the same temperature/
time conditions.
The pH values measured 45 min post mortem of all 6 x 3 slices were
between 5.40 and 5.70. The slices cut at 45 min post mortem from the
muscle weighed about 80 g and were stored in plastic pouches under
atmospheric pressure between 50 min and 5 hours post mortem in water
baths of different temperatures. After 5 hours all samples were stored
in a chiller at 0°C.
Temperature/time conditions are indicated in top part of the Figure.
All temperature data are those of the storage medium.
Experiment 1: 40°C between 50 min and 1.5 hours
              38°C between 1.5 and 2 hours
              32°C between 2 and 3 hours
              25°C between 3 and 4 hours
              20°C between 4 and 5 hours
               0°C after 5 hours post mortem

Experiment 2: 38°C between 50 min and 1.5 hours
           32°C between 1.5 and 2 hours
           25°C between 2 and 3 hours
           20°C between 3 and 4 hours
            0°C after 4 hours post mortem

Experiment 3: 32°C between 50 min and 1.5 hours
           25°C between 1.5 and 2 hours
           20°C between 2 and 3 hours
            0°C after 3 hours post mortem

Experiment 4: 25°C between 50 min and 1.5 hours
           20°C between 1.5 and 2 hours
            0°C after 2 hours post mortem

Experiment 5: 20°C between 50 min and 1.5 hours
            0°C after 1.5 hours

Experiment 6: 0°C after 50 min post mortem

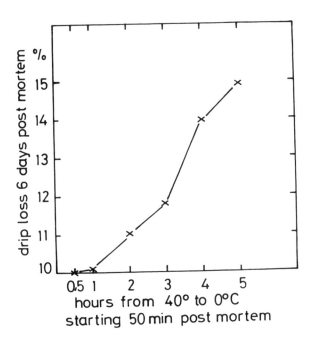

Fig. 3 The relationship between drip loss 6 days post mortem in M.long.dorsi (pH 5.4-5.7) and the mean velocity of temperature fall in the muscles between 40° and 0°C.
The data are derived from Fig. 2
The mean velocity was calculated by measuring the temperature immediately before incubation and the time taken to reach 0°C in the muscle.

Fig. 3 demonstrates that at day 6 the drip loss in PSE-prone muscles after hot boning with fast chilling rates can be reduced from 15 to about 10 %. This reduction of the exudate of PSE muscles could be of considerable economic importance. Present work is aimed at transferring from these experimental conditions to the conditions in slaughter and cutting facilities.

Influence of chilling conditions on meat colour

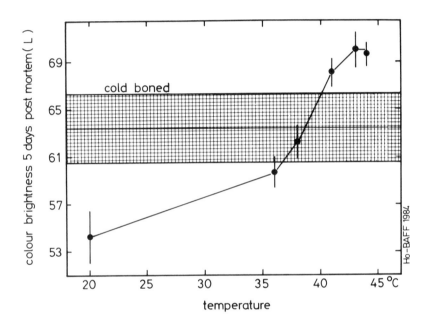

Fig. 4   Influence of temperature of storage between 45 min and 5 hours post mortem on the colour brightness (L) of four hot-boned M.long.dorsi of pork (pH$_1$ 5.4 - 5.8) after 5 days post mortem in comparison with the colour brightness of M.long.dorsi cold-boned after 24 hours.
From the left side of 4 carcasses the M. long. dorsi were hot-boned between 30 and 45 min post mortem and slices were stored between 45 min and 5 hours post mortem at the temperatures indicated. Afterwards they were held at 0°C. The muscles in the right side of carcasses were cold-boned after 24 hours and stored thereafter at 0°C. Colour brightness was measured 5 days post mortem.

Parallel to the work on the effect of fast chilling of PSE-prone muscles on drip loss we also studied its influence on the colour of meat. The paleness of PSE muscles can also be reduced by fast chilling as shown in Fig. 4. In PSE muscles with pH$_1$ values below 5.8, the effect of hot boning from one side of the carcass and storage for 4 hours at temperatures below 35°C

was compared with cold boning from the other side. The colour brightness
(L) values were considerably reduced (the muscles became redder) in the
hot boned material. The difference was observed even at 5 days post
mortem. Drip loss and colour brightness after 5 days show a linear
relationship (Fig. 5).

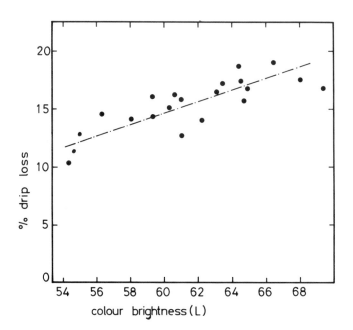

Fig. 5 Relationship between colour brightness (L) and drip of
M. long. dorsi after 5 days of storage.
Data are obtained from the same muscles as in Fig. 4.
Drip loss was determined as described in Methods.

CONCLUSIONS

Fast chilling of fast glycolysing (PSE-prone) muscles reduces the
drip loss and the paleness of the meat. However, the drip loss of normal
glycolysing muscles is still lower. Therefore, fast chilling after hot
boning can diminish the PSE characteristics but cannot solve the PSE
problem. The results reported here will be extended in further experiments
under industrial conditions of slaughter and cutting, with the aim of
evaluating the best temperature/pH/time conditions for normal and fast
glycolysing pork muscles.

REFERENCES

Bendall, J.R. and Wismer-Pedersen, J. 1962. Some properties of
fibrillar proteins of normal and watery pork muscle. J. Food
Sci. 27, 144-159.

Fischer, Chr., Hamm, R. and Honikel, K.O. 1979. Changes in solubility
and enzymic activity of muscle glycogen phosphorylase in PSE
muscles. Meat Sci. 3, 11-19.

Honikel, K.O. 1985. How to measure the water-holding capacity of meat.
Recommendation of standardized methods. Proc. EC Seminar on
Evaluation and Control of Meat Quality in Pigs, Dublin, Nov 1985.

Honikel, K.O. and Kim, C.J. 1985. Über die Ursachen der Entstehung von
PSE-Schweinefleisch. Fleischwirtschaft 65 (in press).

Honikel, K.O., Kim, C.J., Roncalés, P. and Hamm, R. 1986. Sarcomere
shortening of prerigor muscles and its influence on drip loss.
Meat Science (in press).

Honikel, K.O. and Reagan, J.O. 1986. Influence of different chilling
conditions on hot-boned pork. J. Food Sci. (submitted).

Scopes, R.K. 1974. Studies with a reconstituted muscle glycolytic
system. Biochem. J. 142, 79-86.

Stabursvik, E., Fretheim, K. and Fryskin, T. (1984). Myosin denaturation
in pale, soft, and exudative (PSE) porcine muscle tissue as studied
by differential scanning calorimetry. J. Sci. Food Agric. 35,
240-244.

Voyle, C.A. 1971. Sarcomere length and meat quality. Proc. 17th
European Meeting of Meat Research Workers, Bristol, pp. 95-97.

DISCUSSION - Session III

P A Barton-Gade   Dr Warriss gave a very good overall view of how
genotype and the environment interact and then stated at the end of his
paper that we do not know enough at present to be able to give good
directions for preventing PSE. Do we really agree with that after his
talk today, because I do not think I do?

Dr Gregory showed very clearly one of the things which I have always
thought about concerning the advantages and disadvantages of different
stunning methods. It is very difficult to distinguish between the stunning
method itself, and the systems that we apply to get the pig to the point
of slaughter and subsequent sticking. Finally Dr Honikel, you have a lot
of beautiful pictures and I have a great deal of difficulty sometimes
relating what happens in experiments where we take samples from the pig
to what happens in the intact carcass. I am sure that some PSE is caused
by poor chilling, but I am not so sure that the situation is quite as bad
as you make it.

B W Moss   I agree with Dr Barton-Gade that in general we have a great
deal of information on preslaughter handling, however I also believe as
Dr Warriss commented that there is still more detailed information required.
May I illustrate this with examples from some behavioural studies I
carried out in the lairage. Increasing the number of producer groups
mixed in a lairage pen increased aggressive activity and prolonged the
time taken for the pigs to settle down. In a related study it was found
that in single producer lots, those lots transported for longer settled
down more readily in lairage. I would not recommend that pigs should be
transported longer distances as a means of getting them to settle in
lairage. A survey carried out by a colleague Dr W S Campbell indicated
that 20% of the pigs slaughtered in Northern Ireland had longer than 20 h
from last feed to slaughter, and in experiments pigs starved 24 h prior
to slaughter showed greater aggressive activity in lairage. These examples
I believe show some of the problems and interactions that have to be
taken into consideration when making specific recommendations on lairage
design.

A J Kempster   I believe we have considerable information which could
be applied by the industry to good effect. The difficulty is that this
information is fragmented, consisting of comparisons of individual
factors or sometimes a first order interaction. We cannot go to the meat
industry and say that this is the best overall plan and what the total
advantage is of one particular system over others. I think the Danes
are developing a systems approach and we can learn a great deal from their
experience. We need to know what proportion of the overall effect on
meat quality is genetic, or is preslaughter, around the time of slaughter,
or post slaughter. Some of these interactions negate one another. More
consideration should be given to total systems and the demonstration of
overall benefits to the industry.

D Lister   I feel that it is perhaps more important now to establish
principles of operation and management rather than systems. The industry
is still so diverse and fragmented that it needs guidance on how to do
things rather more than systems which it must adhere to! In the longer
term it may well be that the systems approach will be more appropriate.

P A Barton-Gade   The aim in Denmark is to have as low transport
mortality as possible, to have systems of collection which protect the
health of the herd and to have systems which are easy to use for the
personnel, because if they are not easy to use then they just will not be
used. Using these general principles, we worked out a thirteen point
programme which tells the producer what we would like to aim at in the long
term. For example, on transport we have a recommendation that all lorries
have loading devices. We have a certain number of rules concerning
slaughter. Finally, the most important thing is that there must be
somebody in the factory who follows this up, because if there is not then
the system won't work. I am not saying that all factories use all the
recommendations. Certainly they use most of them.

D Wright   We have prepared similar codes of practice, based on studies
in Northern Ireland, to encourage farmers and slaughterers to improve
transport and pre-slaughter handling. Could I ask Dr Gregory about the
need for restraining systems for high voltage stunning. Is there a risk
that such restrainers and enclosed raceways to the stunning system are
creating stress of the type you referred to with the carbon dioxide
tunnel? Is this likely to lead to excessive stimulation prior to stunning
and exacerbate the PSE-condition?

N G Gregory   I do not know of any documented data that answers your
question. We have been looking at the incidence of PSE meat in pigs which
have been put into a 'squeeze box' restrainer. This is slightly different
from the restraining conveyor which most of us are familiar with. The
squeeze box conveyor is not widely used. We introduced it to Britain and
are trying to develop it as a cheap system for restraining the animal
when using head to back stunning. The incidence of PSE meat in the
squeeze box restrainer is identical to that in a conventional low voltage
system. So we are fairly confident that it does not cause undue stress,
at least in stress resistant pigs. I agree that races are subject to bad
handling practice. It is difficult to make generalisations but I am not
convinced that race systems are inheritently associated with PSE meat.

H J Swatland   In planning for the future we have to think in terms of
almost total automation, because labour costs are of such vital importance.
We should be planning for robotic systems from slaughter through to meat
grading. Considerable work is required in the development of automated
abattoir equipment.

G Malmfors   I want Dr Warriss to comment on the following proposal:
Why not send the whole production group (10 to 15 pigs) to slaughter at
the same time? This homogenous group would probably be very resistant to
various stressful handling procedures. The group should, of course, be
handled as a unit throughout the whole pre-slaughter handling chain.

P D Warriss   This would certainly be a solution to many of the problems
associated with preslaughter handling, particularly the effects of
fighting caused by mixing unfamiliar animals. Its major drawbacks would
probably be that a wider range of liveweight at slaughter would have to be
tolerated to allow for the different growth rates likely between members
of the rearing group. The slaughter plant might find this unacceptable
despite the probable benefits of less damaged carcasses and better meat
quality. In addition, it is a very expensive system and requires quite
a large logistical organisation.

D E Hood   What is Dr Gregory's recommendation to industry regarding the best available stunning system, for example to install in a new plant? Do you recommend electrical or carbon dioxide stunning and under what specific conditions?

N G Gregory   Taking account of both pig welfare and commercial requirements we recommend 240 V head to back stunning for 3 sec using a restrainer. There are 4 reasons for this recommendation. Firstly, in our hands this method causes a cardiac arrest in every pig and in so doing it reduces the chance of the animal regaining consciousness. Secondly, inducing a cardiac arrest at stunning will help to eliminate blood splash formation, and this is because it stops the pump which forces blood through the ruptured blood vessels. Thirdly, it should help to reduce susceptibility of the carcass to bruising where the bruise is inflicted shortly before stunning or between stunning and sticking. Lastly, the carcass tends to be more relaxed if it has undergone a cardiac arrest, and it is therefore easier to handle. It happens that the system that we advocate, head to back stunning, is illegal in some countries of the European Community where one is only allowed to stun across the head of the animal.

P A Barton-Gade   I recall that Temple Grandin in the United States was critical of head to back electrical stunning. Also, the welfare aspects of carbon dioxide stunning are presently under investigation in several countries. I believe that we need to investigate a little more before deciding on what the best stunning method is.

J B Ludvigsen   In your list of methods to counteract undesirable influences of transport and lairage you mention magnesium. Do you know what the mechanism of action of magnesium is, Dr Warriss?

P D Warriss   Magnesium has been used intravenously to lower the rate of post mortem glycolysis. It acts as an anaesthetic and neuromuscular blocker. I do not know if the exact mechanism of action is understood. Whether magnesium asparate hydrochloride administered orally works in the same way is unclear.

G Monin   In your paper in Session II, Prof Honikel, you found a shortening of sarcomers in normal muscle that was incubated post mortem at body temperature. In your paper in this session, however, you reported no shortening of the sarcomers in fast glycolysing meat. Can you explain this apparent contradiction? In our own work we have found that muscle from PSE pigs with a very rapid onset of rigor had longer sarcomers than muscle from normal carcasses, even where rigor had developed comparatively rapidly in the normal carcasses.

K O Honikel   Rigor shortening is quite a slow process. In PSE muscles, apparently the onset of rigor mortis is so fast that the muscle is not able to contract. In the paper on water holding capacity methods we stored normal glycolysing muscles under PSE temperature conditions. Normal glycolysing muscles held under high temperature conditioning regimes show rigor shortening (of 10 to 15 per cent) with increased drip loss. PSE prone muscles, however, with their rapid pH fall do not exhibit rigor shortening and have normal sarcomere lengths. Nevertheless they release a high amount of drip. The reasons for the exudation of drip in the latter case are membrane damage and protein denaturation. In normal glycolysing muscles held at high temperatures the shortening increases drip, which is released slowly and is only a fraction of the drip exudating from fast glycolysing PSE meat.

G Eikelenboom   The association of longer sarcomers with more rapid pH fall in pigs, fits very well with our experience with electrical stimulation in beef and veal, in which we find consistently longer sarcomers in ES carcasses associated with a more rapid pH fall.

K Lundström   There is a possible explanation in the Norwegian results (Staborsvik, E., Fretheim, K. and Froystein, T. 1984, J Sci Fd Agric, 35, 240-244) showing denaturation of the myosin heads in PSE muscle. That would prevent contraction of the sarcomeres.

R G Kauffman   Would Prof Honikel comment on the possibility of cold shortening occurring during rapid chilling of pork musculature.

K O Honikel   Pork muscles in situ with a normal rate of glycolysis require three to five hours to enter rigor mortis. At this time and under commercial chilling conditions in Germany the temperature of the muscles will not normally be low enough to cause cold shortening. However, in the case of hot deboned pork I recommend the chilling of pork muscles of normal and PSE disposition to below 20°C within three to four hours, but not below 10°C before five hours post mortem.

P A Barton-Gade   It is very interesting that you have not a cold shortening problem in Germany because we certainly have in Denmark. Using very rapid chilling tunnels the temperature is below 10°C at certain points of the loin at one hour after slaughter. We have found that pigs with high pH_1 values have cold toughening under these conditions.

SESSION IV

CONTROL OF MEAT QUALITY IN BREEDING AND SELECTION

Chairman: P. Sellier

INCLUSION OF MEAT QUALITY CRITERIA IN THE AGGREGATE GENOTYPE TO
PREVENT DETERIORATION OF MEAT QUALITY IN THE DANISH PIG BREEDING
PROGRAMME

T Vestergaard

National Institute of Animal Science
Postboks 39
8833 Orum Sdr. Lyng
Denmark

ABSTRACT
    Meat quality has been part of the Danish pig breeding programme since
1954. From 1954 to 1972 the selection criterion was leanness and there
was almost no selection for meat quality score. During this period the
frequency of PSE carcasses increased markedly. During the period 1972 to
1980 selection included a meat quality index to such a degree that it was
the only trait that showed phenotypic improvement. From 1980 to the
present all breeding animals, including progeny, full and half sibs have
been evaluated by the selection index method. The aggregate genotype
consists of four traits: daily gain, feed conversion ratio, per cent meat
in carcass, all weighted according to their economic value, and meat
quality calculated so as not to lead to a negative genetic response in
meat quality. Changes in parameter estimates including meat quality from
period to period are probably due to change in the preslaughter stunning
method. Use of the selection index has improved genetic merit, however,
genetic parameters must be properly monitored and the index reevaluated
for observed and predicted response to correspond.

INTRODUCTION
    Since the publication of the investigations carried out by Ludvigsen
in the years 1948 to 1950, meat quality has been taken into account in the
Danish pig breeding programme (Ludvigsen, 1954). The well-known
phenomenon later designated pale, soft and exudative (PSE) pig meat was
described as muscle degeneration by Ludvigsen and defined as 'alterations
of the muscle appearing macroscopically as a discolouration, having a
pale or greyish colour resembling that of chicken or fish meat'. Since
these investigations, an evaluation of meat colour and structure has been
part of the carcass evaluation on tested pigs. From 1954 meat quality
has been assessed using different criteria and will therefore be described
separately for the periods 1954 - 1972, 1972 - 1980 and 1980 - 1985.

RESULTS
Period 1954 - 1972: At this time the main selection criterion was
leanness, measured in the carcass as back fat thickness, and from 1967

292

percent meat was also included. The routine carcass evaluation procedure for performance tested pigs included an evaluation of meat colour and meat structure but almost no selection was applied to it. Meat quality was measured on a scale ranging from 0.5 - 5 (Pedersen, 1979) on which 0.5 to 2.0 indicated marked to slight PSE, 2.5 to 3.0 indicated the desired quality and 3.5 to 5.0 indicated slight to marked DFD (dark, firm, dry) pork.

The results from this period showed an increase in the frequency of PSE carcasses. The decrease in the score for meat colour, recorded until 1970, was probably a correlated response to the selection for leaner pigs. The decrease in meat colour score followed the decrease in backfat thickness (Fig 1).

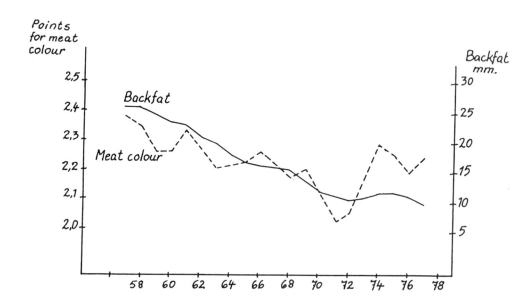

Fig. 1   Trends in meat colour and backfat thickness from 'on stations' results. More than 7000 pig carcasses were measured in each year (Vestergaard, 1979).

Period 1972 - 1980:   During this period meat quality was introduced as a breeding goal together with leanness. In order to improve meat quality a Meat Quality Index (MQ) was introduced as a selection criterion in 1972. The Index was calculated from measurements of pH and reflectance in loin

and ham. The Index was scaled to be in the interval (0; 10), with both
PSE and DFD meat corresponding to low index values, (Barton 1974).

From 1972 until the end of the seventies, meat quality was the only
trait that showed phenotypic improvement. The index value rose steadily
from 6.77 in 1972 to 7.26 in 1980, whereas daily gain, feed conversion
ratio and meat percentage remained unchanged (Table 1; Pedersen 1979).

Period 1980 - 1985: Since autumn 1980, all breeding animals have been
assessed by the selection index method, taking into account information
from their own performance, from progeny and from full- and half-sibs.
The aggregate genotype consists of the four traits listed below, with
their respective economic weights in Danish kroner per unit.

	TRAIT	ECONOMIC WEIGHT
1	Daily gain (g)	0,14 kr
2	Feed conversion ratio (SFU per kg gain)	-120,00 kr
3	Per cent meat in carcasses	7,00 kr
4	Meat quality index (MQ)	10,00 kr

The economic weight of the first three traits is calculated on the
basis of the economic value of the traits, whereas the economic weight
of MQ is calculated so that selection by the selection index may not lead
to a negative genetic response in MQ.

The value of 10 kr per unit of MQ is based on the estimates of
parameters from period 1 in Table 2. In 1983, the estimated parameters
were updated (period 2 in Table 2).

The heritability (h^2) of MQ decreased from 0.59 in period 1 to 0.26
in period 2 and the genetic correlation between MQ and daily gain changed
from -0.17 to 0.09. The much lower h^2 in period 2 may be due partly to
the fact that the method of stunning at most abattoirs changed from
electrical stunning to carbon dioxide anaesthesia. The substantial
decrease in MQ index for the year 1984/85 (Table 1) suggests that the
estimates of parameters deviate strongly from the true parameters.
Estimates of parameters for period 3 show that the genetic correlation
between MQ and daily gain is negative once more and to a greater degree
(-0.22) than before. In addition, the genetic correlation between MQ
and feed conversion ratio has increased from 0.08 to 0.35.

TABLE 1   Means of 4 traits from 'on stations' results for Danish Landrace pigs (adjusted for station x season effect according to Andersen and Vestergaard, 1984).

	Number of test groups	Daily gain (g)	FCR[1]	Meat percentage	Meat quality index (MQ)
1984/85	1086	827	2.74	63.9	7.05
1983/84	1099	795	2.81	64.1	7.22
1982/83	1468	760	2.86	64.0	7.16
1981/82	2423	743	2.93	63.6	7.30
1980/81	2982	727	2.96	63.5	7.26
1979/80	2889	726	2.98	63.3	7.19
1978/79	2654	722	3.00	63.2	7.16

[1]Feed conversion ratio, expressed in Scandanavian Feed Units per kg of gain

TABLE 2   Estimated genetic parameters for the Danish Landrace breed. (Family heritability on the diagonal and genetic correlations on top; standard errors of estimates in brackets).

Traits	Periods[1]	Average daily gain	Gain/ feed intake	Percent meat in carcass	MQ index
ADG	1	.37 (.06)	−.57 (.09)	−.39 (.07)	−.17 (.06)
	2	.62 (.11)	−.56 (.19)	−.27 (.14)	.09 (.19)
	3	.49 (.09)	−.43 (.14)	−.27 (.09)	−.22 (.11)
FCR	1		.43 (.06)	.01 (.06)	.19 (.05)
	2		.49 (.11)	−.34 (.15)	.08 (.20)
	3		.36 (.10)	−.25 (.11)	.35 (.12)
Percent meat	1			.57 (.06)	−.08 (.05)
	2			.58 (.11)	−.10 (.20)
	3			.59 (.09)	−.02 (.11)
MQ	1				.59 (.06)
	2				.26 (.12)
	3				.32 (.09)
Genetic variance	1	2107	.020	2.43	.63
	2	2546	.014	1.97	.20
	3	1954	.008	1.77	.34

[1]Period 1: Based on records of test group means from 1974 to 1976; number of test groups = 6304; number of sires = 2381.
Period 2: Based on records of test group means from May 1981–June 1982; number of test groups = 1612; number of sires = 747.
Period 3: From June 1982 to May 1985; number of test groups = 2609; number of sires = 986.
The parameters were calculated according to Andersen and Vestergaard (1984).

The consequence of applying parameter estimates from period 3 is that the economic weight of MQ will increase from kr 10.00 to kr 21.00 per unit in order to avoid a negative response in MQ when using the selection index.

The estimates for parameters that include MQ vary more from one period to another than estimates not involving MQ, which could be partly due to the change in stunning method in period 2 (Table 2).

The use of the selection index has been successful in the improvement of genetic merit, but the data on meat quality illustrate a common problem: the efficiency of the selection index can be severely affected if genetic parameters are poorly estimated. Sampling errors of the estimates and unaccounted changes of the parameters during the breeding programme are two factors that reduce the efficiency of index selection.

REFERENCES

Andersen, S. and Vestergaard, T. 1984. Estimation of genetic and phenotypic parameters for selection index evaluation in the Danish pig breeding programme. Acta Agric. Scand., 34, 231-243.

Barton, P. 1974. A standardised procedure for the pre-slaughter treatment of pigs to be tested for meat quality. Proc. 20th Europ. Meeting of Meat Res. Workers, Dublin, Ireland, 52-54.

Ludvigsen, J. 1954. Undersøgelser over den såkaldte 'muskeldegeneration' hos svin. 272, ber. fra forsøgslaboratoriet (with an English Summary)

Pedersen, O.K. 1979. Testing of breeding animals for meat production and meat quality. Acta Agric. Scand. Suppl. 21, 122-132.

Vestergaard, T. 1979. Selektionsindeks som redskab i dansk svineavl. Hyologisk Tidsskrift 4, 25-28.

THE HALOTHANE TEST IN IMPROVING MEAT QUALITY

A.J. Webb, O.I. Southwood, S.P. Simpson

AFRC Animal Breeding Research Organisation, West Mains Road,
Edinburgh EH9 3JQ, United Kingdom

ABSTRACT
        Recent research confirms that genetic liability to halothane reaction
and porcine stress syndrome (PSS) is largely controlled by a single locus
(HAL), with evidence of genetic and maternal effects on expression.   It
now appears that the gene causing halothane reaction (n) may not be fully
recessive,either for halothane sensitivity or meat quality.   Nevertheless
annual benefits from producing a heterozygous (Nn) slaughter generation
in the UK could reach £20 million.   The incidence of halothane reaction
in European Landrace strains ranges from 1 to 100%.   Continued selection
for improved lean content is expected to increase the frequency of the
halothane gene (n).   Phenotypic halothane screening programmes would then
only serve to hold gene frequencies at intermediate equilibria.   A long
term case therefore exists for eliminating the gene from maternal lines.
Elimination by test-mating to positive homozygotes (nn) may prove
prohibitively expensive.   One option might be to await cheaper and more
efficient methods of genotyping before attempting elimination.   Immediate
priorities for research are to devise new methods of genotyping, and to
determine the net economic advantage of the heterozygote (Nn).

INTRODUCTION

        It is now some 15 years since the association between porcine stress

syndrome (PSS) and a pig's reaction to halothane anaesthesia was first

observed (Eikelenboom and Minkema, 1974).   Good evidence has accumulated

that the halothane reaction is controlled by a single locus (HAL), situated

within a group of linked blood type loci (Carden et al., 1983;   Andresen,

1985).   Early indications that the halothane gene (HAL^n) was additive

for lean content but recessive for PSS (Minkema et al., 1976) have

encouraged the use of halothane positive (nn) terminal sires to give a

heterozygous (Nn) slaughter generation with an economic advantage.   Many

European breeding programmes have therefore already adopted measures to

control the frequency of the gene in maternal lines (Kallweit, 1985).

This paper reviews the state of knowledge on the halothane test and its

likely role in breeding strategy within EEC countries.

INHERITANCE OF HALOTHANE SENSITIVITY

Major gene

        Empirical observations of a fully recessive gene have now been tested against alternative models of inheritance (Carden et al., 1983). At the Animal Breeding Research Organisation (ABRO), selection for positive (HP) and negative (HN) halothane reaction has been practised in two sets of experimental lines (Figure 1) : Pietrain/Hampshire (PTH) and British Landrace (BLR) (Webb et al., 1985). The lines clearly demonstrate that the HP incidence can be very quickly changed by selection over the intermediate range, but the approach to fixation in the HP BLR line appeared slower than expected from a straightforward recessive gene.

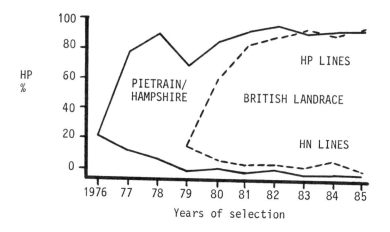

Fig. 1   Observed incidences of halothane positive reaction at 7-8 weeks of age in ABRO Pietrain/Hampshire (———) and Landrace (-----) lines selected for positive (HP) and negative (HN) reaction.   (Generation interval slightly greater than one year in Pietrain/Hampshire;   exactly one year in Landrace).

Comparing single- and two-locus models (Carden et al., 1983), the best fit in PTH was obtained with a fully recessive allele (n), and penetrance values of 0.91 for nn and zero for Nn.   In BLR the existence of a second recessive suppressor locus could not be completely ruled out. However, the best fit for a single-locus model was obtained with penetrance values of 0.91 for nn and 0.22 for Nn.

Penetrance of heterozygote

To investigate the possibility of HP reactions by heterozygotes (Nn), the ABRO BLR lines were test-mated in all combinations (Table 1).   The observed incidence of 12% HP among expected Nn progeny from HP♂ x HN♀

matings can be explained by the presence of the gene (n) in the HN line.
The significantly higher incidence of 28% HP from the reciprocal HNσ x HP♀
matings suggests that Nn progeny may be more likely to react when born out
of HP dams.

TABLE 1   Results of test matings between ABRO positive (HP) and
negative (HN) British Landrace selection lines (Southwood et al.,
1985).

| Selection line | | Expected offspring genotype | Number | | Offspring | |
Sire	Dam		Sires	Litters	No. tested	% HP
HN	HN	NN	20	28	240	2.9
HP	HN	Nn	22	37	287	11.8
HN	HP	Nn	21	36	302	27.5
HP	HP	nn	23	33	267	91.0

P < 0.01 for reciprocal difference : (HN x HP) - (HP x HN), and
for heterozygote versus normal : (Nn - NN).

The reciprocal difference implies some form of maternal inheritance.
This might for example involve the mitochondria, which are already
implicated in PSS (Cheah et al., 1984), contain their own DNA, and are
maternally inherited (Wagner, 1972).   Mitochondrial inheritance has
recently been suggested for Huntingtons Disease and spina bifida in humans
(Merril and Harrington, 1985),

Gene expression
The expression of the halothane gene appears to be affected by both
genetic and environmental factors.   For example, repeat halothane tests
on progeny from HP x HP matings in the ABRO lines gave estimated
probabilities of HP reaction averaging 43, 74 and 77% at 3, 5 and 7 weeks
of age respectively (Carden and Webb, 1984).   This suggests that the gene
is not fully expressed at 3 weeks.   In the same trials, the probabilities
of HP reaction were consistently lower in boars than gilts, averaging 62
and 73% respectively (Carden, 1982).   However, a preliminary analysis of
later generations of the BLR lines suggests that the sex difference may
have disappeared (O.I. Southwood, unpublished).
The time taken for a HP reaction may provide a quantitative measure

of the expression of the halothane gene.   In the ABRO PTH and BLR lines,
reaction times have averaged 84 and 88 seconds respectively, with
phenotypic standard deviations of 40 and 36 seconds.   A preliminary pooled
estimate of heritability was 0.38±0.31 (Webb et al., 1985).   A similar
estimate in Belgian Landrace was 0.37±0.19 (Lampo, 1978), although as in
the ABRO lines other authors have been unable to demonstrate significant
differences among sires (e.g. Reik et al., 1983).   Examples of genetic
modification of expression can be seen for the genes affecting polydactyly
(Bodmer, 1960) and white spotting (Doolittle, 1983) in laboratory mice.

The possible involvement of neighbouring blood group loci in modifying
the expression of the halothane gene has been investigated in the ABRO BLR
lines (Imlah, 1984;   summarised by Webb et al., 1985).   Accumulating
results from four consecutive halothane tests per individual showed a trend
towards earlier age of onset and faster reaction times from haplotypes
normally associated with HAL^n, involving alleles S^s, H^a and PHI^B.
Although the association was not demonstrated beyond doubt, this raises the
possibility of one or more modifier loci within the region, either marked
by or directly involving the blood group loci.

FREQUENCY OF HALOTHANE REACTION
Landrace

Some recent reports of the HP incidence in European Landrace strains
are compared with pre-1980 estimates in Table 2.   Countries such as
Switzerland and the Netherlands have been successful in lowering the
incidence of HP by halothane testing and selection (Kallweit, 1985).   In
countries with intermediate HP incidences, and where little selection
against the gene has taken place, such as the UK, Sweden and France, the
incidence has not changed seriously.   Incidences in the large-hammed
German and Belgian strains appear to be approaching 100%.

Large White

Latest reports confirm that the halothane gene is either absent
altogether or present at a very low frequency in most Large White strains.
Low HP incidences have been reported in Denmark (5%), the Netherlands (3%),
Switzerland (6%), Finland (2%), Austria (2%) and Canada (2%), with an
unexpected higher incidence of 16% in Sweden (see Webb et al., 1982;
D'Allaire and Deroth, 1982;   Kallweit, 1985).

TABLE 2   Comparison of published incidences of halothane positive reaction (%) in European Landrace strains before and after 1980.

Country	1977-80[+]	1982-85	Recent author
Norway	5	2	Kallweit (1985)
Denmark	7	5	Kallweit (1985)
United Kingdom	11	10	Southwood (unpublished)
Switzerland	13	1	Kallweit (1985)
Sweden	15	21	Kallweit (1985)
France	17	16	Kallweit (1985)
Netherlands	22	8	Eikelenboom (1985)
Germany	68	79	Kallweit (1985)
Belgium	86	100	Kallweit (1985)

[+]For 1977-80 references see Webb et al. (1982).

In 1979 a British survey of 764 Large Whites from seven herds showed no HP reactions (Webb, 1980). Since then, a sample of 200 Large Whites have been test-mated to HP reactors (expected nn) from the ABRO PTH and BLR lines. Surprisingly, 42 out of 200 (21%) left at least one HP offspring, implying a national HP incidence in purebred Large White of around 1% (Webb et al., 1985).

Possible reasons for the sudden "appearance" of the gene in British Large White could include : (1) HP reactions in Nn test cross progeny, (2) chance sampling, (3) immigration, (4) some form of genetic suppressor mechanism in the breed. A total of 56 suspected Large White carriers were therefore mated inter se at ABRO, and have so far left a total of 42 out of 233 (18%) purebred Large White HP progeny. This confirms that the gene was present in at least some of the test-mated Large Whites. Examination of the pedigrees of suspect carriers implicates a large number of herds, ruling out a recent immigration. The most plausible explanation seems to be chance sampling. However, there is at least a possibility of some form of suppressor mechanism in which the gene is progressively "uncovered", for example as some threshold of leanness is crossed.

EFFECTS ON PERFORMANCE

Phenotypic effects

Good evidence now exists that, compared with HN, HP reactors show advantages in lean yield and feed efficiency, accompanied by disadvantages in meat quality, postweaning survival, carcass length, appetite and litter productivity (e.g. Webb et al., 1982). However, the difference in performance between phenotypes appears to vary widely over studies. As an example, the phenotypic differences (HP-HN) from the ABRO PTH and BLR lines have been compared (Table 3) with each other, and with those from 20 other published studies. The two sets of lines farrowed on different farms, and their growth performance was measured at Meat and Livestock Commission (MLC) testing stations in different years : PTH fed twice daily to-appetite, and BLR litters split between ad libitum and scale feeding.

TABLE 3   Phenotypic differences (HP-HN) in performance from Pietrain/Hampshire (PTH) and British Landrace (BLR) selection lines at ABRO, compared with published differences in a variety of breeds and countries.

Trait	BLR[+] Diff	se	PTH[+] Diff	se	Published differences[++] Mean	Min	Max
Litter size born	-0.1	0.3	-1.2	0.4	-	-	-
Litter size weaned	-0.3	0.3	-1.8	0.4	-	-	-
Piglet weight at birth (kg)	-0.11	0.02	-0.08	0.04	-	-	-
Piglet weight at weaning (kg)	-0.40	0.20	-0.40	0.77	-	-	-
Daily liveweight gain (g)	3	7	5	19	-2	-47	28
Food conversion ratio	-0.07	0.02	0.00	0.08	-0.06	-0.30	0.02
Carcass length (mm)	-14	7	-12	5	-11	-29	1
Eye-muscle area (cm^2)	0.8	0.3	0.8	0.6	1.1	-2.7	3.4
Killing out (g/kg)	1	3	10	7	10	2	26
Lean proportion (g/kg)	10	7	38	13	26	9	46
PSE (% carcases)	31	3	25	8	46	22	80
Mortality (% pigs)	5	2	10	5	10	5	17

[+] From Carden et al. (1985); Simpson et al. (1985); Webb and Jordan (1978); Webb and Simpson (1986).

[++] For references see Webb et al. (1982).

The phenotypic reductions (HP-HN) in litter size weaned were 1.8 piglets in PTH and 0.3 piglets in BLR, but there was little difference in the effects on piglet weights (Table 3).   Similarly, the phenotypic differences in lean content and mortality were greater in PTH than BLR, although the differences in PSE incidence were not consistent.   Across all studies, there was threefold variation in meat quantity, quality and viability.

Clearly the differences among studies must be treated with caution since they are affected by environment, sampling, and varying degrees of homozygosity.   Nevertheless, it seems both possible that the effects of the gene may vary among breeds, and likely that the genetic background could affect expression.   Thus the effects on litter size, lean content and mortality might be expected to be smaller in BLR than in PTH with its greater muscle development.   If so, the cost-effectiveness of strategies to manipulate the frequency of the gene will need to be assessed separately for each breed combination.

Genotypic effects

Although the halothane gene (n) appears additive for its beneficial effects on lean yield, the extent to which the gene (n) is recessive for its adverse effects on viability and meat quality remains uncertain (see Webb et al., 1982).   Recent genotypic comparisons have shown the heterozygote (Nn) to be intermediate between the two homozygotes (NN and nn) for PSE incidence, pH_1, drip loss, CK activity, EEL reflectance and other meat quality criteria (Schworer, 1982;   Jensen and Barton-Gade, 1985; Lundstrom et al., 1985).

In the ABRO PTH lines, a within-sire comparison of the performance of Nn and NN was conducted by mating NN boars to both nn and NN females. Progeny were performance tested at MLC stations on twice daily to-appetite feeding.   Standard errors of differences (Table 4) were large, since they include an estimate of the variance due to genetic drift which is itself subject to large errors.   However, compared with NN, Nn showed significantly paler meat colour accompanied by evidence of expected improvements in lean yield, but no change in muscle pH.   For all markets, further clarification is needed as to whether the financial penalty from a deterioration in meat quality of Nn is likely to seriously reduce the economic advantage from improved meat quantity.

TABLE 4   Differences between Nn and NN estimated from the
Pietrain/Hampshire selection lines on twice-daily to-appetite
feeding (from Carden, 1982).

Trait	Mean	s.d.	Nn-NN	se[+]	
Daily live-weight gain (g)	662	57	13	17	NS
Daily food consumption (g)	2000	124	25	34	NS
Food conversion ratio	3.03	0.20	-0.02	0.07	NS
Average backfat (mm)	28.6	3.4	-1.3	1.1	NS
Carcass length (mm)	720.0	17.1	-1.8	6.9	NS
Eye-muscle area (cm^2)	32.3	2.5	0.4	0.6	NS
Killing out (g/kg)	781	14	3	4	NS
Lean propn. in rumpback joint (g/kg)	522	37	15	14	NS
pH 90 min post mortem	6.1	0.8	0.0	0.1	NS
Eye-muscle colour (EEL)	43.1	3.5	1.6	0.9	*

[+]Standard errors of Nn-NN include estimated variance due to
   genetic drift.   No. pigs : 108 Nn and 216 NN.

BREEDING STRATEGY

What breeding strategy should be adopted for the halothane gene both
to ensure that meat quality does not deteriorate, and to exploit any
advantage in lean content?   For any one pig population, the options are
to completely eliminate the gene, to simply reduce the frequency to a
lower level, to fix the gene at a frequency of 100%, or to take no action.
Strategies for sire and dam populations are considered separately below.

Halothane positive sire lines

The cost-effectiveness of developing nn terminal sire lines will
depend on:

(1)   the magnitude of the effect of HAL^n on all aspects of
       performance, including any effects of the genetic
       background,

(2)   the extent to which HAL^n is recessive for meat quality
       and PSS,

(3)   the frequency of HAL^n in the corresponding dam line,

(4)   the costs of eliminating HAL^n, or reducing its frequency,
       in the dam line,

(5)   the additional maintenance and selection costs of the
       HP sire line.

What would be the benefit of switching to a Nn slaughter generation
in a country such as the UK?   Consider a national herd of 800 000 F_1
Large White x Landrace sows producing some 16 000 000 slaughter pigs per
year from a backcross to Large White.   Assume that the halothane gene
(n) is fully recessive with initial nn frequencies of 1% in Large White
and 15% in Landrace.   Further assume that the net economic difference
in performance in the dam breeds is £-1.4 for nn-NN and £+0.8 for Nn-NN
(Webb, 1983), with corresponding differences of £-3.9 and £+1.7 for the
terminal sire breed (Webb et al., 1982).   A very approximate calculation
(Table 5) shows that the net benefit from introducing a nn terminal sire
and eliminating the gene from the maternal lines could be of the order of
£20 million per year, after allowing for additional selection costs in
the sire line.   Assuming an instantaneous change, and with an annual
discount rate of 5%, the current value of the improvement would be £173
million after 10 years and £264 million after 20 years.

TABLE 5   Approximate calculation of economic benefit of changing
from a Large White x (Large White x Landrace) backcross to a
heterozygous (Nn) slaughter generation produced by mating a nn
terminal sire to a NN F_1 Large White x Landrace dam.

| | Present | | | Future | | |
Breed or cross	Number (000)	£ per pig	Total £ (000)	Number (000)	£ per pig	Total £ (000)
Large White (LW)	30	0.13	3.9	30	0.00	0.0
Landrace (LR)	30	0.17	5.2	30	0.00	0.0
F_1 (LWxLR)	800	0.28	224.0	800	0.00	0.0
Terminal sire (TS)	0	-	-	50	-3.90	-19.5
Slaughter generation	16 000	0.21	3 280	16 000	1.70	27 000
Less selection costs for TS	-	..	-	16	-200.0	-3 000
Total benefit	-	-	3 513	-	-	23 805

Clearly the predicted benefits depend heavily on the relative
performance of Nn, which is poorly estimated at present.   In addition,

the improvement of £1.7 for Nn progeny amounts to only two years of
selection in the British pig improvement scheme, estimated to be worth
£0.8 per year (Mitchell et al., 1982).   The terminal sire population
would therefore have to be continuously selected to maintain its advantage.
A further complication is that present methods of payment may not adequately
reflect the changes in meat quantity and quality from introducing a
halothane positive (nn) terminal sire.   In Britain, a comparison of the
economic performance of Large White and specialised large-hammed terminal
sires is currently being conducted by MLC (A.J. Kempster, personal
communication).   However, it is unlikely that the halothane gene will be
solely responsible for the changes in meat quantity and quality from the
large-hammed sire lines.

Maternal lines

If nn terminal sires should prove worthwhile, the case for a short
term reduction in frequency of HAL^n in the maternal breeds depends on the
expected change in frequency of nn in the slaughter generation.   For
example, starting from nn frequencies of 1% in maternal Large White and
15% in maternal Landrace, and assuming fully recessive inheritance, the
expected frequency of nn in the slaughter generation would be 25% (Table 6).
Eliminating HAL^n from Large White altogether (Case I) then reduces the nn
frequency in the slaughter generation from 25 to 20%.   Eliminating HAL^n
from Landrace (Case II) reduces the nn frequency in the slaughter
generation from 25 to 5%.   Reducing the frequency of nn in Landrace from
15 to 5% (Case III) reduces the nn frequency in the slaughter generation
from 25 to 16%.   Increasing the frequency of nn in Landrace from 15 to
25% (Case IV) only raises the nn frequency in the slaughter generation from
25 to 30%.   The frequency of nn in the slaughter generation, and more
especially the frequencies of stress deaths (5-10% of nn) and PSE (20-80%
of nn), are therefore relatively insensitive to small changes in frequency
of HAL^n in the maternal breeds.

In the longer term, if no countermeasures are taken in segregating
maternal breeds such as European Landrace, continued selection for leanness
can be expected to increase the frequency of the gene.   For example, the
doubling time for a nn frequency of 15% could be as little as three
generations (Smith, 1982).   Any action to maintain the gene frequency at
a low level will result in an expensive equilibrium, with selection against

halothane sensitivity working in opposition to selection for increased leanness. A strong case will almost certainly exist for eliminating the gene from maternal lines within the next few years.

TABLE 6   Effect of a change in frequency of nn in maternal breeds on the frequency of nn in the slaughter generation when using a nn terminal sire.   Table entries are frequencies (%) of nn assuming that the gene is fully recessive.

	Maternal breeds				
	Large White	Landrace	Terminal sire	F$_1$ dam	Slaughter generation
Case	(LW)	(LR)	(TS)	(LWxLR)	TSx(LWxLR)
Start	1	15	100	3	25
I	0	15	100	0	20
II	1	0	100	0	5
III	1	5	100	2	16
IV	1	25	100	5	30

ELIMINATING THE GENE

Halothane testing

The effectiveness of alternative halothane testing strategies in eliminating the gene can be compared using the approach of Smith and Webb (1981).   Assume for example that pigs are selected on an index of performance for which nn has an advantage over NN of 0.50 phenotypic standard deviation (σ).   Assume also that the gene is additive so that Nn has an advantage over NN of 0.25σ.   In the presence of selection on the index, but with no halothane testing whatsoever, an initial frequency of 15% nn would be expected to increase to 85% in just four generations of selection (Figure 2).   With phenotypic halothane screening of boars only, the initial frequency of 15% nn would still increase to around 45% in four generations.   Halothane screening of both sexes would bring the nn frequency to an equilibrium of around 20%.   The initial nn frequency of 15% would be held roughly constant by selecting only HN reactors (Nn and NN) from 100% HN litters.

In the absence of selection on the index (Figure 2), halothane screening of both sexes would still fail to eliminate the gene after

10 generations. Even with selection from 100% HN litters, elimination would require roughly 8 generations. Clearly the quickest method of elimination is by test-mating both sexes to known nn. However, the occurrence of HP Nn among the progeny (Table 1) would wrongly implicate the parent as a carrier. To overcome this for a true frequency of 20% HP Nn, some 50 halothane-tested progeny (or six litters) would be required to give a 95% probability of correctly genotyping a parent. This could dramatically increase the cost of elimination, although it is not yet clear whether Nn reactions are simply an artefact resulting from a change in genetic background in the ABRO HP BLR line.

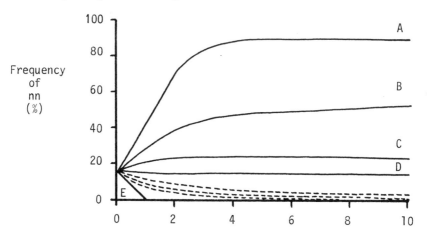

Fig. 2   Effect of different halothane testing strategies on the expected frequency of nn in the presence of selection on a performance index where the differences nn-NN and Nn-NN are 0.50 and 0.25 sd respectively.   A : no halothane screening; B: halothane screening of boars only; C: halothane screening of both sexes; D: rejection of segregating litters; E: test-mating of both sexes to nn.   Dotted lines show strategies B to D in the absence of index selection.

Halothane and succinylcholine

The use of succinylcholine in conjunction with halothane as a means of detecting the heterozygote (Nn) has been investigated at ABRO. Boars from the HP and HN BLR lines were mated to HN F_1 Large WHite x Norwegian Landrace females to give expected Nn and NN classes of progeny. Pigs received intravenous succinylcholine during halothane testing at 6-10 weeks of age. Compared with NN, Nn showed a significant increase in the duration of rigor and muscular spasms, and in a subjective score of the severity of rigor (Table 7).   However,

the accuracy of genotyping an individual pig was poor due to considerable
overlap of the distributions (Figure 3). Nevertheless, prior
succinylcholine testing could be used to reduce the proportion of Nn
in groups of HN (Nn and NN) for test mating. For example, in a herd
with an initial frequency of 20% nn, the number of pigs needed for
test mating to produce a given number of NN could be reduced by up to 50%
(Webb et al., 1986).

TABLE 7   Effect on hind leg musculature of an
intravenous injection of 10-15mg succinylcholine
during halothane anaesthesia in predicted
heterozygotes (Nn) and normal homozygotes (NN)
at the halothane locus (from Webb et al., 1986).

Observation	Nn	NN	s.e. of difference
Number of pigs	54	67	0
Rigor duration (sec)	14.5	8.1	3.1*
Rigor severity score (0-5)	3.6	2.2	0.5*
Spasm duration (sec)	21.3	15.9	1.6**

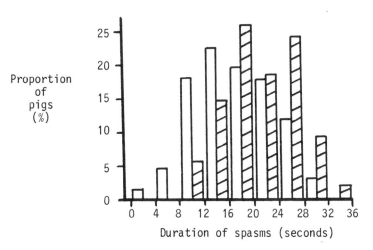

Fig. 3   Distribution of duration of spasms in response to
succinylcholine and halothane (Clear bars = predicted NN;
hatched bars = predicted Nn).

Alternative methods of genotyping

Clearly a direct method of detecting the heterozygote (Nn) at the HAL

TABLE 8 Subjective assessment of the relative merits of alternative methods of genotyping at the HAL locus.

Test	Procedure	Cost	Accuracy	Sample reference
Halothane test	Face mask	Low	Low	Eikelenboom & Minkema (1974)
Creatine kinase activity	Blood sample	Low	Low	Bickhardt (1981)
Linkage with blood type genes	Blood sample	Moderate	Variable	Imlah (1982)
Erythrocyte fragility	Blood sample	Low	Low	King et al. (1976)
Mitochondrial calcium efflux	Post mortem	High	High	Cheah et al. (1984)
Halothane and succinylcholine	Injection	Moderate	Moderate	Webb et al. (1986)
Muscle contracture (eg Caffeine)	Biopsy	High	High	Britt et al. (1978)
Halothane progeny testing	Test mating	High	High	Southwood et al. (1985)

locus is required, and a number of possibilities have been invesitgated. A very subjective appraisal of the tests shows a close association between cost and accuracy (Table 8). So far, only blood typing and CK activity are a practical proposition on a large scale. The most promising, blood typing, is discussed elsewhere in this symposium (Andresen, 1985). Other methods such as mitochondrial calcium efflux may offer a prospect of a cheaper test through a related blood parameter.

PRIORITIES FOR RESEARCH

Short term research priorities appear twofold : first to provide a quick cheap method of genotyping at the HAL locus, and second to determine the relative performance of the heterozygote, particularly for meat quality. In addition to the various tests already under investigation (Table 8), new techniques in molecular biology for detecting DNA polymorphism offer a good prospect of direct methods of genotyping (Archibald, 1985), although the timescale is uncertain. Accurate genotyping will in turn be required for precise estimation of the relative performance of Nn. Homozygous selection lines such as those at ABRO have the disadvantage of accumulating genetic drift, as well as possible systematic changes in genetic background affecting gene expression. Differences among genotypes will therefore best be estimated from current nucleus populations.

A longer term priority will be to assess the possibility of separating the beneficial and harmful effects of the halothane gene within the same strain of pig. Prospects for success depend on whether the effects on performance arise from a single pleiotropic gene, or two or more loci in linkage disequilibrium. If a single gene, background genetic variation may allow the expression for PSS to be reduced while retaining some of the advantage in leanness, for example by selection for slow reaction time. If two or more loci close together, it may be possible to identify favourable recombinants using blood type or DNA polymorphic markers (Archibald, 1985).

New techniques in biotechnology offer the opportunity for further investigation of possible mitochondrial inheritance of PSS. Since mitochondrial DNA is only roughly 16 500 base pairs in length (Watanabe et al., 1985), it would be possible to compare the amino acid sequence from the BLR HP and HN lines at ABRO. It may also be possible to establish whether mitochondria from the HN line protect against halothane reactions in the HP line, for example by transplanting a HP nucleus into the cytoplasm of a fertilized HN egg.

Finally, there is growing concern that the eating quality of pork may be declining. An important question for the future will therefore be the extent to which the unfavourable genetic correlation between meat quantity and quality will remain in the absence of the halothane gene.

CONCLUSIONS

It now appears that the halothane gene (HAL^n) may not be fully recessive for its effects on halothane sensitivity and PSS. In particular, the economic loss due to meat quality for the heterozygote (Nn) is unclear. Nevertheless the net economic benefit of producing a 100% Nn slaughter generation in the UK could be as great as £20 million per year. In the long term a case will almost certainly exist for eliminating HAL^n from maternal lines. However, elimination by test-mating to nn is likely to be prohibitively expensive. Immediate research priorities are therefore to develop direct methods of genotyping at the HAL locus, and to determine the relative economic advantage of Nn.

With rapid advances in molecular genetics, it seems likely that easier and cheaper methods of eliminating the halothane gene could be available within 5 to 10 years. In view of the uncertainty over the benefits of HP sire lines, the relatively slow short-term rise in gene frequency and the expense of existing methods of elimination, the best strategy for the present may be either to take no action or to simply prevent a further serious increase in gene frequency. In the event of an unforeseen rapid rise in frequency or a shift in economic importance, existing halothane screening methods would still be available to reduce the frequency to a low level rather quickly. In the meantime, it may therefore be sufficient to monitor the change in frequency in the parent breeds.

REFERENCES

Andresen, E.   1985.   Selection against PSS by means of blood typing.
    EEC Seminar "Evaluation and Control of Meat Quality in Pigs",
    Dublin.
Archibald, A.L.   1985.   A molecular genetic approach to the porcine
    stress syndrome.   EEC Seminar "Evaluation and Control of Meat
    Quality in Pigs", Dublin.
Bickhardt, K.   1981.   Blood enzyme tests.   In "Porcine Stress and
    Meat Quality" (Ed. T. Froystein, E. Slinde, N. Standal)
    (Agric. Food Res. Soc., As, Norway).   pp. 125-134.
Bodmer, W.F.   1960.   Interaction of modifiers : the effect of pallid
    and fidget on polydactyly in the mouse.   Heredity, 14, 445-448.
Britt, B.A., Kallow, W. and Endrenyi, L.   1978.   Malignant hyper-
    thermia-pattern of inheritance in swine.   In "Second International
    Symposium on Malignant Hyperthermia" (Ed. J.A. Aldrete, B.A. Britt)
    (Grune and Stratton, New York).   pp. 195-211.
Carden, A.E.   1982.   Genetics of Halothane Susceptibility in Pigs.
    Ph.D Thesis, Univ. Edinburgh, 150pp.
Carden, A.E., Hill, W.G. and Webb, A.J.   1983.   The inheritance of
    halothane susceptibility in pigs.   Genet. Sel. Evol., 15.
    65-82.
Carden, A.E., Hill, W.G. and Webb, A.J.   1985.   The effects of
    halothane susceptibility on some economically important traits
    in pigs.   1. Litter productivity.   Anim. Prod., 40, 351-358.
Carden, A.E. and Webb, A.J.   1984.   The effect of age on halothane
    susceptibility in pigs.   Anim. Prod., 38, 469-475.
Cheah, K.S., Cheah, A.M., Crosland, A.R., Casey, J.C. and Webb, A.J.
    1984.   Relationship between Ca^{2+}, glycolysis and meat quality
    in halothane-sensitive and halothane-insensitive pigs.   Meat
    Science, 10, 117-130.
D'Allaire, S. and Deroth, L.   1982.   Incidence of porcine stress
    syndrome susceptibility at the St. Cyrille Record of Performance
    station in Quebec.   Can. Vet. J., 23, 202.
Doolittle, D.P.   1983.   Effect of maternal age and parity on spot
    size in mice.   J. Hered., 70, 390-394.
Eikelenboom, G.   1985.   Ways to improve meat quality in pigs.   In
    "Stress Susceptibility and Meat Quality in Pigs" (Ed. J.B. Ludvigsen).
    EAAP Publn. No. 33, 68-79.
Eikelenboom, G. and Minkema, D.   1974.   Prediction of pale, soft,
    exudative muscle with a non-lethal test for halothane-induced
    porcine malignant hyperthermia syndrome.   Neth. J. Vet. Sci. 99
    : 421-426.
Imlah, P.   1982.   Linkage studies between the halothane (HAL),
    phosphohexose isomerase (PHI) and the S(A-O) and H red blood cell
    loci of Pietrain/Hampshire and Landrace pigs.   Anim. Blood Gps.
    Biochem. Genet., 13, 245-262.
Imlah, P.   1984.   Blood group association with severity and speed of
    the halothane reaction.   Anim. Blood Gps. Biochem. Genet., 15,
    275-284.
Jensen, P. and Barton-Gade, P.A.   1985.   Performance and carcass
    characteristics of pigs with known genotypes for halothane
    susceptibility.   In "Stress Susceptibility and Meat Quality in
    Pigs" (Ed. J.B. Ludvigsen).   EAAP Publn. No. 33, 80-87.

Kallweit, E. 1985. Selection for stress resistance of pigs in various West European countries. In "Stress Susceptibility and Meat Quality in Pigs" (Ed. J.B. Ludvigsen). EAAP Publn. No. 33, 60-67.

King, W.A., Ollivier, L. and Basrur, P.K. 1976. Erythrocyte osmotic response test on malignant hyperthermia susceptible pigs. Ann. Genet. Sel. Anim., 8, 537-540.

Lampo, P. 1978. Quantification of the halothane anaesthesia test for stress susceptibility in pigs. Proc. 5th Wld. Congr. Int. Pig Vet. Soc., Zabreb, Paper KA28.

Lundstrom, K., Rundgren, M., Edfors-Lilja, I., Essen-Gustavsson, B., Nyberg, L. and Gahne, B. 1985. Effect of halothane genotype on immune response, muscle characteristics, meat quality and performance - within litter comparison. 36th Ann. Mtg. EAAP, Greece, 20pp.

Merril, C.R. and Harrington, M.G. 1985. The search for mitochondrial inheritance of human diseases. Trends in Genetics, 1, 140-144.

Minkema, D., Eikelenboom, G. and van Eldik, P. 1976. Inheritance of MHS-susceptibility in pigs. Proc. 3rd Int. Conf. Prod. Disease in Farm Animals, Wageningen. (Pudoc, Wageningen). pp. 203-205.

Mitchell, G., Smith, C., Makower, M. and Bird, P.J.W.N. 1982. An economic appraisal of pig improvement in Great Britain. 1. Genetic and production aspects. Anim. Prod., 35, 215-224.

Reik, T.R., Rempel, W.E., McGarth, C.J. and Addis, P.B. 1983. Further evidence on the inheritance of halothane reaction in pigs. J. Anim. Sci., 57, 826-831.

Schworer, D. 1982. [Study of meat quality and stress resistance in pigs from the MLP-sempach testing scheme]. Doctors Thesis. Univ. Zurich, 138pp.

Simpson, S.P., Carden, A.E., Webb, A.J. and Wilmut, I. 1985. Influence of the halothane gene on reproduction in pigs. Anim. Prod., 40, 529-530. (Abstract).

Smith, C. 1982. Estimates of trends in the halothane gene in pig stocks with selection. Z. Tierzuchtg. Zuchtgsbiol., 99, 232-240.

Smith, C. and Webb, A.J. 1981. Effects of major genes on animal breeding strategies. Z. Tierzuchtg. Zuchtgsbiol., 98, 161-169.

Southwood, O.I., Webb, A.J. and Carden, A.E. 1985. Halothane sensitivity of the heterozygote at the halothane locus in British Landrace pigs. Anim. Prod., 40, 540-541. (Abstract).

Wagner, R.P. 1972. The role of maternal effects in animal breeding. III. Mitochondria and animal inheritance. J. Anim. Sci., 35, 1280-1287.

Watanabe, T., Hayashi, Y., Ogasaware, N. and Tomoita, T. 1985. Polymorphism of mitochondrial DNA in pigs based on restriction endonuclease cleavage patterns. Biochem. Genet., 23, 105-113.

Webb, A.J. 1980. The incidence of halothane sensitivity in British pigs. Anim. Prod., 31, 101-105.

Webb, A.J. 1983. Effect of halothane phenotype on the performance of British Landrace pigs. Anim. Prod., 36, 520. (Abstract).

Webb, A.J., Imlah, P. and Carden, A.E. 1986. Succinylcholine and halothane as a field test for the heterozygote at the halothane locus in pigs. Anim. Prod. (In press).

Webb, A.J., Carden, A.E., Smith, C. and Imlah, P. 1982. Porcine stress syndrome in pig breeding. Proc. 2nd Wld. Congr. Genet. Appl. Livest. Prod., Madrid. (Editorial Garsi, Madrid). Vol. 5, 588-608.

Webb, A.J. and Jordan, C.H.C. 1978. Halothane sensitivity as a field
    test for stress-susceptibility in the pig. Anim. Prod., 26,
    157-168.
Webb, A.J. and Simpson, S.P. Effect of halothane phenotype on the
    performance of British Landrace pigs. 2. Growth and carcass
    traits. (In preparation).
Webb, A.J., Southwood, O.I., Simpson, S.P. and Carden, A.E. 1985.
    Genetics of porcine stress syndrome. In "Porcine Stress and
    Meat Quality" (Ed. J.B. Ludvigsen). EAAP Publn. No. 33, 9-30.

SELECTION AGAINST PSS BY MEANS OF BLOOD TYPING

Erik Andresen

Department of Animal Genetics
The Royal Veterinary and Agricultural University
Bülowsvej 13, DK-1870 Frederiksberg C, (Copenhagen), Denmark

ABSTRACT

The discoveries leading to establishment of the 6-locus linkage group Phi, Hal, S, H, Po2, Pgd are reviewed. The halothane locus, Hal, is a major quantitative trait locus, QTL, affecting the porcine stress syndrome, PSS, and associated meat quality traits. The five loci closely linked to the Hal locus are useful for marker assisted selection, MAS, against the recessive Hal^n gene and, thereby, against PSS. Application of MAS to population data, in contrast to family data, requires linkage disequilibrium between the Hal genes and the marker genes considered. Examples of selection are presented. The usefulness of haplotyping within families for distinguishing between halothane NN and Nn genotypes is emphasized. Attention is called to complications due to heterozygous advantage of the Hal genes. The efficiency of selection based on halothane tests alone versus halothane tests combined with blood typing remains to be evaluated. The need for, and possibility of, exact heterozygosity-diagnosis is discussed. Finally, the view is expressed that animal welfare problems should not be ignored whenever eradication or exploitation of the halothane gene is considered.

EARLY HISTORY

The symptoms observed in pigs affected with stress were initially studied and defined by Ludvigsen during the period 1948-57 (Ludvigsen, 1954, 1955, 1957, see also Wismer-Pedersen, 1959). Incidentally, at the end of this period immunogenetic studies revealed the H blood group system (Andresen, 1957) which is one of the marker systems for the porcine stress syndrome, PSS, a designation introduced by Topel et al. (1968). MHS stands for malignant hyperthermia syndrome (Sybesma and Eikelenboom, 1969). The two designations are synonymous, but malignant hyperthermia is more descriptive of the symptoms which develop in the live animal and leads to PSE (pale, soft, exudative) meat or DFD (dark, firm, dry) meat after slaughter.

Genetic linkage between the H blood group locus and the two enzyme loci Phi (phosphohexose isomerase) and Pgd (phosphogluconate dehydrogenase) was established in 1971 (Andresen, 1971). In 1974 the halothane test was introduced as a diagnostic aid for detecting pigs which are able to develop the MHS (Eikelenboom and Minkema, 1974, Christian, 1974). In several breeds of pigs the abnormal reaction to halothane behaves as a simple autosomal, recessive trait provided the pigs are tested under specified conditions (Ollivier et al., 1975).

In 1976 three associations were discovered which soon proved to be of great significance for an understanding of the causal effects. The first association discovered was between halothane reactions and phenotypes of the H blood group system (Rasmusen and Christian, 1976). The second was the association between H blood groups and meat quality (Jensen et al., 1976), so both halothane reactions and H blood groups are associated with meat quality (cf. Barton et al., 1977). The third was an association between halothane reactions and genotypes of the Phi system (Jørgensen et al., 1976).

The most reasonable deductions to be made from the observations during the period 1974-76 was that (1) the halothane locus is a major locus for meat quality and (2) the observed associations with the H and Phi loci were caused by close linkage between the three loci Hal, H and Phi.

The next step was to confirm the assumption that the Hal locus was a member of the previously established linkage group comprising H, Phi and Pgd. This assumption was soon confirmed by establishing close linkage between Hal and Phi (Andresen and Jensen, 1977).

EXPLOITING THE GENETIC LINKAGE

The halothane locus is a major locus affecting meat quality as the difference between the average effects of alternative homozygotes, NN and nn, exceeds one standard deviation with respect to both meat colour and meat percentage (cf. Andresen et al., 1981a). Thus, the Hal locus is a major quantitative trait locus (QTL) which can be exploited for direct selection as surveyed by Webb et al. (1985b) at this conference.

The genes closely linked with the Hal locus can be used in marker assisted selection (MAS). A number of polymorphic marker loci are available for this purpose (see later section). The linkage between the marker locus and Hal may lead to association between the products of the two loci, i.e. PSS (MHS) and the marker trait, i.e. a blood group factor, an enzyme variant or a variant of a plasma protein; see illustration.

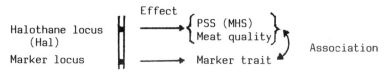

Marker assisted selection can be implemented within families or in population material or at both levels. Association within families is due to joint segregation of the linked genes and their respective products. The outcome depends on the linkage phase (coupling vs. repulsion) of the parents and the relative distance between the two loci. Clearly, the effect summarized over families also depends on the gene frequencies at the two loci.

Whereas marker assisted selection within families is not affected by presence or absence of linkage disequilibrium, MAS within populations is only effective if the degree of linkage disequilibrium, D, is significantly different from zero. This is due to the fact that the association (r) is zero if D is zero as shown by Bodmer and Payne (1965). Thus, search for existence of linkage disequilibrium (D≠0) in various pig populations having PSS problems have been of paramount importance prior to implementation of MAS at the population level. Hill (1974) has described the necessary procedures for estimating D, and the advantages and limitations of MAS have been reviewed critically by Smith and Simpson (1985).

Breeding strategies involving the Hal locus can be aimed at complete eradication of the gene n from the population or at production of a homozygous reactor (nn) sire line and a homozygous normal (NN) dam line to exploit heterozygous advantage in the slaughter generation. Either goal can be approached by halothane test alone or by halothane test combined with blood typing.

COMPLICATIONS DUE TO OVERDOMINANCE

Heterozygous advantage (overdominance) at the Hal locus was originally proposed by Ollivier et al. (1975) and later considered by Andresen et al. (1981a) and economically quantified by Smith and Webb (1981) and Webb et al. (1985a). The relationship between the three halothane genotypes, NN, Nn, nn, and their relative chances of being included in the breeding population: $f_1 = 1-s_1$, $f = 1$ and $f_2 = 1-s_2$, leads to a balanced polymorphism with gene frequencies depending solely on the selection pressures s_1 and s_2, thus, $\bar{p}_{(N)} = s_2/(s_1+s_2)$ and $\bar{q}_{(n)} = s_1/(s_1+s_2)$. Therefore, the gene frequencies depend on the breeding and production goal at any given time. The selection pressure s_1 is mainly due to low meatiness for NN individuals and s_2 to high mortality and inferior meat quality for nn individuals. Whenever meaty pigs are favoured in production there is a tendency for maintaining the n gene in the population. This complicates eradication programmes, cf. last section.

USEFUL GENETIC MARKERS

The linkage group Phi, Hal, H, Pgd (Andresen and Jensen, 1977) includes three polymorphic systems which are applicable in marker assisted selection (MAS) against the recessive halothane gene n. This linkage group was later expanded to include the S locus (Andresen, 1981, Rasmusen, 1981) and the Po2 locus (Juneja et al., 1983). For a short period of time it was erroneously assumed that this 6-locus linkage group was on the same chromosome as the SLA, C, J linkage group established by Hradecký et al. (1982). The latter linkage group is now assigned to chromosome number 7 (Geffrotin et al., 1984) whereas Pgd is not located on chromosome 7 (Christensen et al., 1985). This implies that the "Hal linkage group" is not located on chromosome 7. Some authors have proposed that the "Hal linkage group" belongs to chromosome 15 together with the G blood group locus; however, conclusive evidence for this assignment is not available (Fries et al., 1983).

Various studies have indicated that the likely order of the six loci in the "Hal linkage group" is as follows: Phi, Hal, S, H, Po2, Pgd (Juneja et al., 1983), although some results suggest that the Phi, Hal order should be reversed (Guerin et al., 1982). The latter authors indicate a distance between Phi and Pgd of 1-1.5 cM. Other estimates for the distance between the two loci are 6-8 cM (Andresen, 1971) and 4-6 cM (Gahne and Juneja, 1985b). Two reports indicate a distance exceeding 10 cM (Imlah, 1982, Van Zeveren et al., 1984).

MARKER SELECTION AND MEAT QUALITY

The five genetic loci which are markers for genes at the Hal locus and thereby for stress susceptibility and associated problems can be used in various ways. For example an observed association between a marker gene and inferior meat quality can be exploited directly without any consideration of the causative locus, i.e. the Hal locus. This approach was made in Denmark during the period 1976-79 following the discovery of an association between genotypes of the H blood group system and meat colour score (see e.g. Jensen and Andresen, 1980). In 1979 Imlah and Thomson published results on association between the H blood groups and meat quality (colour) and on the use of blood groups to predict halothane reactors, a topic to be discussed in the next section.

The aim of the selection in Denmark was to improve the population mean of the socalled KK (meat quality) index for Danish Landrace pigs from breed- ing centres. This leads to a reduction of the occurrence of PSE and thus a reduction of the Hal^n gene frequency. This is clear from the average KK in- dices and PSE frequencies for individuals of the three Hal genotypes, NN, Nn and nn (Andresen et al., 1981a, Jensen and Barton-Gade, 1985).

The KK index has a range of 0-10. An index between 0 and 6.5 corre- sponds to inferior meat quality (PSE+DFD), an index between 6.5 and 7.5 in- dicates acceptable meat quality and an index above 7.5 corresponds to good meat quality. Andresen et al. (1979) have summarized the expected results of various methods of marker assisted selection. Some of the initial obser- vations are indicated in Table 1.

TABLE 1   Presence or absence of PSE relative to H blood groups among 274 pigs of the Danish Landrace breed.

PSE present		PSE absent	
H(a+)	H(a-)	H(a+)	H(a-)
38^a(23.2)	23(37.8)	66^a(80.8)	147(132.2)

[a] assumed to be eliminated.

The excess of PSE meat from pigs possessing the blood group gene H^a is highly significant. Thus the initial approach was to eliminate all H(a+) pigs. The average KK index for the total of 274 pigs was 7.11 prior to this selection.   After eliminating the 38 + 66 H(a+) pigs the remaining 170 pigs had an average KK index of 7.33. The difference of 0.22 which represents the maximum attainable improvement, corresponds to $2-2\frac{1}{2}$ years of selection. The results also showed that a smaller improvement of 0.15 (7.26-7.11) KK units was expected if the selection was complete against individuals pos- sessing a combination of the H^a allele and the genotype $Phi^B Phi^B$. The ob- servations within the same material of 274 individuals are indicated in Table 2.

The improvement of 0.15 KK units, corresponding to $1\frac{1}{2}$ years of selec- tion, was less than that attainable by complete elimination of all indivi- duals possessing the H^a allele. Nevertheless, this approach was more accep- table to the breeders due to the smaller risk of culling H(a+) individuals having a high breeding value. Notice that in this case only 52 (23+29) H(a+) individuals are eliminated in contrast to 104 (38+66) in the first situation.

With the present knowledge it is clear that the reason for the observ-

TABLE 2 Presence or absence of PSE relative to combinations of
H blood groups and Phi genotypes among 274 pigs of the Danish Landrace breed.

PSE present $H(a+)$, $Phi^B Phi^B$		PSE absent $H(a+)$, $Phi^B Phi^B$	
present	absent	present	absent
$23^b(11.6)$	$38(49.4)$	$29^b(40.4)$	$184(172.6)$

[b) assumed to be eliminated.

ed associations was linkage disequilibrium caused by an excess of the Phi^B, H^a,
Hal^n haplotype in the Danish Landrace breed. This was confirmed in a separate study (Andresen et al., 1981b).

The examples just considered indicate the expected improvement of meat
quality (KK index) by a one-step complete elimination of marker genes associated with inferior meat quality. However, marker assisted selection usually implies marker selection in conjunction with conventional selection
procedures. The expected results under Danish conditions were considered by
Andresen et al. (1979). The conclusion was that the information on known
marker loci add relatively little to the rate of improvement of the population mean when normal selection is effective. This was in agreement with
results of other studies (Neimann-Sørensen and Robertson, 1961, Smith,
1967).

The described selection in Denmark was stopped in 1979 because the evidence indicated that the previously observed association between H blood
types and meat quality had disappeared. The KK index had reached an acceptable level. A new selection index, which included the KK index, was introduced in 1980/81 and since that time marker selection with or without
halothane testing has been used in "problem herds" only.

Although exact frequencies are not available, selection for improved
meat quality is bound to be accompanied by selection against the Hal^n gene
within breeds possessing this gene. Other methods of selecting against
Hal^n are considered in the following.

SELECTION AGAINST Hal^n USING MARKERS

Population data

Within a population with random mating the frequency of a recessive
gene, in casu Hal^n, will decrease in a predictable way if complete selection is against the recessive genotype, in casu $Hal^n Hal^n$. The rate of

change from one generation to the following can be increased in some cases by blood typing alone. The effect of selection using blood typing is predictable provided the various linkage disequilibria are known. The expected rate of change can be increased by using a combination of halothane testing and blood typing. Sellier (1985) has exemplified such predictions using data on French Landrace pigs. Doubtless the effect could be improved further if all parents of halothane reactors were culled.

Vögeli et al. (1984) have studied the association between marker genes and halothane reactivity within a line of 129 Swiss Landrace pigs. The following results were obtained (Table 3).

TABLE 3 Distribution of genotypes at three loci relative to halothane reactions among 129 pigs of the Swiss Landrace breed. (From Vögeli et al., 1984).

	H^aH^a, H^a/		Phi^BPhi^B		S^sS^s	
	present	absent	present	absent	present	absent
Hal+	53	1	52	2	51	3
Hal−	48	27	50	25	9	66

In this material the S^sS^s is the best marker-genotype to be used in selection against the Hal^n gene because almost all Hal+ individuals (51/54), but only 9 Hal− individuals would be removed, thereby leaving 69 (66+3) individuals for further breeding. In contrast, by using H^a or Phi^BPhi^B only 28 or 27, respectively, would be available for further breeding. Clearly, these results reflect the gene frequencies and the linkage disequilibria present in this particular line.

FAMILY DATA

Rasmusen et al. (1980), Jørgensen (1981), and Juneja et al. (1983) have described and exemplified the procedure involved in detecting the parental origin of haplotypes containing the halothane gene. The designation haplotype, derived by elision of haploid genotype, refers in casu to a chromosome possessing a sequence of genes from the linkage group Phi, Hal, S, H, Po2, Pgd.

Gahne et al. (1983) applied the method of haplotyping to distinguish between halothane genotypes by using the three marker loci Phi, Po2, and Pgd in conjunction with halothane testing. The material comprised 2047 offspring within 257 litters of the Swedish Landrace and Yorkshire breeds. The offspring were from Nn x Nn or Nn x nn matings. About 90% of the offspring could be assigned Hal genotype.

Čepica et al. (1984) applied a similar strategy, but they used another combination of marker loci, namely S, H, and Phi, for predicting Hal genotypes. Among 270 offspring within 42 litters of the Czech Landrace breed the Hal genotypes of 94% of these offspring were correctly predicted.

Comprehensive reviews of continued Swedish investigations have been published by Gahne and Juneja (1985a, b). In 1983-84 blood samples from more than 19.000 pigs were analysed for the Phi, Po2, and Pgd markers. Halothane testing was performed simultaneously. The predicted Hal genotypes were compared with the observed Hal phenotypes among approx. 4000 offspring of the Swedish Landrace and Yorkshire breeds. The offspring were from Nn x Nn or Nn x nn matings. The authors concluded that 90-95% of the offspring had been correctly genotyped, thus the prediction was in agreement with the initial results published in 1983.

CONCLUDING REMARKS

In most countries the harmful effects of the Hal^n gene are regarded to overshadow the beneficial effects. In addition to economical considerations one ought to consider the animal welfare problems. Malignant hyperthermia is an extremely unpleasant syndrome for humans and doubtless also for pigs. Whatever the main reasons are, the aim in most countries is to decrease the frequency of the Hal^n gene. At the present time this can be achieved by halothane testing, by blood typing or by a combination of both. In Switzerland the frequency of halothane reactors decreased from 17.7% in 1978 to 1.1% in 1983 by using the halothane test alone (Schwörer et al., 1985). The combined use of halothane testing and blood typing appears also to have been effective in the large scale Swedish investigation, but a comparison of the achievements in the two countries is not yet available.

Selection directly by means of a quantitative trait locus (QTL), in casu Hal, or by using marker assisted selection (MAS) is usually accompanied by conventional selection procedures as exemplified in this review. As meatiness continues to be a goal for selection the heterozygous advantage at the Hal locus will be maintained. Thus, application of the various methods for selection against Hal^n is counteracted by regression of the Hal^n gene frequency towards gene frequency balance. This implies that surveillance of population is required at appropriate intervals if the breeding and production goal is not changed. Alternatively, the Hal^n gene could be removed completely, but this would require a method for exact diagnosis of heterozygosity. The best approach would be to reveal the primary function of the Hal locus. This problem should have high priority among biochemists. However, animal geneticists can also contribute further to improve heterozygosity diagnosis, for example by detecting restriction fragment length polymorphism, RFLP, in the Hal region. This possibility is definitely within sight and the necessary studies should be pursued with vigour.

REFERENCES

Andresen, E. 1957. Investigations on blood groups of the pig. Nord. Vet. Med., 9, 274-284.

Andresen, E. 1971. Linear sequence of the autosomal loci PHI, H and 6-PGD in pigs. Animal Blood Groups Biochem. Genet., 2, 119-120.

Andresen, E. 1981. Evidence for a five-locus linkage group involving direct and associative interactions with the A-0 blood group locus in pigs. In "Papers dedicated to Professor Johannes Moustgaard on the occasion of his seventieth birthday" (Ed. E. Brummerstedt). (The Royal Danish Agricultural Society, Copenhagen). pp. 208-212.

Andresen, E., Barton-Gade, P., Hyldgaard-Jensen, J., Jørgensen, P.F., Nielsen, P.B. and Moustgaard, J. 1979. Selection for improved meat quality with the aid of genetic markers in pigs of the Danish Landrace breed. Acta Agric. Scand., 29, 291-294.

Andresen, E. and Jensen, P. 1977. Close linkage established between the HAL locus for halothane sensitivity and the PHI (phosphohexose isomerase) locus in pigs of the Danish Landrace breed. Nord. Vet.-Med., 30, 37-38.

Andresen, E., Jensen, P. and Barton-Gade, P. 1981a. The porcine Hal locus: A major locus exhibiting overdominance. Z. Tierzüchtg. Züchtgsbiol., 98, 170-175.

Andresen, E., Jensen, P. and Jonsson, P. 1981b. Population studies of Phi, Hal, H haplotype frequencies and linkage disequilibria in Danish Landrace pigs. Z. Tierzüchtg. Züchtgsbiol.,98, 45-54.

Barton, P., Jørgensen, P.F., Nielsen, P.B. and Moustgaard, J. 1977. Blodtypesystemer hos svin. En undersøgelse over sammenhængen mellem et enzymsystem, et blodtypesystem og kødkvalitet hos Dansk Landrace. Årsberetn. Inst. Sterilitetsforsk., 20, 93-100.

Bodmer, W.F. and Payne, R. 1965. Theoretical consideration of leukocyte grouping using multispecific sera. In "Histocompatibility Testing" (Munksgaard, Copenhagen). Series Haemat., No. 11, pp. 141-149.

Čepica, S., Hojný, J., Pazdera, J., Hradecký, J. and Králová, D. 1984. Determination of genotypes for Hal locus by means of simultaneous halothane testing and typing of linked marker loci. Z. Tierzüchtg. Züchtgsbiol., 101, 291-297.

Christensen, K., Kaufmann, U. and Avery, B. 1985. Chromosome mapping in domestic pigs (Sus scrofa): MPI and NP located to chromosome 7. Hereditas, 102, 231-235.

Christian, L.L. 1974. Halothane test for PSS-field application. Proc. Am. Ass. Swine Practitioners Conf. Des Moines, Iowa, 6 p.

Eikelenboom, G. and Minkema, D. 1974. Prediction of pale, soft, exudative muscle with non-lethal test for the halothane induced porcine malignant hyperthermia syndrome. Tijdschr. Diergeneesk., 99, 421-426.

Fries, R., Stranzinger, G. and Vogeli, P. 1983. Provisional assignment of the G blood-group locus to chromosome 15 in swine. Gene mapping in swine using natural and induced marker chromosomes. J. Hered., 74, 426-430.

Gahne, B. and Juneja, R.K. 1985a. Use of blood typing for prediction of halothane genotypes of pigs. In J.B. Ludvigsen, "Stress susceptibility and meat quality in pigs", EAAP publication, No. 33, pp. 31-42.

Gahne, B. and Juneja, R.K. 1985b. Prediction of the halothane (Hal) genotypes of pigs by deducing Hal, Phi, Po2, Pgd haplotypes of parents and offspring: Results from a large-scale practice in Swedish breeds. Animal Blood Groups Biochem. Genet. (In press).

Gahne, B., Juneja, R.K. and Petterson, H. 1983. Genotypes at the locus for halothane sensitivity (Hal) in pigs revealed by Hal-linked genetic markers in blood. Proc. 5th Internat. Conf. on Production Disease in Farm Animals. Uppsala, Sweden, pp. 100-103.

Geffrotin, C., Popescu, C.P., Criblie, E.P., Boscher, J., Renard, Ch., Chardon, P. and Vaiman, M. 1984. Assignment of MHC in swine to chromosome 7 by in situ hybridization and serological typing. Ann. Génét., 27, 213-219.

Guérin, G., Ollivier, L. and Sellier, P. 1982. Segregation study of Hal, Phi and 6-Pgd in the pig. Abstract 28. Internat. Conf. Anim. Bloodgroups and Biochem. Polymorphisms, Ottawa, Canada, p. 4.

Hill, W.G. 1974. Estimation of linkage disequilibrium in randomly mating populations. Heredity, 33, 229-239.

Hradecký, J., Hruban, V., Pazdera, J. and Klandy, J. 1982. Map arrangement of the SLA chomosomal region and the J and C blood group loci in the pig. Animal Blood Groups Biochem. Genet., 13, 223-224.

Imlah, P. 1982. Linkage studies between the halothane (Hal), phosphohexose isomerase (Phi) and the S (A-O) and H red blood cell loci of Pietrain/ Hampshire and Landrace pigs. Animal Blood Groups Biochem. Genet., 13, 245-262.

Imlah, P. and Thomson, S.R.M. 1979. The H blood group locus and meat colour, and using blood groups to predict halothane reactors. Acta Agric. Scand., suppl., 21, 403-410.

Jensen, P. and Andresen, E. 1980. Testing methods for PSE syndrome: Current research in Denmark. Livestock Prod. Sci., 7, 325-335.

Jensen, P. and Barton-Gade, P. 1985. Performance and carcass characteristics of pigs with known genotypes for halothane susceptibility. In J.B. Ludvigsen, "Stress susceptibility and meat quality in pigs", EAAP publication, No. 33, pp. 80-87.

Jensen, P., Staun, H., Nielsen, P.B. and Moustgaard, J. 1976. Undersøgelse over sammenhængen mellem blodtypesystem H og points for kødfarve hos svin. Medd. Statens Husdyrbrugsforsøg, No. 83.

Juneja, R.K., Gahne, B., Edfors-Lilja, I. and Andresen, E. 1983. Genetic variation at a pig serum locus, Po-2 and its assignment to the Phi, Hal, S, H, Pgd linkage group. Animal Blood Groups Biochem. Genet., 14, 27-36.

Jørgensen, P.F. 1981. Blood types and other biochemical markers for stress-susceptibility and meat quality in pigs. In "Porcine stress and meat quality" (Ed. T. Frøystein, E. Slinde and N. Standal) (Agricultural Food Research Society, Ås, Norway). pp. 146-159.

Jørgensen, P.F., Hyldgaard-Jensen, J. and Moustgaard 1976. Phosphohexose isomerase (PHI) and porcine halothane sensitivity. Acta. Vet. Scand., 17, 370-372.

Ludvigsen, J. 1954. Investigations on the socalled "Muscle Degeneration" in pigs. 272. Report from Institute of Animal Science, Copenhagen.

Ludvigsen, J. 1955. Investigations on the socalled "Muscle Degeneration" in pigs. 284. Report from Institute of animal Science, Copenhagen.

Ludvigsen, J. 1957. On the hormonal regulation of vasomotor reactions during exercise with special reference to and the action of adrenal cortisol steroids. Acta Endocrin., Copenhagen, 26, 406-416.

Neimann-Sørensen, A. and Robertson, A. 1961. The association between blood groups and several production characters in three Danish cattle breeds. Acta Agr. Scand., 11, 163-196.

Ollivier, L., Sellier, P. and Monin, G. 1975. Determinisme génétique du syndrome d'hyperthermie maligne chez le porc de Piétrain. Ann. Génét. Sél. Anim., 7, 159-166.

Rasmusen, B.A. 1981. Linkage of genes for PHI, halothane sensitivity, A-O inhibition, H red blood cell antigens and 6-PGD variants in pigs. Anim. Blood Groups Biochem. Genet., 12, 207-209.

Rasmusen, B.A., Beece, C.K. and Christian, L.L. 1980. Halothane sensitivity and linkage of genes for H red cell antigens, phosphohexose isomerase (PHI) and 6-phosphogluconate dehydrogenase (6-PGD) variants in pigs. Anim. Blood Groups Biochem. Genet., 11, 93-107.

Rasmusen, B.A. and Christian, L.L. 1976. H blood types in pigs as predictors of stress susceptibility. Science, 191, 947-948.

Schwörer, D., Blum, J.K., Vögeli, P. and Rebsamen, A. 1985. Stressresistance and meat quality in Swiss pig breeds in the years 1977-1985. 36. Ann. Meeting of the Europ. Ass. Anim. Prod., Kallithea, Halkidiki, Greece. Session VI, Poster session.

Sellier, P. 1985. The use of blood markers as an aid for selecting against the halothane-sensitivity gene in the French landrace pig breed. 36. Ann. Meeting of the Europ. Ass. Anim. Prod., Kallithea, Halkidiki, Greece. Vol. 1, p. 128.

Smith, C. 1967. Improvement of metric traits through specific genetic loci. Anim. Prod., 9, 349-358.

Smith, C. and Simpson, S.P. 1985. The use of genetic polymorphisms in livestock improvement. Main paper presented at the 36. Ann. Meeting of the Europ. Ass. Anim. Prod., Kallithea, Halkidiki, Greece. Abstract 64, 1.

Smith, C. and Webb, A.J. 1981. Effects of major genes on animal breeding strategies. Z. Tierzüchtg. Züchtgsbiol., 98, 161-169.

Sybesma, W. and Eikelenboom, G. 1969. Malignant Hyperthermia Syndrome in pigs. Neth. J. Vet. Sci., 2, 155-160.

Topel, D.G., Bicknell, E.J., Preston, K.S., Christian, L.L. and Matsushima, C.Y. 1968. Porcine stress syndrome. Med. Vet. Pract., 49, 40-41,59-60.

Van Zeveren, A., Van de Weghe, A. and Bouquet, Y. 1984. The use of markersystems of blood substances in an improved diagnosis of stress-susceptibility in Belgian Landrace pigs. In "Proc. 8th Internat. Pig Veterinary Society" (Eds. M. Pensaert et al.). (Ghent, Belgium). p. 249.

Vögeli, P., Stranzinger, G., Schneebeli, H., Hagger, C., Künzi, N. and Gerwig, C. 1984. Relationships between the H and A-O blood types, phosphohexose isomerase and 6-phosphogluconate dehydrogenase red cell enzyme systems and halothane sensitivity, and economic traits in a superior and an inferior selection line of Swiss Landrace pigs. J. Anim. Sci., 59, 1440-1450.

Webb, A.J., Southwood, O.I., Simpson, S.P. and Carden, A.E. 1985a. Genetics of porcine stress syndrome. In "J.B. Ludvigsen, Stress susceptibility and meat quality in pigs". EAAP publication, No. 33, pp. 9-30.

Webb, A.J., Southwood, O.I. and Simpson, S.P. 1985b. The Halothane test in improving meat quality. European Community Seminar, Evaluation and Control of Meat Quality in Pigs. Dublin, Ireland, 21 -22 Nov.

Wismer-Pedersen, J. 1959. Quality of pork in relation to rate of pH change postmortem. Fd. Res., 24, 711.

CROSSBREEDING AND MEAT QUALITY IN PIGS

P. Sellier

Institut National de la Recherche Agronomique
Station de Génétique quantitative et appliquée
78350 Jouy-en-Josas, France.

ABSTRACT

A review is presented on heterosis for meat quality. The classical view, i.e. no significant heterosis, is to be revised to some extent when considering specific meat quality traits in specific crosses. Evidence is presented for two main deviations from additive inheritance of breed differences in crossing. Abnormally fast rate of fall in pH in the early post-mortem period, an indicator of typical PSE meat condition, appears to be inherited as a recessive or partially recessive trait in Pietrain crosses. The halothane locus is likely to play the major role in this respect. On the other hand, it may be hypothesized that the comparatively larger extent of fall in pH (i.e. lower ultimate pH) shown by Hampshire purebreds is inherited as a dominant or partially dominant trait in Hampshire crosses. Recent results suggest that a dominant major gene could be involved in this "Hampshire effect" on meat quality.

INTRODUCTION

Over the last 25 years, a great deal of research work has been done on genetic variation in pig meat quality, mostly in relation to Porcine Stress Syndrome (see e.g. McGloughlin, 1980). As far as variation among populations is concerned, it has been established that marked differences exist between certain breeds with regard to meat quality traits. In the same period, the use of breed crosses has steadily increased in practical breeding programmes in most European countries.

This article is an attempt to summarise our present state of knowledge on the question of how breed differences in meat quality behave in crosses. In other words, to what extent are there heterosis effects on meat quality and, if they actually exist, in what direction these effects operate ?

CLASSICAL VIEWS ON HETEROSIS IN MEAT QUALITY

Meat quality traits were considered in several comparisons between contemporary pigs from F_1 crosses and parent breeds, allowing the estimation of individual heterosis effects on these traits. More often, the breed crosses involved were among the following ones :
- Yorkshire x Landrace : e.g. Skarman (1965), Walstra et al. (1971) and

Minkema et al. (1974) ;

- Pietrain x Landrace : e.g. Jonsson (1962), Kirsch et al. (1963), Schmidt (1964), MacDougall and Disney (1967) and Lean et al. (1972);
- Pietrain x Large White : e.g. Knoertzer (1961) and Jacquet and Ollivier (1971) ;
- Pietrain x Minnesota No. 1 : Elizondo et al. (1976, 1977) and McKay et al. (1982) ;
- various crosses between U.S. breeds (Duroc, American Yorkshire, Hampshire and Chester White) : e.g. Young et al. (1976), Schneider et al. (1982) and review by Johnson (1981).

From most of these studies, it was concluded that F_1 crosses do not differ from expectation, assuming additive inheritance, and meat quality traits are not affected to any significant extent by heterosis. The same conclusion was expressed in several review papers on crossbreeding and heterosis in pigs : e.g. Bichard and Smith (1972), Sellier (1970, 1974, 1976) and McGloughlin (1980). For instance, the latter author stated that " in crossbreds meat quality appears to be intermediate between the two parent breeds (...), suggesting that there is no beneficial effect to be gained from heterosis ".

NEED FOR REANALYSING THE PROBLEM

The lack of heterosis effects on meat quality is indeed a safe statement as far as available information is "averaged" over various crosses and various meat quality traits. However, a more useful approach in this area is likely to consider specific traits in specific crosses, and the purpose of this paper is to draw some tentative conclusions from literature data in this respect.

As regards pig meat quality, until now three specific defects have been recognized, namely :

    (1) pale, soft, exudative (PSE) meat ;

    (2) dark, firm, dry (DFD) meat ;

    (3) meat of "Hampshire type".

The DFD condition, associated with a high ultimate pH of meat, is most often of environmental origin (e.g. long-term stress in the pre-slaughter period leading to low energy content in the muscle). Genetic proneness to DFD is probably of minor importance, as pointed out by Nielsen (1981), and this abnormality will not be considered further.

Typical PSE meat stems from the very fast muscle glycolysis and consequently rapid post-mortem drop in pH. This induces a denaturation of muscle proteins, leading to pale colour and decreased water holding capacity. An indicator commonly used to determine PSE condition is the pH at 45 minutes or less after slaughter (known as pH_1), which is markedly lower in PSE than in not-PSE meat. It is well-documented that PSE condition is closely related to Porcine Stress Syndrome and halothane sensitivity (e.g. Eikelenboom and Minkema, 1974 and review by Webb, 1981). Regarding breed variation of PSE incidence, the highest incidence is found in the Pietrain breed in connection with the high frequency of halothane reactors in this breed (e.g. Hanset et al., 1983). Pietrain crosses are therefore of particular interest for studying heterosis effect on pH_1 and PSE condition.

In typical meat of "Hampshire type", the fall in muscle pH in the immediate post-mortem period takes place at a normal rate, but the total extent of fall in pH is abnormally large resulting in a very low ultimate pH of meat (measured at 24 h after slaughter and known as pH_2). The most likely explanation for this low pH_2 lies in the high "glycolytic potential" of the muscle at the time of slaughter (Monin and Sellier, 1985).

The rate as well as the extent of drop in pH appear as the two distinct phenomena of primary importance which are to be taken into account when analysing between-breed variation and consequently heterosis for meat quality. Consideration will be successively given to pH_1 and pH_2 in the following sections, with special emphasis on Pietrain crosses and Hampshire crosses respectively.

HETEROSIS FOR RATE OF FALL IN pH

As the halothane locus (Hal) plays a prominent role in determining rate of fall in pH, crosses among breeds largely differing in gene frequencies at this locus are the most informative for assessing heterosis in this trait. Crosses between the Pietrain and breeds where the halothane sensitivity gene ($\underline{Hal^n} = \underline{n}$) is absent or at a low frequency will be considered.

Pietrain crosses

Estimates of heterosis effect on pH_1 in 2-way crosses involving the Pietrain breed are few to our knowledge, and some of them come from studies of rather small scale (Table 1). As a general rule, F_1 pigs are intermediate between parental breeds for pH_1. However, it is to be emphasized that in 3 out of 4 experiments reported in Table 1, F_1's are much closer to the "stress-resistant" breed than to the "stress-susceptible" Pietrain, indicating some recessiveness of the Pietrain trait in crossing. These results can be interpreted in terms of Hal genotypes, assuming that all Pietrains used are nn and pigs from the partner breed are NN or predominantly NN (N = Hal^N = normal allele). It should be mentioned that in such Pietrain crosses, an "apparent" trend towards additivity can be found when the partner breed is not completely free from the gene n ($q \neq 0$). The F_1 group then contains a proportion q of nn pigs, which lowers the average pH_1 value in F_1's (for instance, q is of the order of 0.3 in the British Landrace according to Webb, 1981).

TABLE 1 Heterosis effect on pH_1 in F_1 Pietrain crosses.

Reference	Means (no. of pigs)			Heterosis
	Pietrain	F_1	partner breed (a)	
MacDougall and Disney (1967)	5.70 (31)	6.16 (32)	6.33 (31)	0.14 (2.4%)
Lean et al. (1972) (b)	5.68 (43)	5.95 (46)	6.00 (47)	0.11 (1.9%)
Elizondo et al. (1976)	5.7 (8)	5.9 (8)	6.3 (8)	-0.10 (-1.7%)
Elizondo et al. (1977) (c)	5.9 (4)	6.3 (8)	6.4 (8)	0.15 (2.4%)

(a) British Landrace in the two first studies ;
Minnesota No.1 (a "stress-resistant" inbred line) in the two others.
(b) Means of values found at 4 different liveweights.
(c) Means approximately estimated from a figure (crossbreds are from a F_2xF_2 cross, and theoretically only half of heterosis is expressed).

In addition to pH_1, other traits affected by PSE condition were investigated in several studies with Pietrain crosses. Muscle protein solubility is usually regarded as one of the best criteria for PSE evaluation. In the study by MacDougall and Disney (1967), denaturation of sarcoplasmic and myofibrillar proteins (as measured by loss of solubility) was considerably greater in Pietrain than in Pietrain x Landrace and Landrace pigs. The two latter groups were very similar in this respect. The same pattern was found for transmission value, a trait closely related to muscle protein solubility, in the Pietrain x Minnesota No.1 cross (Elizondo et al., 1977). As regards meat colour, assessed visually or objectively, results are somewhat less consistent. Pietrain F_1 crosses exhibited the same meat colour as pigs from the "stress-resistant" parent breed in some studies (Jacquet and Ollivier, 1971 ; Elizondo et al., 1977 ; McKay et al., 1982), whereas they were intermediate between parent breeds in other studies (Jonsson, 1962 ; Kirsch et al., 1963 ; Lean et al., 1972 ; Elizondo et al., 1976). In fact, there is some evidence that the Pietrain breed shows pale colour independently of halothane locus effect. For example, halothane-positive (HP) Pietrain pigs were found to give significantly paler meat than HP pigs from French or Belgian Landrace breeds (Sellier et al., 1984 ; Monin and Sellier, at this meeting). This "Pietrain effect" on meat colour, which acts in addition to that of Hal genotype , is likely to be additively inherited and could explain the intermediate position of some Pietrain crosses in meat colour.

Within-population comparisons between halothane genotypes

As stated above, the halothane locus is a major locus with respect to rate of fall in pH and other indicators of PSE meat condition. In the 5 last years, a number of studies dealing with within-population comparison between halothane genotypes (nn, nN and NN) for meat quality have been published. These data are of special interest in the present context, since they make it possible to accurately assess the heterosis effect at the Hal locus level, without the above mentioned "disturbing" influence of breed effects.

Results available are reported in Table 2. Accuracy of the different studies is variable, depending on the number of animals involved and the design of the comparisons (either direct or indirect). This could explain why results of Table 2 do not follow a consistent pattern. As far as pH_1 is concerned, heterozygotes (Nn) are significantly intermediate between the two homozygotes according to Schneider et al. (1980) and Christian and Rothschild (1981), and intermedia-

TABLE 2 Within-population comparisons between halothane genotypes in meat quality traits.

Reference	Population (1)	No. of animals(2)	Method (3)	Traits studied (4)	Main results, with special emphasis on pH_1 (when available)
Eikelenboom et al.(1980)	DuL	338,728,325	I	MQS	NN and Nn very close to each other, and both markedly better than nn.
Jensen and Andresen (1980)	DaL	376,169,19	I	MQI	Nn (6.7) intermediate between nn (6.4) and NN (7.5).
Schneider et al. (1980)	SL	21,48,54	I(?)	pH_1,REF, MQS	Nn significantly intermediate for the 3 traits :e.g. for pH_1 5.95 (NN), 5.79 (Nn) and 5.69 (nn).
Christian and Rothschild (1981)	cross	28,32,23	II	pH_1,REF, TV	Nn significantly intermediate for the 3 traits : e.g. for pH_1 6.42 (NN), 6.15 (Nn) and 5.73 (nn) ; Nn close to NN for TV.
Webb (1981) (trial 2)	cross	126 (NN+Nn)	II	REF,%PSE	no significant difference between NN and Nn.
Carden (1982) (5)	PTH	216,108,0	III	pH_1, REF	no difference in pH_1 between NN and Nn, but Nn showing significantly paler meat than NN.
Jensen and Barton-Gade (1985)	DaL	61,167,111	IV	pH_1,rigor, CS,WHC,MQI, %PSE,pH_2	Nn generally intermediate between NN and nn. However, no significant difference between NN and Nn in CS and MQI, and Nn closer to NN than to nn in pH_1 and %PSE.
Lundström et al. (1985)	cross	23,68,23	II	REF,FOP, drip loss, DLF, pH_2	Nn significantly intermediate between NN and nn (except for pH_2, unaffected by Hal genotype).

(1) DuL=Dutch Landrace ; DaL=Danish Landrace ; SL=Swiss Landrace ; PTH=Pietrain/Hampshire composite lines.
(2) No. of NN, Nn and nn individuals, respectively (approximate numbers calculated for the study by Eikelenboom et al., 1980).
(3) I=indirect estimation derived from comparisons between groups differing in expected genotypic frequencies : see Andresen and Jensen (1978) and Eikelenboom et al. (1980).
II=within-litter comparison, with the aid of blood-typing.
III=within-sire comparison.
IV=pooling NN vs Nn and Nn vs nn comparisons (crosses between homozygous lines).
(4) MQS=meat quality score ; MQI=meat quality index (Danish "KK index") ; REF=reflectance ; TV=transmission value ; CS=colour/structure score ; WHC=water holding capacity ; FOP=fibre optic probe value ; DLF=dielectric loss factor (Testron).
(5) as quoted by Webb et al. (1985) ; pH_1 was measured at 90 minutes post-mortem.

te but closer to <u>NN</u> than to <u>nn</u> in the study by Jensen and Barton-Gade (1985). No difference was found between <u>NN</u> and <u>Nn</u> by Carden (1982). From examining all of these results in detail, it may be reasonably concluded that some positive (i.e. favourable) heterosis is found at the Hal locus for traits which are indicators of PSE condition. Comparing <u>Nn</u> to <u>NN</u> animals is of primary interest regarding breed crossing strategy. That <u>Nn</u> individuals are not prone to PSE is indeed among the arguments presented by several authors who have recommended use of <u>nn</u> sire lines and <u>NN</u> dam lines in the terminal cross. In our present state of knowledge, this argument may be considered as acceptable, in spite of "contrary" results presented by Lundström et al. (1985) and to a lesser extent by Christian and Rothschild (1981). It may be argued that a threshold in pH_1 is likely to exist for the occurrence of typical PSE condition. Even though <u>Nn</u> individuals actually show lower average pH_1 than <u>NN</u>, one may hypothesize that only a small proportion of them fall below the pH_1 threshold and exhibit fast development of the PSE meat condition. This view is partly supported by data presented by Jensen and Barton-Gade (1985).

HETEROSIS FOR EXTENT OF FALL IN pH

General heterosis effect on ultimate pH (pH_2) will be first considered before paying attention to the specific case represented by crosses involving the Hampshire breed.

General heterosis

As a general rule, it has been found that pH_2 is not affected to any significant extent by heterosis (e.g. Mac Dougall and Disney, 1967 ; Jacquet and Ollivier, 1971 ; Elizondo et al., 1977). In the study by Jacquet and Ollivier (1971), Pietrain x Large White pigs were also intermediate between parent breeds for technological yield of Paris ham processing, a trait more closely related to ultimate pH of meat (correlation is around 0.7) than to rate of fall in pH (e.g. Jacquet et al., 1984). As far as the halothane locus is concerned, neither effect of halothane genotype nor heterosis effect were found on pH_2 by Jensen and Barton-Gade (1985) and Lundström et al. (1985).

Hampshire crosses

Comparatively low pH_2 in purebred Hampshires has been reported by several authors (see Monin and Sellier (1985) and earlier studies cited in that article). The difference in pH_2 between pigs sired by Hampshire x Pietrain boars and pigs sired by Pietrain boars was found by Sellier (1975) to be approximately half of that found by Sellier and Jacquet (1973) and Sellier (1981) between pigs sired by Hampshire boars and pigs sired by Pietrain boars (e.g. 0.10 ± 0.03 vs 0.19 ± 0.03 and 0.24 ± 0.03 for M. adductor femoris). On the basis of these results it was suggested by Sellier (1982) that the low pH_2 shown by the Hampshire breed is transmitted in an additive manner in crossing. In fact, this hypothesis needs to be revised to some extent as indicated by several other results given below.

Pale colour of meat is associated with low pH_2 in the Hampshire breed, and Hampshire crosses follow a specific pattern for colour score, as pointed out by Johnson (1981) in a review of Iowa and Oklahoma crossbreeding experiments. Summarized results given by Johnson (1981) are shown in Table 3. For Hampshire crosses, 4 out of 5 estimates of heterosis effect on colour score are significantly negative (i.e. unfavourable) whereas none of the 4 other heterosis estimates relative to crosses between Duroc, Yorkshire and Chester White was significant (though often negative also). From detailed data given on these crosses by Young et al. (1976) and Schneider et al. (1982), it appears that 50% Hamsphire crossbreds are, on average, similar to Hampshire purebreds as regards meat colour, suggesting some dominance of the "Hampshire effect" on meat quality.

Concerning ultimate pH, the above dominance hypothesis appears to be supported by combining the results from a series of experiments carried out in France on purebred and crossbred pigs (Table 4). The difference in pH_2 between 50% Hampshire and 50% Pietrain crosses is of the same magnitude as that found between Hampshire and Pietrain purebreds. Considering the unweighted mean of pH_2 values reported in Table 4 for ham muscles, the pH_2 of Hampshire pigs, as compared to Pietrain pigs, is lower by 0.21, whereas the pH_2 of 50% Hampshire pigs, as compared to 50% Pietrain pigs, is lower by 0.20. In contrast, the difference in pH_2 between Large White and Pietrain crosses is about half of the difference between Large White and Pietrain purebreds (respectively 0.04 and 0.10 for the unweighted mean of pH_2 values in hams muscles). Such an additive inheritance is in accordance with the lack of heterosis effect

TABLE 3   Specific heterosis estimates for meat colour score
(from Johnson, 1981).

Breed cross (a)	Individual heterosis (%)	
	Oklahoma experiment (b)	Iowa experiment (c)
D x Y	-0.4	-2.9
C x D		-4.0
C x Y		1.3
H x D	0.6	-4.7*
H x Y	-11.0*	-6.5*
H x C		-10.2*

(a) C = Chester White, D = Duroc, H = Hampshire, Y = Yorkshire.
(b) Young et al. (1976).
(c) Schneider (1978) ; see also Schneider et al. (1982).
 * P <0.05

on pH_2 in the crosses other than those involving the Hampshire breed.

The fact that the unfavourable effect of Hampshire on pH_2 and colour of meat could be inherited as a more or less completely dominant trait in crossing could afford an explanation for a result repeatedly found in the French "commercial product evaluation" programme. This result is the surprisingly large disadvantage in meat quality shown by pigs from crossbreeding schemes using the Hampshire breed as a component of the sire line (Runavot and Sellier, 1984).

A dominant major gene responsible for inferior technological quality of meat ?

Results from a recent research work carried out in a French breeding company (S.C.A. Pen Ar Lan) merit attention in connection with the above considerations on meat quality in Hampshire crosses. These results, to be presented at the "18èmes Journées de la Recherche Porcine en France" (Naveau, 1986), may be summarized as follows. A measurement of meat quality, the so-called "Napole" technological yield, has been studied on a large sample of animals from two composite sire lines (Penshire =P66 and Pen Ar Lan =P77). The "Napole" technique consists of curing (for 24 h) and cooking (for 10 minutes) a 100g sample of M. semimembranosus. The sample of fresh muscle is taken on the slaughterline, kept at 4°C for 24 h and then cured and cooked. The

TABLE 4   A summary of French results on differences in pH$_2$ between Pietrain (P) Large White (LW) and Hampshire (H) pure breeds and crosses.

Muscle	Pure breeds				Crosses (a)			
	reference (b)	P (control)	LW (c)	H (c)	reference (b)	PX (control)	LWX (d)	HX (d)
Adductor femoris	(1)	5.90	0.09		(5)	6.27		-0.19
	(2)	5.70	0.05		(6)	5.88	0.05	
	(3)	5.81	0.16		(7)	5.55		-0.24
	(4)	5.80	0.08	-0.25				
Biceps femoris	(2)	5.52	0.09		(5)	5.99		-0.20
	(4)	5.74	0.09	-0.16	(6)	5.69	0.06	
					(7)	5.38		-0.17
Gluteus superficialis	(2)	5.66	0.15	-0.22	(6)	5.73	0.02	-0.17
					(7)	5.44		
Semi-membranosus	(3)	5.75	0.05		(5)	6.21		-0.24
	(4)	5.69	0.12					
Gluteus profondus	(2)	5.90	0.16	-0.21				
Longissimus dorsi	(2)	5.43	0.10		(6)	5.61	0.06	-0.07
	(3)	5.49	0.06	-0.03	(7)	5.34		
	(4)	5.75	0.08					

(a) P, LW and H were used as terminal sire breeds with Large White dams in (5) and French Landrace x Large White dams in (6) and (7).
(b) (1) Sellier et al. (1984) ; (2) Monin et Sellier (1985) ; (3) experiment 3 from Monin and Sellier (this meeting) ; (4) Monin et al. (unpublished results) ; (5) Sellier and Jacquet (1973) ; (6) Sellier (1977) ; (7) Sellier (1981).
(c) expressed as deviations from contemporary P control.
(d) expressed as deviations from contemporary PX control.

"Napole" yield is defined as the ratio of cooked weight over fresh weight. Bimodal distribution of the trait in both lines and the positive relationship between sire progeny means and within-sire progeny variances have led Naveau (1986) to postulate that a dominant unfavourable gene is segregating in both lines, whose effect is to cause the "Napole" yield to be lower than normal (i.e. less than 91%). A point of special interest is that both lines are partially of Hampshire breeding. Penshire contains Hampshire (around 50%), Large White and Duroc genes, whereas Pen Ar Lan is a composite line established 10 years ago with equal proportions of Hampshire, Pietrain and Large White genes. As suggested by Naveau (1986), an attractive hypothesis is that the postulated major gene, called RN⁻ (RN = "Rendement Napole"), was brought by Hampshire founders into the two lines, and that the gene RN⁻ would be responsible, at least partially, for the "Hampshire effect" on meat quality described by Monin and Sellier (1985). It can be added that the correlation between "Napole" yield and ultimate pH is of the order of 0.6-0.7, whereas that between "Napole" yield and technological yield of Paris ham processing is about 0.75 (Naveau et al., 1986).

CONCLUSION

This review of literature data on heterosis for specific meat quality traits suggests that there is a need to revise to some extent the classical view in this field, i.e. the lack of heterosis effect on meat quality. Two main deviations from additive inheritance of breed differences in crossing appear to exist.

As regards the abnormally fast rate of fall in pH (associated with typical PSE meat condition), it appears that this trait is inherited as a recessive or partially recessive trait in Pietrain crosses. The halothane locus probably plays the major role in this respect. On the other hand, there are several pieces of evidence for hypothesizing that the comparatively large extent of fall in pH (i.e. low ultimate pH) characterizing the Hampshire breed is inherited as a dominant or partially dominant trait in Hampshire crosses. Recent results suggest that a dominant major gene could be involved in the "Hampshire effect" on meat quality.

Further research is however needed on these two points in order to confirm the above tentative statements.

REFERENCES

Andresen, E. and Jensen, P. 1978. Evidence of an additive effect of alleles of the Hal locus on the KK index for porcine meat quality. Nord. Vet. Med., 30, 286-288.

Bichard, M. and Smith, W.C. 1972. Crossbreeding and genetic improvement. In "Pig Production" (Ed. D.J.A. Cole), pp. 37-52. Butterworths, London.

Carden, A.E. 1982. Genetics of halothane susceptibility in pigs. Ph.D. Thesis, University of Edinburgh.

Christian, L.L. and Rothschild, M.F., 1981. Performance and carcass characteristics of normal, stress-carrier and stress-susceptible swine. Pig Producers Day Reports, Iowa State Experiment Station, 3p.

Eikelenboom, G. and Minkema, D. 1974. Prediction of pale, soft, exudative muscle with a non-lethal test for the halothane-induced porcine malignant hyperthermia syndrome. Tijdschr. Diergeneesk., 99, 421-426.

Eikelenboom, G., Minkema, D., van Eldik, P. and Sybesma, W. 1980. Performance of Dutch Landrace pigs with different genotypes for the halothane-induced malignant hyperthermia syndrome. Livest. Prod. Sci., 7, 317-324.

Elizondo, G., Addis, P.B., Rempel, W.E., Madero, C. and Antonik, A. 1977. An approach to determine the relationships among breed composition, skeletal muscle properties and carcass quantitative traits. J. Anim. Sci., 45, 1272-1279.

Elizondo, G., Addis, P.B., Rempel, W.E., Madero, C., Martin, F.B., Anderson, D.B. and Marple, D.N. 1976. Stress response and muscle properties in Pietrain (P), Minnesota No. 1 (M) and PxM pigs. J. Anim. Sci., 43, 1004-1014.

Hanset, R., Leroy, P., Michaux, C. and Kintaba, K.N. 1983. The Hal locus in the Belgian Pietrain pig breed. Z. Tierzücht. ZüchtBiol., 100, 123-133.

Jacquet, B. and Ollivier, L. 1971. Résultats d'une expérience de croisement Piétrain x Large White. II. Aptitude du jambon à la transformation en jambon de Paris. In Journées de la Recherche Porcine en France 1971, pp. 23-33. Institut Technique du Porc, Paris.

Jacquet, B., Sellier, P., Runavot, J.P., Brault, D., Houix, Y., Perrocheau, C., Gogué, J. and Boulard, J. 1984. Prediction of the technological yield of "Paris ham" processing by using measurements at the abattoir. In "Biophysical PSE-muscle analysis" (Ed. H. Pfützner), pp. 143-153. Vienna Technical University.

Jensen, P. and Andresen, E. 1980. Testing methods for PSE syndrome: current research in Denmark. Livest. Prod. Sci., 7, 325-335.

Jensen, P. and Barton-Gade, P.A. 1985. Performance and carcass characteristics of pigs with known genotypes for halothane susceptibility. In "Stress susceptibility and meat quality in pigs" (Ed. J.B. Ludvigsen), pp. 80-87, E.A.A.P. Publication No. 33.

Johnson, R.K. 1981. Crossbreeding in swine : experimental results. J. Anim. Sci., 52, 906-923.

Jonsson, P. 1962. Experiments with Pietrain pigs. In Forsog med swin, pp. 273-283. Forsogslab-Arb., Copenhagen.

Kirsch, W., Fender, M., Rabold, K., Fewson, D. and Schoen, P. 1963. Vergleichende Zucht-, Mast- und Ausschlachtungversuche mit veredelten Landschweinen, Pietrain-Schweinen und F1-Kreuzungstieren. Zuchtungskunde, 35, 254-264.

Knoertzer, E. 1961. Le croisement porcin Piétrain x Large White. Bull. tech. Ingrs Serv. agric., 165, 1021-1042..

Lean, I.J., Curran, M.K., Duckworth, J.E. and Holmes, W. 1972. Studies on Belgian Pietrain pigs. 1- A comparison of Pietrain, Landrace and Pietrain Landrace crosses in growth, carcass characteristics and meat quality. Anim. Prod., 15, 1-9.

Lundström, K., Rundgren, M., Edfors-Lilja, I., Essen-Gustavsson, B., Nyberg, L. and Gahne, B. 1985. Effect of halothane genotype on immune response, muscle characteristics, meat quality and performance. A within-litter comparison. 36th E.A.A.P. Meeting, Kallithea, Greece, Paper MP5.18.

MacDougall, D.B. and Disney, J.G. 1967. Quality characteristics of pork with special reference to Pietrain, Pietrain x Landrace and Landrace pigs at different weights. J. Fd Technol., 2, 285-297.

McGloughlin, P. 1980. Genetic aspects of pigmeat quality. Pig News and Information, 1, 5-9.

McKay, R.M., Rempel, W.E., McGrath, C.J., Addis, P.B. and Boylan, W.J. 1982. Performance characteristics of crossbred pigs with graded percentages of Pietrain. J. Anim. Sci., 55, 274-279.

Minkema, D., Cöp, W.A.G., Buiting, G.A.J. and van de Pas, J.G.C. 1974. Pure breeding compared with reciprocal crossbreeding of Dutch Landrace (NL) and Dutch Yorkshire (GY) pigs. In "Proceedings of the Working Symposium on Breed Evaluation and Crossing Experiments with Farm Animals", pp. 297-312. "Schoonoord", Zeist.

Monin, G. and Sellier, P. 1985. Pork of low technological quality with a normal rate of muscle pH fall in the immediate post-mortem period: the case of the Hampshire breed. Meat Sci., 13, 49-63.

Naveau, J. 1986. Contribution à l'étude du déterminisme génétique de la qualité de viande porcine. Héritabilité du rendement technologique "Napole". In 18èmes Journées de la Recherche Porcine en France, in press.

Naveau, J., Pommeret, P. and Lechaux, P. 1986. Proposition d'une méthode de mesure du rendement technologique : la méthode "Napole". TechniPorc, 9, in press.

Nielsen, N.J. 1981. The effect of environmental factors on meat quality and on deaths during transportation and lairage before slaughter. In "Porcine stress and meat quality" (Ed. T. Froystein, E. Slinde and N. Standal), pp. 287-297. Agricultural Food Research Society, As, Norway.

Runavot, J.P. and Sellier, P. 1984. Bilan des dix tests de contrôle des produits terminaux réalisés en France de 1970 à 1983. In 16èmes Journées de la Recherche Porcine en France, pp. 425-438. Institut Technique du Porc, Paris.

Schmidt, L. 1964. Kreuzungsversuche mit Pietrain- und veredelten Landschweinen. Bayer. landw. Jb., 41, 905-919.

Schneider, J.F. 1978. Individual and maternal heterosis estimated from single crosses and backcrosses of swine. Ph.D. Thesis, Iowa State University, Ames.

Schneider, J.F., Christian, L.L. and Kuhlers, D.L. 1982. Crossbreeding in swine : genetic effects on pig growth and carcass merit. J. Anim. Sci., 54, 747-756.

Schneider, A., Schwörer, D. and Blum, J. 1980. Beziehung des HalothanGenotyps zu den Produktions- und Reproduktionsmerkmalen der Schweizerischen Landrasse. 31st E.A.A.P. Meeting, München, Paper GP3.9.

342

Sellier, P. 1970. Hétérosis et croisement chez le Porc. Ann. Génét. Sél. anim., 2, 145-207.

Sellier, P. 1974. Le croisement dans l'espèce porcine. In 1st World Congress on Genetics applied to Livestock Production, Vol.1, 859-871. Editorial Garsi, Madrid.

Sellier, P. 1975. Valeur en croisement de verrats Piétrain et Hamsphire x Piétrain. In Journées de la Recherche Porcine en France 1975, pp. 253-258. Institut Technique du Porc, Paris.

Sellier, P. 1976. The basis of crossbreeding in pigs ; a review. Livest. Prod. Sci., 3, 203-226.

Sellier, P. 1977. Valeur en croisement de verrats Large White et Piétrain : influence du poids d'abattage. In Journées de la Recherche Porcine en France 1977, pp. 85-89. Institut Technique du Porc, Paris.

Sellier, P. 1981. Une première évaluation de la race Duroc. In 13èmes Journées de la Recherche Porcine en France, pp. 299-306. Institut Technique du Porc, Paris.

Sellier, P. 1982. Le choix de la lignée mâle du croisement terminal chez le Porc. In 14èmes Journées de la Recherche Porcine en France, pp. 159-182. Institut Technique du Porc, Paris.

Sellier, P. and Jacquet, B. 1973. Comparaison de porcs Hampshire x Large White et Piétrain x Large White. In Journées de la Recherche Porcine en France 1973, pp. 173-180. Institut Technique du Porc, Paris.

Sellier, P., Monin, G., Houix, Y. and Dando, P. 1984. Qualité de la viande de quatre races porcines : relations avec la sensibilité à l'halothane et l'activité créatine phosphokinase plasmatique. In 16èmes Journées de la Recherche Porcine en France, pp. 65-74. Institut Technique du Porc, Paris.

Skarman, S. 1965. Crossbreeding experiments with swine. LantbrHögsk. Annlr, 31, 3-92.

Walstra, P., Minkema, D., Sybesma, W. and van de Pas, J.G.C. 1971. Genetic aspects of meat quality and stress resistance in experiments with various breeds and breed crosses. 22nd E.A.A.P. Meeting, Versailles, 13p.

Webb, A.J. 1981. The halothane sensitivity test. In "Porcine stress and meat quality" (Ed. T. Froystein, E. Slinde and N. Standal), pp. 105-124. Agricultural Food Research Society, Ås, Norway.

Webb, A.J., Southwood, O.I., Simpson, S.P. and Carden, A.E. 1985. Genetics of porcine stress syndrome. In "Stress susceptibility and meat quality in pigs" (Ed. J.B. Ludvigsen), pp. 9-30, E.A.A.P. Publication No. 33.

Young, L.D., Johnson, R.K., Omtvedt, I.T. and Walters, L.E. 1976. Postweaning performance and carcass merit of purebred and two-breed cross pigs. J. Anim. Sci., 42, 1124-1132.

A MOLECULAR GENETIC APPROACH TO THE PORCINE STRESS SYNDROME

A.L. Archibald

AFRC Animal Breeding Research Organisation, West Mains Road,
Edinburgh EH9 3JQ, United Kingdom

ABSTRACT

Halothane-induced malignant hyperthermia is inherited as a single autosomal recessive gene (Hal n) in pigs. The Hal locus lies within the S - Phi - Hal - H - Po2 - Pgd linkage group. However, the nature of the Hal lesion remains unknown. The techniques of genetic engineering now make it possible to study this genetic disorder in greater detail. In this paper proposals are presented for a molecular genetic approach to the development of an improved method for determining the Hal genotypes of pigs as well as for identifying the Hal gene itself.

INTRODUCTION

The stress syndromes (porcine stress syndrome PSS, malignant

hyperthermia syndrome MHS and the pale soft exudative meat syndrome PSE)

have been the subject of considerable study over the past twenty years.

These syndromes are determined by segregation at one or several closely

linked loci including a locus where segregation affects sensitivity to

the anaesthetic halothane. This research has been comprehensively

reviewed by Harrison (1979), Mitchell and Heffron (1982) and Webb et al.

(1982). Although it is almost twenty years since the halothane-induced

malignant hyperthermia syndrome was first recognised in pigs, (Hall

et al., 1966; Harrison et al., 1968) the nature of the primary lesion

remains unknown. Many different aspects of the physiology and

biochemistry of stress susceptible and resistant pigs have been compared

without any definitive conclusion being drawn as to the identity of the

defect. There is perhaps some agreement that a membrane defect is the

underlying cause and that the membrane may have a calcium-related

function.

It has been clear for many years that susceptibility to the various

porcine stress syndromes is genetically controlled. The inheritance of

halothane-induced stress has been shown to be controlled by a recessive

gene at a single autosomal locus (Hal) (Ollivier, Sellier and Monin,

1975; Minkema, Eikelenboom and van Eldik, 1977; Smith and Bampton, 1977

and Mabry, Christian and Kuhlers, 1981). The incomplete penetrance

observed by Ollivier et al. (1975) and Smith and Bampton (1977) has led

some groups to test alternative hypotheses for the inheritance of porcine
stress (Carden, Hill and Webb, 1983 and Grashorn and Muller, 1985).
Both of these groups have tested two locus models with a susceptibility
locus and a suppressor locus.   Carden et al. (1983) concluded that a
strictly recessive single locus mode of inheritance may not be adequate
to explain their observations in the British Landrace.   In both studies
the putative suppressor locus did provide a genetic basis for the
variation in penetrance, but alternatives to the single locus model
have not been adequately tested in most populations.

Although the genetic control of porcine stress may still not be
fully resolved there is a strong body of evidence which places the Hal
locus within a group of linked genes most of whose products can be assayed
in blood.   The linkage relationships of the halothane sensitivity locus
have recently been reviewed by Archibald and Imlah (1985).

Halothane sensitivity is only one manifestation of the general
condition of porcine stress syndrome (PSS).   When compared to normal
(Hal N/N) contemporaries, reactor animals (Hal n/n) show advantages in
lean content accompanied by adverse effects on stress susceptibility
(sudden deaths), meat quality (PSE) and litter productivity.   From
worldwide comparisons of reactors and non-reactors the advantages in lean
content were outweighed by the disadvantages to give an average net loss
of £3.90 per reactor slaughtered at 90kg (Webb et al., 1982).   There is
evidence that the Hal n allele may have additive effects on lean content
but behave as a recessive for the other deleterious performance traits
thereby giving the heterozygote a selective advantage.   This advantage
may amount to about £1.70 for the heterozygote over the normal homozygote
(Webb et al., 1982).

The porcine stress syndrome therefore is comparable to some of the
hereditary disorders which afflict Man in as much as the underlying lesion
is unknown but the genetics of the disorder indicate a simple mode of
inheritance probably controlled by a single locus.   The application of
molecular genetics to the study of some of the genetic diseases of Man has
already yielded significant advances, particularly in the realms of
prenatal diagnosis and genetic counselling (Gusella et al., 1983;
Monaco et al., 1985; Murray et al., 1982 and Reeders et al., 1985).
The halothane sensitivity locus also represents one of the few examples
in domestic animals of a single locus with significant effects on

economically important traits.   This combination of circumstances makes
it desirable, as well as possible, to use the techniques of molecular
biology and genetic engineering to further our understanding of the
porcine stress syndrome.

THE USE OF GENETIC MARKERS

As the halothane sensitivity gene (Hal n) is largely recessive the
standard halothane test only detects a small proportion of the
heterozygotes (Southwood et al., 1985).   The alternative diagnostic
tests are no better and are indeed generally less effective for genotyping
(Webb et al. 1985).   Progeny testing which offers the only reliable way
of genotyping animals is both expensive and time consuming.   Therefore
there is a need for a cheap and accurate test which would facilitate the
unambiguous identification of all three genotypes - N/N, N/n and n/n.

One possible approach is to identify genetic markers tightly linked
to the Hal locus.   Linkage has already been established between the Hal
locus and the following loci - Phi phosphohexose isomerase, H
erythrocyte antigen, Pgd 6-phosphogluconate dehydrogenase, Po2
postalbumin-2, S suppressor locus for the A-O blood groups and possibly
G erythrocyte antigen (for a review see Archibald and Imlah, 1985).   The
structure of the linkage group is summarised in Figure 1.   These loci
meet the requirements of genetic markers as they are polymorphic and
linked to the locus of interest.   Indeed, Gahne and Juneja (1985) have
shown that a combination of the halothane test and haplotyping for Phi,
Po2 and Pgd allows Hal genotypes to be predicted with an accuracy of
90-95% in Swedish Landrace and Yorkshire pigs.   This approach is being
used in marker assisted selection in Sweden to effect a breeding policy
of significantly reducing the frequency of the Hal n allele.   Given the
economic advantages of the heterozygote this is not the only legitimate
breeding policy with respect to the Hal locus.   However, the gene
frequencies at these marker loci are not ideal for marker assisted
selection in all pig populations.   Furthermore, there are technical
difficulties associated with the accurate determination of genotypes at
the S and H loci in addition to the problems of incomplete penetrance of
the Hal locus as discussed by Archibald and Imlah (1985).

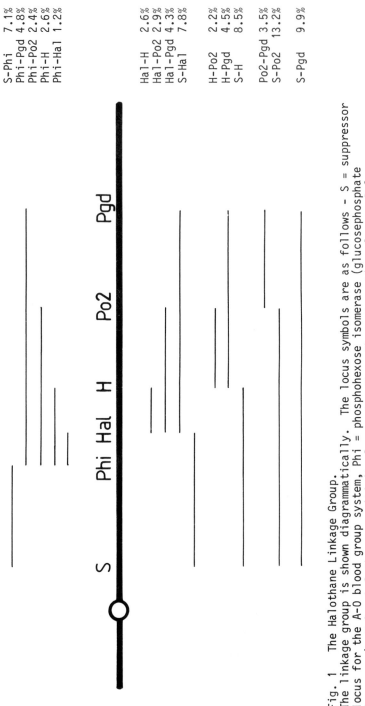

Fig. 1   The Halothane Linkage Group.
The linkage group is shown diagrammatically. The locus symbols are as follows - S = suppressor locus for the A-O blood group system, Phi = phosphohexose isomerase (glucosephosphate isomerase), Hal = halothane sensitivity locus, H = H erythrocyte antigen locus, Po2 = postalbumin-2, and Pgd = 6-phosphogluconate dehydrogenase. The estimates of the recombination frequencies given here are those calculated by pooling most of the published linkage data (see Archibald & Imlah, 1985).

Restriction fragment length polymorphisms

      The big breakthrough in the use of genetic markers in medical science was the recognition of a new class of marker - restriction fragment length polymorphisms (RFLPs). The nature and use of RFLPs has been thoroughly reviewed by Botstein et al. (1980). Briefly, restriction endonucleases cleave DNA reproducibly at specific sites to yield a characteristic series of (restriction) fragments. The fragments can be separated according to their size by agarose gel electrophoresis. The fragments are then transferred to an inert support such as a nitrocellulose or nylon filter (Southern, 1975). The filters are probed with radioactively labelled DNA sequences which will hybridise specifically to any fragments which contain sequences complimentary to the probe sequence. The unhybridised probe is washed away and the filter is placed with X-ray film in order to visualise the fragments which have hybridised to the probe. Base changes can alter the sequences that are recognised by the restriction enzymes, abolishing or creating sites and consequently changing the sizes of the corresponding restriction fragments. Deletion, insertion or transposition of larger elements will make simultaneous changes in the restriction fragment pattern for several enzymes. An example of this type of polymorphism is given in Figure 2. RFLPs are expected to behave as simple Mendellian characters with a codominant mode of inheritance. Almost 1000 RFLP loci have been identified in Man (de la Chapelle, 1985). Furthermore, linkage has been established between RFLP marker loci and the loci responsible for Duchenne muscular dystrophy (Murray et al., 1982), Huntington's disease (Gusella et al., 1983), and adult polycystic kidney disease (Reeders et al., 1985). The use of RFLPs in animal improvement has been extensively discussed by Soller and Beckman (1982, 1983) and Beckman and Soller (1983).

Cloning the phosphoglucose isomerase gene Gpi

      The initial requirement in the study of RFLPs is for a probe - a cloned DNA sequence which will hybridise to the chromosomal region of interest. Any of the genes already known to be linked to the Hal locus would fulfill this requirement. The most tightly linked of the known markers is fortuitously also the gene for which a well tried cloning strategy already exists - phosphohexose isomerase (Phi) or more correctly glucosephosphate isomerase (Gpi). The enzyme glucose phosphate

A.

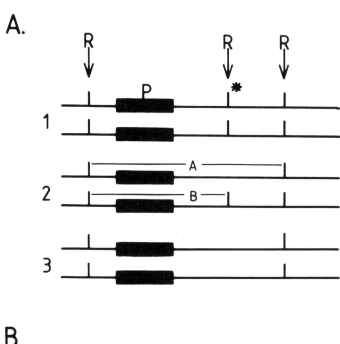

B.

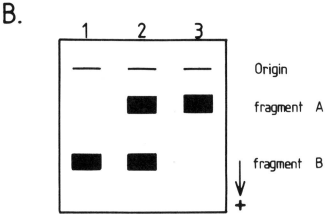

Fig. 2   Illustration of Restriction Fragment Length Polymorphism.
A.   Chromosomal DNA from individuals (1, 2 & 3) are digested with
restriction endonuclease R.   One pair of homologous chromosomes is
shown for each individual.   The polymorphic restriction endonuclease
recognition site is marked with an asterisk (*).   This digestion
produces two different sized fragments (A & B) which can be detected
with a probe (P) for the area boxed in the figure.
B.   The digested fragments are separated by agarose gel electrophoresis
and detected after transfer to am inert filter with labelled probe (P).
These individuals correspond to the two homozygotes and the
heterozygote for this RFLP locus.

isomerase (EC 5.3.1.9) has been isolated from pig muscle and purified to homogeneity (Gee et al. 1979).   There is also a limited amount of amino acid sequence data available for this enzyme (Achari et al., 1981). This amino acid sequence data can be used to design an oligonucleotide which would hybridise to the corresponding DNA sequence (Lathe, 1985). Messenger RNA isolated from pig muscle (GPI is known to represent 1% of the extractable protein in rabbit muscle (Noltmann, 1964)) would be copied in vitro by reverse transcriptase to yield copy DNA (cDNA) which would be used to generate a cDNA library (Okayama and Berg, 1982).   The cDNA library would be screened with the synthetic oligonucleotide probe in order to identify those clones containing the Gpi gene.   This approach has been used to clone a number of genes including those for porcine aspartate aminotransferase and porcine gastrin (Joh et al., 1985 and Noyes et al., 1979).

The cloned Gpi gene would be used as a probe to look for RFLPs linked to the Gpi locus and hence to the Hal locus.   The linkage relationships between any such RFLP loci and Hal would need to be checked carefully as the objective is to find markers for the Hal locus rather than for Gpi.

Chromosomal localisation, 'walking' and chromosome-specific gene banks

By in situ hybridisation of labelled Gpi sequences to metaphase chromosomes it should be possible to make a chromosomal assignment for the Gpi locus and hence for the Hal locus also (Harper and Saunders, 1981).   There is preliminary evidence from linkage studies with chromosomes marked by translocations and centromere size polymorphisms that the Hal linkage group is located on chromosome 15.   (Tikhonov et al., 1984, Fries et al., 1983, 1984).

The next logical step would be to study this chromosome in greater detail and to use the knowledge of the chromosomal localisation to narrow the search for genetic markers for Hal in the same way as those studying X-linked disorders of humans have done.   As porcine chromosomes are diverse in size (Hsu and Benirschke, 1967) it should be possible to fractionate the chromosomes in a Fluorescence Activated Cell Sorter (Griffiths et al., 1984).   DNA isolated from such flow-sorted chromosomes could be used to establish a chromosome-specific gene bank of the appropriate chromosome.   This gene bank would be used as a source of probes to look for RFLPs tightly linked to the Hal locus.

It should also be possible to use the gene bank to explore the chromosomal region adjacent to the Gpi locus by the technique of 'chromosome walking' (Hadfield, 1983 and Steinmetz et al., 1982). Briefly, a cloned sequence is used to identify a series of genomic clones which carry at least part of the original clone. Fragments within these clones which are at the greatest distance from the original sequence are then used in turn to identify a second series of genomic clones, which contain the second sequence, and are therefore linked to the original sequence. By a repeated series of steps of this nature it is possible to 'walk' along the chromosome from the initial cloned gene (see Figure 3). The walk would be initiated from the Gpi gene and the chromosome specific gene bank would be used as the source of genomic clones. At each stage the isolated clones could be tested as probes for RFLPs linked to Hal. The distance between Gpi and Hal is between 1 and 5 centimorgans (or 1000 to 5000 kilobases (kb)) (Archibald and Imlah, 1985). The walk from Gpi to Hal therefore is a major undertaking, as present walking techniques operate in steps of 50kb. Collins and Weissman (1984) and Lehrach and his colleagues at the European Molecular Biology Laboratory, Heidelberg have been developing techniques for increasing the size of each step to 100kb, 200kb or greater. Steps of this size would bridge the gap between the molecular analyses of linkage and the more established methods using somatic cell genetics and segregation studies.

Other markers

In addition to this logical, but probably time consuming, method of identifying genetic markers for porcine stress, it may also be possible to identify additional markers using the increasingly sophisticated ways of looking for protein polymorphisms. The comparison of one and two dimensional gel electrophoresis patterns of proteins from halothane negative and positive pigs may reveal differences, which could be used diagnostically. Initially, one would examine integral membrane proteins from muscle, as the halothane lesion appears to be expressed in muscle and is generally assumed to a membrane defect. No differences were observed between the skeletal muscle proteins of stress susceptible and normal pigs by Lorkin and Lehmann (1983) or

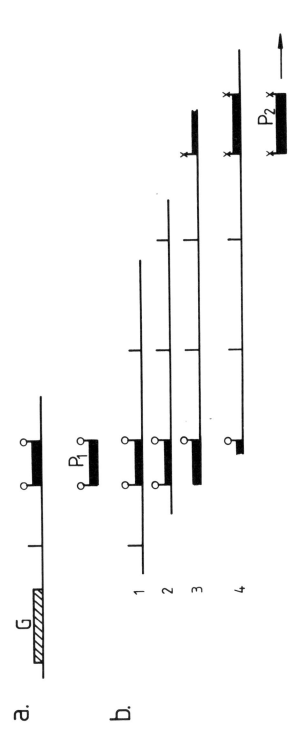

Fig. 3   Illustration of chromosomal walking.
From a genomic clone (a.) containing gene G, a fragment distal to G is subcloned and used as a
probe (P_1).   The probe P_1 is used to identify a series of overlapping random genomic clones
(b. 1-4).   A fragment distal to P_1 in clone 4, is in turn subcloned and used as a probe (P_1).
The probe P_2 is then used to identify a further series of overlapping genomic clones, and so on.
The sequences present in the series identified by P_2 represent sequences linked to those present in
the P_1 series and so to G.   By repeating this procedure it is possible to 'walk' along the
chromosome from G.   Restriction enzyme sites are shown as vertical bars.

Marjanen and Denborough (1984).   A recent study of malignant
hyperthermic muscle in Man, however, revealed two low molecular weight
proteins present in malignant hyperthermia patients, but absent in the
control individuals (Blanck et al., 1984).   The protein extraction
procedures used by this latter group differed from those employed to
examine the pig muscle proteins.   Given the similarity between PSS
and malignant hyperthermia, it would be worthwhile to attempt to
reproduce the study of Blanck and his coworkers on pigs.   Although any
protein polymorphisms detected in this way could be used as genetic
markers in their own right, it is now possible to isolate the corresponding
genes.   First antisera would be raised against the cognate protein by
excising the band from a polyacrylamide gel and directly immunising rabbits
- the polyacrylamide in fact acts as a potent adjuvent.   The antisera
could then be used to screen cDNA libraries established in expression
vectors such as that described by Stanley and Luzio (1984).   Expression
vectors are designed so that the cloned gene sequences are transcribed
and translated to produce the corresponding protein.   The range of
experiments discussed above for Gpi would then be possible for any clones
isolated in this way.

Another source of markers, if not the Hal gene itself, would be the
cDNA library created to identify the Gpi gene.   Muscle from a halothane
resistant pig (Hal N/N) would be chosen as the source of mRNA from
which the cDNA would be made.   Labelled total muscle mRNA from HAL N/N
and Hal n/n individuals could be used to screen the cDNA library.
Clones, which exhibited different hybridisation characteristics, either
presence/absence or quantitative variation, when screened with these
two types of mRNA would be isolated and examined further.   Initially,
one would look for linkage between the Hal locus and any genes isolated
in this way.

LONGER TERM DEVELOPMENTS

Molecular genetics offers, not only the very probable development
of a better method for genotyping the Hal locus, but also the possibility
of isolating the halothane sensitivity gene itself and thereby identifying
and understanding the lesion.

The chromosome specific gene bank could be used to as a source of
probes to examine differences in expression between Hal+ and Hal- animals.

The chromosome specific sequences could be used to screen a muscle cDNA library and thereby identify the genes from this chromosome, which are expressed in muscle. These cloned genes could in turn be used to monitor differences in expression between the stress phenotypes. RNA isolated from stress susceptible and normal pigs would be separated by denaturing agarose gel electrophoresis, transferred to an inert filter and probed with the chromosome-specific sequences shown to be expressed in muscle. Alternatively, the chromosome-specific gene bank could be screened with labelled total mRNA from the muscles of Hal N/N and Hal n/n individuals. In this way, it might be possible to identify genes, whose expression correlates with the halothane sensitivity phenotypes, and so identify the Hal gene. Information gleaned from the chromosomal 'walk' described above would be useful in confirming any such putative halothane sensitivity gene.

The identification and cloning of the halothane gene itself would open up a number of exciting possibilities. For example, the halothane resistant allele could be introduced by microinjection into a fertilised porcine oocyte known to be homozygous for the halothane sensitive allele (Hammer et al., 1985). It is now possible to generate 'transgenic' animals in which the introduced gene is present as a single copy. The 'transgenic' pigs created in this way would be homozygous for Hal n at the original locus and heterozygous for Hal N at another chromosomal location. By a conventional breeding programme the new locus would be made homozygous for Hal N. The products of this breeding programme would be pigs, homozygous for both Hal alleles, and which could be described as stably heterozygous for Hal. It would be hoped that such pigs would have the same phenotype and performance as conventional Hal heterozygotes. The introduction of a cloned putative Hal N gene into the genome of a Hal n/n pig/oocyte would also provide the ultimate proof that the gene was indeed Hal N, if the transgenic animal was halothane resistant and also capable of transmitting this character to its offspring.

A cloned Hal gene could also be used to identify the homologous gene(s) in other species. The gene might be worth introducing into other livestock species as a means of improving lean content. If the Hal n gene, however, represents an absence of function, it is difficult to imagine that it would be of any value when transferred to another species.   However, it is possible that, by making a molecular genetic analysis of the Hal chromosomal region, the good and bad effects associated with Hal n could be assigned to different genetic elements, which could then be separated.   The gene may also be valuable in the study of the malignant hyperthermia syndrome in Man.

The cloned gene could be used to identify the corresponding gene product.   The gene could be used to drive in vitro transcription and translation systems to produce the gene product (Stueber et al., 1984). Alternatively the cDNA version of the gene could be introduced into a bacterial expression vector to yield a bacterial-hal fusion protein which could be used to raise antisera against the hal protein (Stanley and Luzio, 1984).   Antisera against the hal protein could then be used to purify the hal protein itself.   Knowledge of the nature of the Hal gene product would be invaluable in understanding the physiological trigger for the porcine stress syndrome.

CONCLUSION

The application of molecular biology skills to the study of the porcine stress syndrome will very probably provide an alternative means of determining Hal genotypes within the next three years.   In the longer term it may be possible to identify the underlying genetic lesion and thereby understand the phenomenom which we currently call porcine stress syndrome.   It seems very likely that molecular genetics will make as big an impact on pig breeding as the halothane test has already done.

ACKNOWLEDGEMENTS

I would like to thank Drs. John Webb, John Clark, John Bishop and Richard Lathe for many valuable discussions.   I also acknowledge the generosity of the Ministry of Agriculture, Fisheries and Food for providing financial support to initiate the research programme outlined in this paper.

REFERENCES

Achari, A., Marshall, S.E., Muirhead, H., Palmieri, R.H. and Noltmann, E.A. 1981. Glucose-6-phosphate isomerase. Philosophical Transactions of the Royal Society (London) B, 293, 145-157.

Archibald, A.L. and Imlah, P. 1985. The halothane sensitivity locus and its linkage relationships. Animal Blood Groups and Biochemical Genetics 16, (in press).

Beckman, J.S. and Soller, M. 1983. Restriction fragment length polymorphisms in genetic improvement: methodologies, mapping and costs. Theoretical and Applied Genetics 67, 35-43.

Blanck, T.J.J., Fisher, Y.I., Thompson, M. and Muldoon, S. 1984. Low molecular weight proteins in human malignant hyperthermic muscle. Anesthesiology 61, 589-592.

Botstein, D., White, R.L., Skolnick, M. and Davis, R.W. 1980. Construction of a genetic linkage map in Man using restriction fragment length polymorphisms. American Journal of Human Genetics 32, 314-331.

Carden, A.E., Hill, W.G. and Webb, A.J. 1983. The inheritance of halothane susceptibility in pigs. Genetique Selection et Evolution 15, 65-82.

Collins, F.S. and Weissman, S.M. 1984. Directional cloning of DNA fragments at a large distance from an initial probe: A circularization method. Proceedings of the National Academy of Sciences 81, 6812-6816.

de la Chapelle, A. 1985. Mapping hereditary disorders. Nature 317, 472-473.

Fries, R., Stranzinger, G. and Vogeli, P. 1983. Provisional assignment of the G blood group locus to chromosome 15 in swine. Journal of Heredity 74, 426-430.

Fries, R., Rasmusen, B.A., Jarrel, V.L. and Maurer, R.R. 1984. Mapping of the gene for G blood group antigens to chromosome 15 in swine. Animal Blood Groups and Biochemical Genetics 15, 251-258.

Gahne, B. and Juneja, R.K. 1985. Prediction of the halothane (Hal) genotypes of pigs by deducing Hal, Phi, Po2, Pgd haplotypes of parents and offspring: results from a large-scale practice in Swedish breeds. Animal Blood Groups and Biochemical Genetics 16, (in press).

Gee, D.M., Palmieri, R.H., Porter, G.G. and Noltmann, E.A. 1979. Large scale isolation of pig muscle phosphoglucose isomerase. Preparative Biochemistry 9, 441-455.

Grashorn, M. and Müller, E. 1985. Relations between blood groups, isozymes and halothane reaction in pigs from a selection experiment. Animal Blood Groups and Biochemical Genetics 16, (in press).

Griffiths, J.K., Cram, L.S., Crawford, B.D., Jackson, P.J., Schilling, J., Schimke, R.T., Walters, R.A., Wilder, M.E. and Jett, J.H. 1984. Construction and analysis of DNA sequence libraries from flow-sorted chromosomes: practical and theoretical considerations. Nucleic Acids Research 12, 4019-4034.

Gusella, J.F., Wexler, N.S., Conneally, P.M., Naylor, S.L., Anderson, M.A., Tanzi, R.E., Watkins, P.C., Ottina, K., Wallace, M.R., Sakaguchi, A.Y., Young, A.B., Shoulson, I., Bonilla, E. and Martin, J.B. 1983. A polymorphic DNA marker genetically linked to Huntington's disease. Nature 306, 324-328.

Hadfield, C. 1983. Chromosome walking. Focus (Bethesda Research Laboratories) 5, 1-5.

Hall, L.W., Woolf, M., Bradley, J.W.P. and Jolly, D.W. 1966. Unusual reaction to succinylcholine. British Medical Journal 2, 1305.

Hammer, R.E., Pursel, V.G., Rexroad, C.E., Wall, R.J., Bolt, D.J., Ebert, K.M., Palmiter, R.D. and Brinster, R.L. 1985. Production of transgenic rabbits, sheep and pigs by microinjection. Nature 315, 680-683.

Harper, M.E. and Saunders, G.F. 1981. Localization of single copy DNA sequences on G-banded human chromosomes by in situ hybridisation. Chromosoma 83, 431-439.

Harrison, G.G. 1979. Porcine malignant hyperthermia. International Anesthesiology Clinics 17, 25-61.

Harrison, G.G., Biebuyck, J.F., Terblanche, J., Dent, D.M., Hickman, R. and Saunders, J. 1968. Hyperpyrexia during anaesthesia. British Medical Journal 3, 594-595.

Hsu, T.C. and Benirschke, K. 1967. An atlas of mammalian chromosomes. Volume 1, Folio 38, Springer-Verlag, Berlin, Heidelberg and New York.

Joh, T., Nomiyama, H., Maeda, S., Shimada, K. and Morino, Y. 1985. Cloning and sequence analysis of a cDNA encoding porcine mitochondrial aspartate aminotransferase precursor. Proceedings of the National Academy of Sciences 82, 6065-6069.

Lathe, R. 1985. Synthetic oligonucleotide probes deduced from amino acid sequence data. Theoretical and practical considerations. Journal of Molecular Biology 183, 1-12.

Lorkin, P.A. and Lehmann, H. 1983. Investigation of malignant hyperthermia: analysis of skeletal muscle proteins from normal and halothane sensitive pigs by two dimensional gel electrophoresis. Journal of Medical Genetics 20, 18-24.

Mabry, J.W., Christian, L.L. and Kuhlers, D.L. 1981. Inheritance of porcine stress syndrome. Journal of Heredity 72, 429-430.

Marjanen, L.A. and Denborough, M.A. 1984. Electrophoretic analysis of proteins in malignant hyperpyrexia susceptible muscle. International Journal of Biochemistry 16, 919-929.

Minkema, D., Eikelenboom, G. and van Eldik, P. 1977. Inheritance of M.H.S.-susceptibility in pigs. Proceedings of the Third International Conference on Production Disease in Farm Animals, Wageningen, The Netherlands, 1976, pp 203-207.

Mitchell, G. and Heffron, J.J.A. 1982. Porcine stress syndrome. Advances in Food Research 28, 167-230.

Monaco, A.P., Bertelson, C.J., Middlesworth, W., Colletti, C.-A., Aldridge, J., Fischbeck, K.H., Bartlett, R., Pericak-Vance, M.A., Roses, A.D. and Kunkel, L.M. 1985. Detection of deletions spanning the Duchenne muscular dystrophy locus using a tightly linked DNA segment. Nature 316, 842-845.

Murray, J.M., Davies, K.E., Harper, P.S., Meredith, L., Mueller, C.R. and Williamson, R. 1982. Linkage relationship of a cloned DNA sequence on the short arm of the X chromosome to Duchenne muscular dystrophy. Nature 300, 69-71.

Noltmann, E.A. 1964. Isolation of crystalline phosphoglucose isomerase from rabbit muscle. Journal of Biological Chemistry 239, 1545-1550.

Noyes, B.E., Mevarech, M., Stein, R. and Agarwal, K.L. 1979. Detection and partial sequence analysis of gastrin mRNA by using an oligonucleotide probe. Proceedings of the National Academy of Science 76, 1770-1774.

Okayama, H. and Berg, P. 1982. High-efficiency cloning of full-length cDNA. Molecular and Cellular Biology 2, 161-170.

Ollivier, L., Sellier, P. and Monin, G. 1975. Determinisme genetique du syndrome d'hyperthermie maligne chez le porc de Pietrain. Annales de Genetique et de Selection Animale 7, 159-166.

Reeders, S.T., Breuning, M.H., Davies, K.E., Nicholls, R.D., Jarman, A.P., Higgs, D.R., Pearson, P.L. and Weatherhall, D.J. 1985. A highly polymorphic DNA marker linked to adult polycystic kidney disease on chromosome 16. Nature 317, 542-544.

Smith, C. and Bampton, P.R. 1977. Inheritance of reaction to halothane anaesthesia in pigs. Genetical Research 29, 287-292.

Soller, M. and Beckman, J.S. 1982. Restriction fragment length polymorphisms and genetic improvement. Second World Congress on Genetics Applied to Animal Production, Madrid, October 1982. Volume 6: 396-404.

Soller, M. and Beckman, J.S. 1983. Genetic polymorphism in varietal identification and genetic improvement. Theoretical and Applied Genetics, 67, 25-33.

Southern, E.M. 1975. Detection of specific sequences among DNA fragments separated by gel electrophoresis. Journal of Molecular Biology 98, 503-517.

Southwood, O.I., Webb, A.J. and Carden, A.E. 1985. Halothane sensitivity of the heterozygote at the halothane locus in British Landrace pigs. Animal Production 40, 540-541 (Abstract).

Stanley, K.K. and Luzio, J.P. 1984. Construction of a new family of high efficiency bacterial expression vectors: identification of cDNA clones coding for human liver proteins. The EMBO Journal 3, 1429-1434.

Steinmetz, M., Minard, K., Horvath, S., McNicholas, J., Srelinger, J., Wake, C., Long, E., Mach, B. and Hood, L. 1982. A molecular map of the immune response region from the major histocompatability complex of the mouse. Nature 300, 35-42.

Stueber, D., Ibrahimi, I., Cutler, D., Dobberstein, B. and Bujard, H. 1984. A novel in vitro transcription-translation system: accurate and efficient synthesis of single proteins from cloned DNA sequences. The EMBO Journal 3, 3143-3148.

Tikhonov, V.N., Nikitin, S.V., Gorelov, I.G., Troshina, A.I., Bobovinch, V.E. and Astakhova, N.M. 1984. Mapping of the loci of the G and H blood group systems on the chromosome 15 of domestic and wild pigs. Genetika 20, 662-671.

Webb, A.J., Carden, A.E., Smith, C and Imlah, P. 1982. Porcine stress syndrome in pig breeding. Proceedings of the 2nd International Congress on Genetics Applied to Livestock Production, Madrid, 5: 588-608. Editorial Garsi, Madrid.

Webb, A.J., Southwood, O.I., Simpson, S.P. and Carden, A.E. 1985. Genetics of porcine stress syndrome. In 'Stress susceptibility and meat quality in pigs', Proceedings of commission on animal management and health and commission on pig production, joint session. Halkidiki, Greece, 30 Sept - 5 Oct, 1985. EAAP Publication No. 33,9-30.

A NATIONAL PROGRAMME ON FACTORS AFFECTING

PIGMEAT QUALITY

A. J. Kempster[1], J. D. Wood[2]

[1]Meat and Livestock Commission, P.O. Box 44,
Bletchley, Milton Keynes, MK2 2EF. U.K.
[2]AFRC Food Research Institute-Bristol,
Langford, Bristol, BS18 7DY. U.K.

ABSTRACT

The national programme on factors affecting pigmeat quality in Britain
is discussed. The programme was set up in 1983 in response to increasing
industry concern about the possible deterioration in pigmeat quality
associated with the trend towards leaner carcasses. It extends over five
years and includes surveys to establish the incidence of problem conditions,
ad hoc trials to examine the effects of specific factors on quality and a
major trial at MLC's Pig Development Unit, Stotfold to compare breed types,
sexes, feeding regimens and slaughter weights. This trial is effectively
the British pig industry in microcosm and allows the interaction of many
different factors to be examined. Results available to date are reported.
These indicate that industry concern about poorer cutting and presentational
characteristics of very lean meat is well founded but that there is little
consumer dissatisfaction with its eating quality. Butchers and consumers
have very different opinions about the eating quality of lean meat and the
most desirable lean to fat ratio.

INTRODUCTION

Like the pig industries in most EEC countries, the British pig
industry has responded successfully to the consumer's increasing demand for
leaner meat. As a result both of substantial genetic improvements and
altered management practices, the national mean P_2 fat thickness (taken
over M. longissimus dorsi at the last rib) has been reduced from 19mm in 1972
to a current (1984) value of 13.5mm. This is estimated to represent a
reduction of 30g/kg in carcass lipid content (Kempster, Cook and Grantley-
Smith, 1986a). The reduction in the mean fat thickness has produced a
substantial increase in the number of very lean pigs: the proportion of the
national kill with P_2 values less than 8mm increased from 3 per cent to
7 per cent over the period. The mean carcass weight of these pigs is also
light: 53kg in 1984 compared with a national mean of 64kg (Kempster et al,
1986a). It is very light in comparison with carcass weights in other EEC
countries.

Associated with the trend towards leaner meat has been increasing
comment by the meat industry, especially retail butchers about the
incidence of meat quality problems. They complain that the meat from very

lean carcasses does not 'set' (is floppy), that the fat is soft and that the tissues separate making cutting or slicing difficult. They also say that the meat looks unattractive to the consumer and is likely to lack succulence and flavour. Entire males are the subject of much of this criticism influenced by the fact that they tend to be slaughtered at lighter weights and typically have 0.5 to 1.0mm lower P_2 fat thickness than the normal mix of castrates and gilts at these weights (Meat and Livestock Commission, 1985). Since 1975, entire males have increased from virtually nothing to 25 per cent of the national slaughter population.

The criticisms are not always restricted to light weight carcasses. The number of complaints about soft fat appears to have increased even with the carcasses of moderate to high levels of fatness. However, we believe that this reflects to some extent the increasing awareness of quality deficiencies and variation in meat quality generally because of improved quality control procedures, in particular the setting up of the Charter Quality Bacon Scheme (Kempster, 1984).

Pale, soft, exudative (PSE) muscle also continues to be of concern because of its major influence on commercial value through reduced water-holding capacity. However, in the case of this condition, substantial information is available on the level of incidence and causal factors (Chadwick and Kempster, 1983; Kempster, Evans and Chadwick, 1984). A measurement of water-holding capacity has been introduced recently to the MLC Supernucleus Breeding Scheme.

The scientific information on the meat quality problems associated with very lean carcasses was reviewed by the Meat and Livestock Commission (1983). Although limited, there is evidence which indicates that the fat quality of lean pigs is poorer than that of fatter pigs. The fatty tissue in lean carcasses is less developed than that in fat carcasses. Thus, it contains higher concentrations of water and connective tissue and less lipid. The lipid is also more unsaturated and all these factors cause the tissue to feel softer. (Factors affecting fat quality were reviewed by Wood (1984)).

Information on the eating quality of lean pigs was found to be especially limited. Early studies in pigs covering a wide range of values for backfat thickness have shown little or no effect of fatness on eating quality (Rhodes, 1972). However, the lower limit to fatness did not include very lean carcasses. A more recent, small-scale study (Wood, Mottram and Brown, 1981) examined eating quality in gilts weighing 48 to 61kg carcass weight and with two distinct fatness categories: 12mm P_2 and very lean carcasses with 6mm P_2. There was no difference in the tenderness and flavour of roasted loins, as judged by a trained taste panel, but the very lean group had slightly less juicy meat ($P<0.05$).

Information on the cutting and eating quality characteristics of meat from entire males was also found to be limited, the main focus of attention of research continuing to be the problem of boar taint. Recent developments and the current state of knowledge on ths use of entire males were reviewed by a European Association for Animal Production Working Group in 1984 (Lundstrom, Malmfors, Vahlun, Kempster, Andresen and Hagelsø, 1985).

DESIGN OF A NATIONAL PROGRAMME

In response to the industry concern and against this background of limited scientific information, a comprehensive five-year programme has been set up in 1983 to obtain the facts necessary to provide guidance to industry. This involves:

(1) surveys in abattoirs to establish the incidence of the problems,

(2) ad hoc trials to examine the effects of different factors on pigmeat quality and to evaluate methods of pigmeat quality measurement,

(3) a detailed comparison of pigs of different types at MLC's Pig Development Unit at Stotfold,

This is a co-ordinated programme involving the Meat and Livestock Commission, AFRC Food Research Institute-Bristol and several other organisations.

SURVEYS

pH surveys have been carried out over the years to examine the incidence of PSE and DFD (the latest published by Chadwick and Kempster, 1983). The series is being continued to provide more up to date information and the pH measurements are being complemented by light reflectance measurements taken by the Fibre Optic Probe developed by D. MacDougal at FRI-Bristol. Results to date suggest there has been some increase in PSE incidence over the last 10 years but the levels are not sufficiently high to give serious cause for concern. DFD levels are insignificant.

In response to concern about soft fat (highlighted as the most serious problem from the level of industry complaints and queries), the

Penetrometer has been developed at the FRI-Bristol by E. Dransfield and colleagues to measure the condition. This is a device which measures the resistance of the fat to penetration by a small probe. It is now being used in a national survey involving 30 abattoirs which was begun in November 1985 and runs for 12 months. Penetrometer measurements are being taken on the split carcass in the shoulder at the dorsal mid line. Samples of fat from the same carcasses are being sent to the FRI-Bristol for more detailed texture measurements and chemical analysis.

Boar taint has not been a serious problem in Britain (Rhodes, 1972; Kempster and Cuthbertson, 1982). However, in view of the uncertainty which now exists about the principal causal agent of boar taint (androstenone or skatole) (Lundstrom et al, 1985), the fat samples from the soft fat survey (referred to above) will also be subjected to analysis for these compounds to provide information on national levels. It is also hoped that collaborative work will be carried out with research institutes in other countries.

One of the main problems of entire males is skin damage caused by increased levels of fighting in transit and lairage. This leads to unsightly bacon rashers and, under more extreme circumstances, condemnations of meat. Following a pilot study in one large abattoir, a national survey has begun to examine the national incidence of skin damage and the relative importance of causal factors. The survey involves a visual assessment of skin damage using a photographic scale for all pigs classified nationally on 12 selected days over a year (beginning November 1985).

AD HOC TRIALS: THE EFFECTS OF FAT THICKNESS AND SEX

Results are available from an interim trial carried out to examine fatness levels and sex effects on meat quality, and to evaluate the various techniques available for measuring meat quality (Kempster, Dilworth, Evans and Fisher, 1986b; Wood, Jones, Dransfield and Francombe, 1986). The trial involved 300 carcasses selected from 10 abattoirs. Equal numbers of gilts and entire males were selected from known producer groups within each abattoir to show the typical range of fatness. Average carcass weights and fat measurements are shown in Table 1. The two sexes were compared at similar P_2 fat thicknesses and were represented equally in the three fatness groups.

TABLE 1   Mean carcass weights and P_2 fat thickness measurements

	Sex		Fatness		
	Gilt	Entire male	Lean	Ave	Fat
Carcass weight (kg)	58.2	58.1	57.0	58.2	59.2
P_2 (mm)	12.4	11.9	8.6	11.7	16.2

From Kempster et al (1986b).

Leg and loin joints were assessed, for cutting characteristics and presentation, by a total of 45 butchers from independent and multiple retailers around the country. Key results for the loin joints are shown in Table 2; results for the leg joints were similar. The butchers judged the fat from leaner loins to be softer and the meat to be wetter and floppier with a greater degree of tissue separation than that from fatter loins. Meat from entire males was also judged to be poorer than that from gilts in these respects, but the differences were much smaller than those found between fatness levels. When asked how they judged the meat in terms of fatness or leanness, the butchers were critical of the lean carcasses. For example, 45 per cent of the 'lean' loins were judged rather lean and 15 per cent much too lean. The butchers also considered that 30 per cent of the lean loins would cook and eat poorly or very poorly against 10 per cent for the average and 15 per cent for the fat loins.

TABLE 2   Key results from the butcher panel and consumer panel assessments of loins (%)

BUTCHER PANEL

Firmness of fat	Firm or very firm	Slightly soft	Soft or very soft
Entire male	55	24	22
Gilt	66	20	14
Lean	32	35	32
Average	64	20	15
Fat	85	9	6

TABLE 2 CONTINUED   Key results from the butcher panel and consumer panel assessments of loins (%)

Level of tissue separation	None or hardly any	Slight	Excessive
Entire male	36	36	29
Gilt	43	37	21
Lean	16	38	46
Average	39	43	18
Fat	62	26	11

CONSUMER PANEL

Juiciness	Extremely or very juicy	Moderately or slightly juicy	Dry
Lean	16	69	16
Fat	23	67	9

Overall acceptability	Excellent to good	Fair or average	Poor or very poor
Lean	69	28	4
Fat	71	26	3

From Kempster et al (1986b).

A loin cut and a sample of fat from the shoulder of each carcass were evaluated for cutting characteristics using laboratory assessments at the AFRC Food Research Institute-Bristol (FRI-B). The results from these tests supported the butchers' panel results in that leaner carcasses and entire males had softer fat and more tissue separation (see Table 3).

The occurrence of softer fat in leaner carcasses is in line with other detailed studies of fat composition. Although several factors contribute to the softer fat, such as the lower lipid concentration in smaller tightly-packed fat cells and higher water concentration, the most important factor seems to be the higher concentration of linoleic acid (C18:2) (Wood and Enser, 1982).

Loin chops and shoulder and leg joints were assessed by a total of 500 families eating the meat under normal conditions in their homes: 200

families were given shoulder and leg joints and 300 families were given
loin chops. Each family received lean and fat cuts from the same producer
group. Cuts from the intermediate fatness group were not evaluated by
consumer panels. Key results for loin joints are given in Table 2. The
results for leg and shoulder joints were similar to those for the loins.
The consumers were much less critical than the butchers about the 'lean'
carcasses, judging only four per cent of the loins (seven per cent of the
legs) to be rather lean or too lean.

TABLE 3   Key results from the assessment carried out at AFRC Food
Research Institute-Bristol

	Sex		Fatness		
	Entire male	Gilt	Lean	Ave	Fat
Sensory assessment of fat firmness (loin) [1]	3.7	4.4	2.8	3.9	5.3
Stevens compression test of the firmness of fat (g)	581	740	432	637	913
Trained taste panel					
Texture of lean [2]	1.2	0.9	1.0	–	1.1
Juiciness of lean [3]	1.2	1.2	1.1	–	1.3
Overall eating quality [4]	0.8	1.0	0.7	–	1.0

Joints of average fatness were not panelled.

(1) 8-point scale with 1=very soft to 8=very hard.

(2) -7 (extremely tough) to +7 (extremely tender).

(3) 0 (dry) to +4 (extremely juicy).

(4) -7 (extremely unacceptable) to +7 (extremely acceptable).

From Wood et al (1986).

Consumers found the joints from leaner carcasses to be less juicy on average with a slight tendency towards toughness and poorer flavour, there being five per cent more critical comments for leaner carcasses. However, joints of different fatness were judged to be the same in terms of overall acceptability (including assessment of leanness/fatness). Consumers found no differences between entire males and gilts in eating quality or overall acceptability.

Meat from a loin joint was assessed by a trained taste panel at FRI-Bristol. This involved a total of 12 assessors. The results were in general agreement with those from the consumer panel. The meat from leaner carcasses was found to be less juicy, and there was a tendency for it to be less tender and of poorer flavour (see Table 3).

The results confirm that complaints from butchers about the poorer cutting and handling properties of meat from leaner pigs are well founded and that the meat from entire males is also slightly poorer in these respects, even at the same level of fatness.

The butchers held strong views about the unacceptability of the very lean carcasses and considered that the meat would not cook and eat as well as that from fatter carcasses. However, these views were not held by the consumers. Although the leaner joints had small disadvantages in eating quality characteristics, the consumers considered them to be as acceptable overall as fatter joints. In contrast to the view held by many people in the industry, the meat from entire males and gilts did not differ significantly in eating quality.

AD HOC TRIALS: THE DUROC CONNECTION

Speculation about the eating quality of Duroc pigs has become an important feature of the meat quality debate in Britain. Imported Danish pigmeat from Duroc crosses is claimed to have a particularly high intra-muscular fat content and have superior eating quality to British pigmeat for this reason. It is certainly true that the intramuscular fat content has been shown to be low in some circumstances: in the trial referred to above to examine fatness and sex effects at light carcass weights, the intramuscular fat contents were low (5.5, 6.6 and 9.6 mg lipid/g fresh weight in M. longissimus dorsi for lean, average and fat pigs respectively). However more information is required before sensible decision can be made because the meat from the Danish Duroc crosses comes from relatively heavy pigs with high fatness levels and breed per se may not be the important factor. The meat may also be uneconomical to produce at such weights/fatness levels.

Several trials are now underway in Britain to examine the growth performance and carcass quality (including eating quality) of purebred Durocs and F1 Duroc crosses in comparison with conventional White breeds and crosses. Results from these trials will be available by the end of 1986. In addition, consideration is being given to the possibility of of running a major trial (beginning in 1987) at MLC's Pig Development Unit to provide an overall economic appraisal of the Duroc as a possible 'third breed' used in various breed combinations with the British Large White and Landrace.

MAJOR TRIAL AT MLC'S PIG DEVELOPMENT UNIT, STOTFOLD

The trial has three main objectives:

(1)   to compare the production efficiency of meat-type sires with conventional White sires.

(2)   to examine the influence of genotype, sex, feeding level and slaughter weight on carcass and meat quality.

(3)   to establish a genetic baseline for measuring future changes in production efficiency and carcass characteristics.

Specialist meat-type sires, incorporating breeds such as the Pietrain and Hampshire with emphasis on carcass meat content, are beginning to find an important use in British pigmeat production, contributing about 10 per cent of the terminal sires used. At the moment it is not clear whether their use is in the best overall interests of the industry, in particular whether they will lead to increased stress susceptibility and poorer meat quality. The trial is designed in such a way that an overall assessment of their performance can be made.

The four major breeding companies in Britain are providing meat-type sires and conventional White-bred sires for crossing on their respective hybrid female populations. A genetic control population is also included.

A total of 12 intakes of breeding stock are involved in the trial, at ten week intervals. Each intake involves 12 hybrid gilts, two meat-type boars and two purebred (Large White or Landrace) boars from each breeding company. It also includes a control population intake comprising 12 Landrace gilts (from Wye College) and four Large White boars (from the AFRC Animal Breeding Research Organisation, Edinburgh). The aim is to obtain two litters from each gilt.

The pigs are being grown on one of two feeding regimens (ad libitum and with restriction to a time-based scale), and slaughtered at one of two carcass wieghts (52.5 and 72.5 kg). Three sexes (entire males, castrates and gilts) are involved. Carcass and meat quality evaluation includes tissue separation, detailed chemical analysis, consumer and trained taste panel assessments, butcher panel assessments, and assessments of cutting and processing characteristics.

The trial is not comprehensive in that it does not involve pigs slaughtered at heavy carcass weights (greater than 80kg). However, a three-year study co-sponsored by MLC is being carried out at the University of Newcastle under the direction of Dr M. Ellis to provide information on the effect of genotype and sex on pigs taken to these weights.

GENERAL CONSIDERATIONS

Information available from the programme to date suggests that the trend towards the production of leaner carcasses can continue without any serious effect on eating quality characteristics, However, ways of dealing with the cutting and presentational problems associated with the meat from very lean carcasses are needed.

Improved temperature control and cutting methods may help but are unlikely to completely overcome the problems since they appear to be associated with the physiological immaturity of pigs at light carcass weights (Wood et al, 1986). A better strategy might be to slaughter pigs at heavier weights when the tissues are more mature. The carcasses would be fatter but recent MLC economic studies show that the overall efficiency of producing, marketing and distribution of the lean meat to the point of consumption is greatest at heavier carcass weights, in the range 60 to 75kg (A. J. Kempster and A. S. Monk, unpublished data).

REFERENCES

Chadwick, J.P. and Kempster, A.J. 1983. A repeat national survey (ten years on) of muscle pH values in commercial bacon carcasses. Meat Sci., 9, 101-111.

Kempster, A.J. 1984. Production and slaughter practices. Proceedings of the 30th European Meeting of Meat Research Workers, Bristol, September 1984 pp 1-3.

Kempster, A.J., Cook, G.L. and Grantley-Smith, M. 1986a. National estimates of the body composition of British cattle, sheep and pigs with special reference to trends in fatness. Meat Sci., (in press).

Kempster, A.J. and Cuthbertson, A. 1982. Meat industry attitudes to entire male pigs in Great Britain. Paper given at the 33rd Annual Meeting of the European Association for Animal Production, Leningrad, 1982.

Kempster, A.J., Dilworth, A.W., Evans, D.G. and Fisher, K.D. 1986b. The effects of fat thickness and sex on pigmeat quality with special reference to the problems associated with overleanness. 1. Butcher and consumer panel results. Anim. Prod., (in press).

Kempster, A.J., Evans, D.G. and Chadwick, J.P. 1984. The effects of source population, feeding regimen, sex and day of slaughter on the muscle quality characteristics of British crossbred pigs. Anim. Prod., 39, 455-464.

Lundstrom, K. Malmfors, B., Vahlun, S., Kempster, A.J., Andresen, O. and Hagleso, A.M. 1985. Recent research on the use of boars for meat production - report from the EAAP Working Group Meeting in Denmark, 1984. Livest. Prod. Sci., 13, 303-309.

Meat and Livestock Commission, 1983. Very lean pigs. Report of an MLC Planning and Development Team, MLC, Bletchley, Bucks.

Meat and Livestock Commission, 1985. Pig Yearbook, April 1985. MLC, Bletchley, Bucks.

Rhodes, D.N. 1972. Consumer testing of pork from boar and gilt pigs. J. Sci. Fd. Agric., 23, 1483-1491.

Wood, J.D. 1984. Fat deposition and the quality of fat tissue in meat animals. In 'Fats in Animal Nutrition' (ed. Wiseman, J.) Butterworths, London, pp 407-435.

Wood, J.D. and Enser, M. 1982. Comparison of boars and castrates for bacon production. 2. Composition of muscle and subcutaneous fat changes in side weight during curing. Anim. Prod., 35, 65-74.

Wood, J.D., Jones, R.C.D., Dransfield, E. and Francombe, M.A. 1986. The effects of fat thickness and sex on pigmeat quality with special reference to the problems associated with overleanness. 2. Laboratory and taste panel results. Anim. Prod. (in press).

Wood, J.D., Mottram, D.S. and Brown, A.J. 1981. A note on the eating quality of pork from lean pigs. Anim. Prod., 32, 117-120.

PORCINE MALIGNANT HYPERTHERMIA: DOSE/RESPONSE TO HALOTHANE

C.P. Ahern[+], W.E. Rempel,[++] J.H. Milde,[+++] G.A. Gronert[+++]

[+]Department of Veterinary Physiology and Biochemistry,
University College Dublin, Veterinary College of Ireland,
Dublin 4, Ireland.
[++]Department of Animal Science, University of Minnesota,
St. Paul, Minnesota 55108.
[+++]Department of Anaesthesiology, Mayo Clinic, Rochester,
Minnesota 55905.

ABSTRACT

Sixteen intact pigs and twenty five isolated perfused caudal porcine preparations were exposed to varying concentrations of halothane to trigger hypermetabolic episodes.   These were detected by hind limb rigidity in intact pigs and by increased O_2 consumption and CO_2 and lactate production in the caudal preparations.

In intact pigs, lower concentrations of halothane provided a slower onset of malignant hyperthermia.   For 3% halothane, mean onset time was 1.4 ± 0.2 min; for 2% halothane 2.5 ± 0.5 min, for 1% halothane 5.4 ± 0.9 min and for 0.5% halothane 11.8 ± 1.7 min (mean, SEM).   All intact pigs received all concentrations of halothane, over a period of several weeks.   In the caudal preparation the onset of malignant hyperthermia by halothane sensitised the preparation to subsequent triggering by a lower concentration of halothane.

The data suggest that clinical maintenance concentrations of halothane are in the less sensitive portion of the dose/response curve for triggering of malignant hyperthermia.

INTRODUCTION

The halothane test as developed by Eikelenboom and Minkema (1974) is the most widely used method for identifying stress-susceptible pigs.   The methods employed in halothane testing vary among investigators particularly in regard to duration of the test and the concentration of halothane used (Eikelenboom and Minkema, 1974; McGloughlin et al, 1980; Webb, 1980; McGrath et al, 1984).   Recent experiments (Gregory and Wilkins, 1984) have shown that the response to halothane in stress-susceptible pigs was related to the dose of halothane administered and not to a triggering mechanism which provoked an all or none response.   Halothane was given intravenously through an indwelling catheter to avoid restraint stress.   In another study (McGrath et al, 1984) the concentration of halothane affected both the number of reactions and reaction times.   False negative reactions were obtained when pigs were given less than 3% halothane for five minutes.   One of the problems encountered when investigating dose/response effects of halothane is that this anaesthetic, by depressing the heart and lowering blood pressure, may cause inadequate tissue perfusion.   In the present study this problem was overcome by using an isolated perfused caudal muscle preparation in which blood flow was maintained constant.   It was possible to study both the magnitude of the metabolic response to halothane and the reaction time in this preparation.   The dose/response effects of halothane in intact pigs were also investigated.

MATERIALS AND METHODS

Intact pigs

Purebred Pietrain pigs were obtained from a herd in which the incidence of malignant hyperthermia (MH) had been greater than 95%. Sixteen pigs from two litters, weighing 25-35 kg, were exposed to varying concentrations of halothane.   On a test day, each litter received a different concentration of halothane but all pigs in a specific litter received the same concentration.   The test was performed by administering the various concentrations of halothane in oxygen (3 l/min) by a face mask using a semi-closed circle system. The animal was restrained manually and halothane was administered until hind limb rigidity occurred.   Four or more days later the tests were repeated and the concentrations to each litter rotated until all the pigs had received 3%, 2%, 1% and 0.5% halothane.

Caudal preparation

A separate group of 25 Pietrain pigs, weighing 30-50 kg, was confirmed as susceptible to MH by the development of hind limb rigidity within three minutes of exposure to 3% halothane by mask. At least one week later these pigs were studied in the laboratory using the isolated perfused caudal preparation as described in earlier studies (Gronert et al, 1980).   The preparation involves corporal transection at the level of the first lumbar vertebra with cannulation of the distal aorta and inferior vena cava and perfusion by a roller pump-oxygenator-heat exchanger.   The process of acute cord section produced marked muscle contractions in the hind limbs and was frequently associated with increased oxygen consumption ($\dot{V}_{O_2}$) of the preparation, a sign that muscle depolarisation had triggered a

hypermetabolic episode.   In one experiment two groups of pigs with
four animals in each group were used to study the effects of 1% and
2% halothane.   The muscle relaxant pancuronium (0.5 mg/kg) was
administered before cord section to prevent muscle contraction and the
associated increase in metabolism.   In another experiment pancuronium
was not administered and when $\dot{V}o_2$ had decreased to control levels
of less than 8ml/min^{-1}/kg^{-1} after cord section, varying concentrations
of halothane were introduced until metabolism increased.   These
concentrations included four preparations exposed to 1% halothane and
thirteen preparations exposed to 0.3%, 0.5% and/or 0.75% halothane.
At this time halothane was discontinued and, when metabolism returned
to control levels, smaller concentrations were introduced to determine
if the prior episodes had sensitised the muscle and lowered the threshold.   (Concentrations greater than 1% were not used as previous
studies had demonstrated that higher concentrations triggered the
preparation very rapidly without recovery to control values).

Isolated preparations were perfused at constant blood flow,
control being that flow that resulted in mixed venous Po_2 levels
greater than 50 mm/Hg and unchanged thereafter.   Baseline $Paco_2$ was
40 mm Hg, set by adjusting oxygenator gas inflow and concentrations of
N_2, O_2 and CO_2; this was also unchanged thereafter.   The onset of
MH was indicated by a rise in $P\bar{v}co_2$ of 5mm Hg; halothane was
discontinued when $P\bar{v}co_2$ exceeded 70mm Hg.   Increases in $\dot{V}o_2$ and
metabolic acidosis, ranging to 20m mol/l negative base excess,confirmed
MH episodes.   Arterial and venous blood gases and reduced/
oxyhaemoglobin levels (co-oximeter) provided data for calculation of
O_2 content.   The Fick formula enabled calculation of preparation
$\dot{V}o_2$.   Base excess estimated degree of metabolic acidosis.   Control
preparation temperature was 37.5°C.   Data are expressed as mean $\pm$
SEM.   Comparisons between groups utilised the unpaired t test,
$P < 0.05$ considered significant.

RESULTS

These are preliminary results and further studies are planned.

Intact Pigs

The dose/response effects for halothane in concentrations varying
from 3% to 0.5% are demonstrated in Figure 1.   Mean response times

374

are different among all four concentrations, P < 0.05.

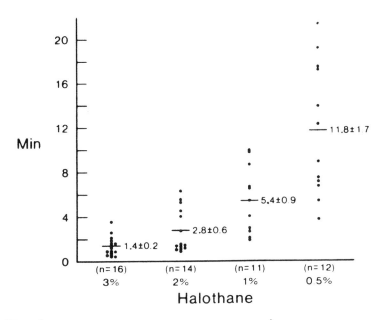

Fig. 1   Reaction times, including mean ($\pm$SEM) reaction times,
          of intact pigs subjected to various concentrations of
          halothane.

For all pigs, 3% halothane produced hind limb rigidity in 3.5 min
or less, 2% halothane in 6.2 min or less, 1% halothane in 10.0 min or
less and 0.5% halothane in 22 min or less.   In the latter group, the
end-point was sometimes obscure because of increased muscle tone
during light anaesthesia and restraint.

Caudal preparation
        When preparations were exposed to 1% and 2% halothane (Figure 2)
the time of onset of MH was faster and the response (as measured by
$P\bar{v}_{CO_2}$) was greater in those preparations which received 2% halothane.
In a separate experiment involving 17 pigs, metabolism increased after
cord section, with $\dot{V}_{O_2}$ ranging from 5-15 ml/min^{-1}/kg^{-1}.   After
conditions were stable and $\dot{V}_{O_2}$ was less than 8 ml/min^{-1}/kg^{-1} halothane
was introduced.   For this study 4 preparations were exposed to 1%
halothane - these triggered in 20, 20, 35 and 50 min but only one
recovered to control levels;   it retriggered when exposed to 0.3%

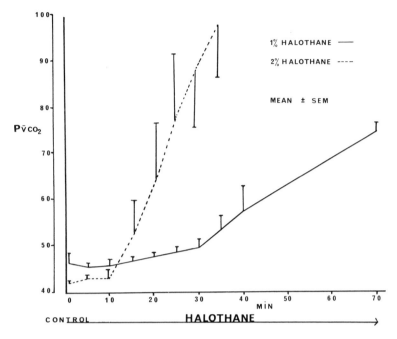

Fig. 2   Carbon dioxide production of isolated perfused
caudal muscle preparations exposed to 1% or 2%
halothane.

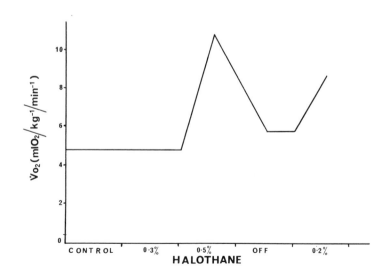

Fig. 3   Example of the oxygen consumption of one isolated
perfused porcine muscle preparation exposed to
different doses of halothane.

halothane.   Results for the remaining 13 preparations are shown in figures 3 and 4.

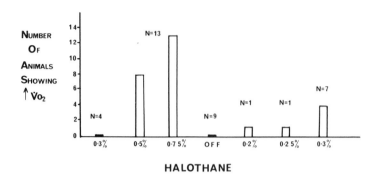

Fig. 4   Number of isolated perfused caudal muscle
          preparations showing increased oxygen consumption
          when exposed to various concentrations of halothane.

An example of the $\dot{V}o_2$ responses to various doses of halothane is shown for one isolated perfused muscle preparation in Figure 3. Four of the preparations were exposed to 0.3% halothane for up to 60 min and none triggered (Figure 4);   all thirteen were then exposed to 0.5% halothane for up to 60 min and eight triggered in 15, 25, 35, 35, 40, 55, 60 and 60 min;   the final five preparations were then exposed to 0.75% halothane and triggered in 10, 15, 20, 20, and 60 min.   Of these   thirteen preparations, nine recovered to control values and were re-exposed to 0.2 - 0.3% halothane;   six retriggered in 10, 10, 15, 15, 20 and 45 min.

DISCUSSION

Halothane triggered MH responses in both intact pigs and the isolated perfused caudal preparation.   The concentration of halothane influenced the time of onset and the magnitude of the MH response. We expected that the muscle depolarisation which was associated with cord section might sensitise muscle in a manner similar to prior exercise (van den Hende et al., 1976) or succinylcholine (Gronert & Milde, 1981) and speed up onset time.   This was apparently not so and perhaps can be explained by the following.   Acute cord section is generally followed immediately by acute spinal shock, with flaccidity and areflexia. Despite the increased $\dot{V}_{O_2}$ observed in most preparations soon after cord section, the sudden neuronal inactivity may have subsequently diminished intrafibrillar ionised calcium levels, resulting in a relatively slower onset of MH upon exposure to halothane. On the other hand, triggering of the caudal preparation by halothane lowered the threshold for halothane, so that lower concentrations now triggered hypermetabolism.

Central depressants delay the onset of MH (Hall et al., 1972; Ahern et al., 1980; Gronert & Milde, 1981). While it is not known how these compounds work they do reduce the level of restraint stress and they also reduce the concentration of halothane required for anaesthesia to maintenance levels. The present results suggest that these levels of 0.5% - 1% halothane are in the dose/response range in which MH is triggered slowly.

The pigs used in the present studies were from a herd with a high incidence of MH. Some other breeds are less susceptible. The incidence of MH in Canadian Landrace pigs was 1.5% when the animals were subjected to the halothane test but the incidence increased to 18% when succinylcholine was used in conjunction with halothane (Seeler et al., 1983). In another Landrace herd a short period of exercise before the administration of halothane increased the incidence of MH (van den Hende et al., 1976). Some animals in a group of Irish Landrace gave a positive MH reaction only after prolonged anaesthesia with halothane (Ahern et al., 1979). The halothane test may be an inadequate screening test for MH in these herds and another stimulus, in addition to halothane, may be required to determine the true incidence of the syndrome. When the halothane test alone is used in herds with a low incidence of MH the duration of the test should be not less than 5 min, anaesthesia should be induced quickly with 5% - 7% halothane and surgical anaesthesia should then be maintained with the appropriate concentration of halothane. This procedure will ensure that the animal is exposed to the highest possible dose of halothane for the duration of the test.

Thus, several factors influence the time of onset and severity of MH. These include drugs used in premedication, stressors such as restraint, environmental temperature and humidity, the susceptibility of the individual animal and the concentration of halothane.

378

ACKNOWLEDGEMENTS

C.P.A. received a travel grant from the Wellcome Trust.

REFERENCES

Ahern, C.P., Somers, C.J. Wilson, P. and McLoughlin,J.V. 1979.
Porcine malignant hyperthermia syndrome: high-energy phosphates
and glycolysis in muscle of pigs showing different responses in
halothane testing. Acta Agric. Scand., Suppl. 21, 457-462.

Ahern, C.P., Somers, C.J., Wilson, P. and McLoughlin, J.V. 1980.
Halothane-induced malignant hyperthermia: creatine phosphate
concentration in skeletal muscle as an early indicator of the
onset of the syndrome. J. Comp. Path., 90, 177-186.

Eikelenboom, G. and Minkema, D. 1974. Prediction of pale, soft,
exudative muscle with a non-lethal test for the halothane-
induced porcine malignant hyperthermia syndrome. Neth.
J. Vet. Sci., 99, 421-426.

Gregory, N.G. and Wilkins, L.J. 1984. The intravenous halothane
test as an experimental method for quantifying stress sensitivity
in pigs. J. Sci. Food Agric., 35, 147-153.

Gronert, G.A. and Milde, J.H. 1981. Variations in onset of porcine
malignant hyperthermia. Anesth. Analg., 60, 499-503.

Gronert, G.A., Milde, J.H. and Taylor, S.R. 1980. Porcine muscle
responses to carbachol, α and β-adrenoceptor agonists, halothane
or hyperthermia. J. Physiol., 307, 319-333.

Hall, L.W., Trim, C.M. and Woolf, N. 1972. Further studies of
porcine malignant hyperthermia. Br. Med. J., 2, 145-148.

McGloughlin, P., Ahern, C.P., Butler, M. and McLoughlin, J.V. 1980.
Halothane-induced malignant hyperthermia in Irish pig breeds.
Livest. Prod. Sci., 7, 147-154.

McGrath, C.J., Lee, J.C. and Rempel, W.E. 1984. Halothane testing
for malignant hyperthermia in swine: dose-response effects. Am.
J. Vet. Res., 45 (9), 1734-1736.

Seeler, D.C., McDonnell, W.N. and Basrur, P.K. 1983. Halothane
and halothane/succinylcholine induced malignant hyperthermia
(porcine stress syndrome) in a population of Ontario boars.
Can. J. Comp. Med., 47 (3), 284-290.

Van den Hende, C. Lister, D., Muylle, E., Ooms, L. and Oyaert,
W. 1976. Malignant hyperthermia in Belgian Landrace pigs
rested or exercised before exposure to halothane. Br. J.
Anaesth. 48, 821-829.

Webb, A.J. 1980. The halothane test: a practical method of
eliminating porcine stress syndrome. Vet. Rec., 106, 410-412.

THE IMPORTANCE OF REACTION INTENSITY IN THE HALOTHANE TEST

K. Fischer

Federal Meat Research Centre
Institute of Meat Production and Marketing
E.C.Baumann-Straße 20, D-8650 Kulmbach,
Federal Republic of Germany

ABSTRACT

The importance of different reaction intensities to the halothane test was examined in 282 piglets of the German Landrace. The reaction intensity was evaluated by the stiffness of the hind legs using a 5 point scale. The subjective evaluation was supported by physiological parameters. With increasing rigidity the criteria measured showed more distinctly the complex of symptoms of malignant hyperthermia. However the increase was not a continuous one. The mean values of the stages 1 and 2 (none or very low rigidity) and those of 3 - 5 (medium to very strong rigidity) were close together. Similar results were found for meat quality traits. In repeated tests the grouping into rigidity stages was not entirely consistent. The shifting however took place generally within groups 1 and 2 or 3 to 5 respectively. The prediction of meat quality was improved a little by including some physiological responses to the halothane test. The results indicated that the observed differences in reaction intensity within the negative or positive reaction groups were not genetically determined; therefore grouping in a 2 stage system (H-, H+) is   sufficient during routine testing. Halothane negative animals show none,   very slight or reversible stiffness during the test.

INTRODUCTION

It is well known that during the halothane test there are differences in the intensity of response over and above the typical positive or negative reaction (Eikelenboom et al., 1977; Kittler and Dzapo, 1978; Augustini et al., 1979; Eiffert, 1984; Haugwitz and v. Lengerken, 1985).

The present investigation examined the following aspects of reaction intensity,

- whether subjectively observed differences in the reaction intensity
  can be quantified,
- whether subjectively observed differences are repeatable,
- whether differences in reaction intensity are important for the prediction
  of meat quality,
- whether physiological responses can be used to predict meat quality.

MATERIAL AND METHODS

Two hundred and eighty two piglets of the German Landrace breed (20 - 30 kg) were halothane tested up to four times at intervals of 2 or 3 days. The tests were carried out in a semi-open system (4 % halothane) with a maximum duration of 4 minutes. The test was interrupted only when irreversible cramps of the legs appeared. The reaction was described subjectively at the end of narcosis using a 5 point scale to describe the rigidity of the musculature of the hind legs. A score of 1 was given for a complete relaxation and a score of 5 for a very strong cramp. After finishing the halothane test, physiological parameters were recorded for 6,5 minutes at intervals of 30 seconds (heart rate, rectal temperature, respiratory rate, minute volume, CO_2 concentration in the expired air, CO_2 output). Blood was taken from an ear vein immediately after halothane application and 6,5 minutes afterwards (pO_2, pCO_2, pH, acid-base values, lactate, glucose). After a normal fattening period the animals were slaughtered at a liveweight of approximately 110 kg. Meat quality was evaluated in the usual manner (pH, WHC according to the Grau/Hamm method, rigor value) supported by measurements of postmortem glycogenolysis. Analysis of variance was carried out after correcting the data for the effect of day of test, liveweight on the day of test and liveweight on the day of slaughter.

RESULTS AND DISCUSSIONS

Results for the most important times of measurement are presented. Table 1 shows that the rectal temperature did not clearly differentiate between the stages of stiffness during the first few minutes after the end of the test. With increasing time after halothane application however a distinct increase in rectal temperature was observed for stages 3 to 5. This was contrary to the decrease observed for stages 1 and 2.

Heart rate values were already different only 0,5 min after the end of the test. With increasing rigidity the LSQ means also increased, however not always significantly.

The respiratory measurements showed especially clear differences between stages of stiffness (Table 2) corresponding to their role in compensating for acidosis, which occured in varying degrees after halothane application. This can clearly be seen in the differences of CO_2 output. In violently reacting animals (stage 5) values were sometimes three times higher than in relaxed animals (stage 1). Forced CO_2 output, necessary for the regulation of pH, depends upon a corresponding increase in respiratory rate and minute volume.

At the first measurement the blood parameters (Table 3) showed only small changes with increasing rigidity. An acidosis was observed in all reaction types caused by the breath-depressive action of the halothane.

TABLE 1   LSQ mean values of rectal temperature and heart rate in pigs grouped according to the degree of hind limb rigidity, at selected times after the halothane test   *

Time of measurement (min after end of test)	1	2	3	4	5	F-Test
	Rigidity stages					
n =	84	24	10	61	72	
	Rectal temperature (°C)					
0	39,04	39,14	39,07	39,04	39,06	
2,5	$38,93^a$	$39,10^{ac}$	$39,55^{bc}$	$39,38^{bc}$	$39,50^b$	***
6,5	$38,75^a$	$38,96^a$	$39,61^b$	$39,67^b$	$39,74^b$	***
	Heart rate (beats/min)					
0,5	$154,9^a$	$159,6^a$	$190,8^b$	$212,2^b$	$224,6^c$	***
2,5	$162,8^a$	$169,0^a$	$211,4^b$	$202,1^b$	$207,2^b$	***
6,5	$147,5^a$	$147,7^a$	$181,2^b$	$169,6^b$	$172,9^b$	***

TABLE 2   LSQ mean values of respiratory measurements for pigs displaying different degrees of hind limb rigidity, measured at selected times after the end of the halothane test

Time of measurement (min after end of test)	1	2	3	4	5	F-Test
	Rigidity stages					
n =	87	25	11	70	78	
	Respiratory rate (breaths/min)					
1,0	$26,4^a$	$26,3^a$	$22,0^a$	$33,3^b$	$33,7^b$	***
2,5	$31,8^a$	$30,4^a$	$33,6^a$	$41,9^b$	$42,9^b$	***
6,5	$33,8^a$	$36,2^a$	$38,1$	$44,7^b$	$43,4^b$	***
	Minute volume - $\dot{V}_E$ (l/min)					
1,0	$8,1^a$	$8,0^a$	$9,6^a$	$16,7^b$	$18,8^b$	***
2,5	$10,2^a$	$12,4^a$	$18,2^b$	$23,4^c$	$24,3^c$	***
6,5	$10,6^a$	$11,4^a$	$17,3^b$	$23,4^b$	$20,3^b$	***
	CO_2 output - $\dot{V}_{CO_2}$ (l/min)					
1,0	$0,28^a$	$0,30^a$	$0,55^b$	$0,85^c$	$0,99^d$	***
2,5	$0,32^a$	$0,44^a$	$0,84^b$	$0,96^b$	$0,97^b$	***
6,5	$0,31^a$	$0,33^a$	$0,50^b$	$0,57^b$	$0,56^b$	***

* Table 1 to 4: LSQ means within the same line marked by different letters are significantly different ($p \leqslant 0,05$)

The test narcosis was interrupted earlier when the subjectively observed reaction appeared more violent. This explains why relatively normal values can still be found in stage 5 compared to stages 1 and 2. The situation was changed at the second measurement. The values for stages 1 and 2 were better while stages 3, 4 and 5 proved to have a strong metabolic acidosis despite forced CO_2 breathing (Table 3).

The physiological measurements for the individual stages of stiffness show that an increasing rigidity of the hind legs was associated with an unfavorable reaction. However, this association was not continuous. In most cases the mean values for stages 1 and 2 were similar, as were those for stages 3, 4 and 5. The values found in stages 3 to 5 correspond with a typical malignant hyperthermia and are in agreement with results in the literature (Berman et al.,1970; Lucke et al., 1976; van den Hende, 1976; Baumgartner et al., 1980; Henning, 1982). Animals in stage 3 had especially unfavorable blood values which indicate slowly developing but distinct malignant hyperthermia. The low values at the beginning (one minute after the test, Table 2) for respiratory measurements shows a limited ability for compensation.

In Table 4 some characteristics of meat quality are grouped according to stages of stiffness. The same tendency was observed as with the physiological parameters, with similar results found for animals of stages 1 and 2 and for those of stages 3, 4 and 5, respectively. This suggests that a subdivision of the subjectively determined halothane reaction gives no advantage for the prediction of meat quality. The somewhat unfavorable mean values of stage 2 compared to stage 1 may have been caused by some halothane positive animals which did not clearly react and were grouped in stage 2.

A further important criterion for judging the importance of intensity of reaction is the stability of the grouping over several tests. The corrected contingent coefficients ($CC_{corr.}$) for conformity to the rigidity stages over 4 repeated tests lay between 0,80 and 0,85. However only the grouping for stage 1 was very stable. In all other stages changes were found, but more regularly within groupings 1 and 2 and 3 to 5 respectively. The constancy of the intermediate stages 2 and 3 is of special interest. Table 5 shows in which stages animals were grouped at the second, third and fourth repeat test which had been placed in stage2 or 3 at the first test. Of the 23 piglets which were classified as stage 2 in the first test ten of them showed constantly in the following tests the reaction behavior of stage 1, six of the animals shifted between stage 1 and 2, three animals remained constant in stage 2 and one shifted between

TABLE 3 LSQ mean values of selected blood measurements in pigs with different levels of hind limb rigidity after the halothane test

Time of measurement (min after end of test)	Rigidity stages					F-Test
	1	2	3	4	5	
n =	46	15	10	53	55	
Lactate (μmol/ml)						
0	4,6	5,6	6,0	5,2	4,4	–
6,5	5,0 [a]	6,8 [b]	11,6 [c]	10,1 [c]	9,6 [c]	***
pCO$_2$ (mm Hg)						
0	69,6	73,0	73,3	70,0	68,4	–
6,5	44,1 [a]	44,5 [ac]	57,3 [bc]	51,9 [bc]	55,5 [b]	***
pH-value						
0	7,164[b]	7,098[ac]	7,030[a]	7,157[bc]	7,187[b]	***
6,5	7,291[a]	7,238[a]	7,034[b]	7,102[b]	7,091[b]	***
Base excess (μmol/l)						
0	-5,7 [b]	-9,8 [ac]	-12,3 [a]	-6,3 [bc]	-4,3 [b]	***
6,5	-5,1 [a]	-8,4 [a]	-16,7 [b]	-14,1 [b]	-13,7 [b]	***

TABLE 4 LSQ mean values of meat quality traits, measured at 45 min post mortem in the carcasses of pigs grouped according to the degree of rigidity displayed after halothane administration

		Rigidity stages					F-Test
		1	2	3	4	5	
	n =	73	23	11	70	73	
pH	(M.l.d.)	5,92[a]	5,88[a]	5,56[b]	5,51[b]	5,53[b]	***
pH	(M.g.m.)	6,03[a]	6,18[a]	5,64[b]	5,58[b]	5,66[b]	***
pH	(M.sm.)	6,14[a]	6,04[a]	5,71[b]	5,69[b]	5,75[b]	***
ATP[1]	(M.sm.)	4,04[a]	3,37[a]	1,37[b]	1,69[b]	1,67[b]	***
Lactate[1]	"	52,4 [a]	59,5 [a]	91,2 [b]	88,5 [b]	86,6 [b]	***
Glycogen[1]	"	28,7 [a]	22,1 [a]	9,2 [b]	13,7 [b]	13,5 [b]	***
R-value	"	0,93[a]	0,94[a]	1,20[b]	1,15[b]	1,15[b]	***
Rigor value(mm)	"	5,94[a]	6,32[a]	9,06[b]	8,66[b]	8,32[b]	***
Liquid area(cm²)	"	4,81[a]	5,29[ac]	6,13[bc]	6,32[b]	6,06[b]	***

[1] μmol/g

TABLE 5 Repeatability of judging for hind limb rigidity in pigs. Animals that were assigned to rigidity stages 2 or 3 in the first test were then subjected to a second, third and fourth test

1st test stage 2 (n = 23)				1st test stage 3 (n = 11)			
repeated test			absolute frequency	repeated test			absolute frequency
2	3	4	of combination	2	3	4	of combination
1	1	1	10	2	1	2	1
1	1	-	2	3	-	-	1
1	1	2	3	3	3	-	1
2	1	1	1	4	3	4	1
2	2	1	1	3	4	5	1
2	1	2	1	4	4	3	1
2	2	2	3	4	4	-	1
3	4	-	1	5	4	4	2
2	3	2	1	4	5	5	1
			23				11

TABLE 6 Optimization of the accuracy of prediction of characteristics of muscle tissue at 1 and 24 hours post mortem by including several halothane reaction criteria in a multiple regression analysis

Dependent variable		A 1st reaction criterion	r^2	B A combined with 2nd reaction criterion	r^2
Glycogen$_1$	(M.sm.)	Rigidity stage	,314***	(Rigidity stage)^3	,349***
Lactate$_1$	"	Rigidity stage	,313***	$\dot{V}_{CO_2}$ (3,5 min)	,353***
ATP$_1$	"	$\dot{V}_{CO_2}$ (3,5 min)	,280***	Rigidity stage	,315***
R-value$_1$	"	Rigidity stage	,274***	$\dot{V}_{CO_2}$ (3,5 min)	,306***
Rigor value$_1$	"	Rigidity stage	,186***	--	--
Liquid area$_1$	"	$\dot{V}_{CO_2}$ (2,5 min)	,164***	--	--
Liquid area$_{24}$	"	Heart rate (1,5 min)	,184***	--	--
pH$_1$	(M.sm.)	$\dot{V}_{CO_2}$ (2,5 min)	,305***	Rigidity stage	,332***
pH$_1$	(M.g.m.)	Rigidity stage	,272***	Heart rate (0,5 min)	,303***
pH$_1$	(M.l.d.)	Rigidity stage	,363***	$\dot{V}_{CO_2}$ (3,5 min)	,408***

[3] rigidity stage in third power

stage 2 and 3. Only one of the 23 animals scored in stage 2 at the first test shifted in the following tests to stages 3 and 4. It is evident that reaction stages 1 and 2 can be considered to be similar and that stage 2 (very low rigidity) can be considered to be a halothane negative reaction. Animals scored into stage 3 at the first test were also unstable in the following tests. Shifts happened mostly into higher stages of reaction. One can therefore conclude that animals of stage 3 (medium rigidity) have to be classified with the positively reacting animals. The corrected contingent coefficients $(CC_{corr.})$ for the conformity judging of hind limb rigidity into either halothane negative (rigidity stage 1 and 2) or halothanepositive (rigidity stages 3-5) lay between 0,95 and 0,98.

Multiple correlations were used to examine whether the physiological measurements made in connection with the halothane test can be used to make a more precise prediction of meat quality. Besides the rigidity judgements, selected physiological criteria at appropriate times of measurement were included into the calculations.

Listed under column A in Table 6 are the reaction criteria which give the best prediction (r^2) for the corresponding meat quality characteristics. When r^2 could be increased by at least 0,02 by combination with a further criterion then it is listed in column B. In six out of ten meat quality characteristics the rigidity stage alone gave the highest r^2 value. Of seven characteristics in which a second calculation may have increased the correlation by at least 0,02 the rigidity stage was always included. This shows the usefulness of a subjective judgement of hind limb stiffness for the prediction of meat quality. The most important physiological measurement was the output of carbon dioxide at 2,5 to 3,5 min. after the end of the test. This measurement however is very expensive and the amount of information obtained was relatively low. Furthermore the optimum combination of physiological measurements changes for each meat quality characteristic. It is therefore not possible to deduce a generally valid combination of measurements. This leads to the conclusion that during routine testing recording of auxiliary parameters does not give decisive advantages.

CONCLUSIONS

The differences observed in reaction intensity in halothane negative and positive animals suggest that the relative stiffness of the hind limbs is not genetically founded and it is sufficient to group animals into two reaction stages (H-, H+) during routine testing. Under sufficiently strong test conditions, halothane negative reactions are only those with none, very low grade or reversible stiffening of the musculature of the limbs.

ACKNOWLEDGEMENT

We wish to express our thanks to Dr. A. Festerling, Institut für Tierzucht-wissenschaft, Universität Bonn, for the statistical programming work.

REFERENCES

Augustini, Chr., Fischer, K. und Scheper, J. 1979. Halothan-Test - Prüfverfahren und Reaktionsverhalten bei Mastprüfungstieren. Fleischwirtschaft 59,1268-1273

Baumgartner, W., Schlerka, G., Schöpf-Grey, Johanna und Köfer, J. 1980. Das Verhalten des Säure-Basen-Haushaltes, des Kohlendioxidpartialdruckes und der inneren Körpertemperatur während der Halothan-Narkose beim Schwein. Berl.Münch.Tierärztl.Wschr. 93, 227-232

Berman, M.C., Harrison, G.G., Bull, A.B. and Kench, J.E. 1970. Changes under-lying halothane induced malignant hyperpyrexia in Landrace pigs. Nature 225, 653-655

Eikelenboom, G., Minkema, D. and van Eldik, P. 1977. The application of the halothane-test. Differences in production characteristics between pigs quali-fied as reactors (MHS-susceptible) and non reactors. Proc. 3rd Int.Conf. Prod. Disease Farm Animals, Wageningen, 13-16 Sept.,1976,Pudoc,Wageningen

Eiffert, L. 1984. Vererbung der Halothanreaktion und Beziehungen zu Marker-genen in verschieden stressanfälligen Schweinepopulationen. Diss.Göttingen

Haugwitz, T. und von Lengerken, G. 1985. Untersuchungen zur methodischen Durchführung des Halothantests beim Schwein. Arch.Tierz. 28,147-155

Henning, M. 1982. Untersuchungen zur Methodik und Anwendung des Halothan-tests beim Schwein. Diss. Göttingen

Kittler, L. und Dzapo, V. 1978. Zur Anwendung der Halothannarkose und Creatin-Phospho-Kinase-(CPK) Bestimmung im Blut als indirektes Kriterium der Vitalität beim Schwein. Dtsch.Tierärztl.Wschr. 85, 467-470

Lucke, J.N., Hall, G.M. and Lister, D. 1976. Porcine malignant hyperthermia. I: Metabolic and physiological changes. Br.J.Anaesth. 48, 297-304

Van den Hende, C., Lister, D., Muylle, E., Ooms, L. and Oyaert, W. 1976. Malignant hyperthermia in Belgian Landrace pigs rested or exercised before exposure to halothane. Br. J.Anaesth. 821-829

RESULTS OF CROSSBREEDING EXPERIMENTS USING MEAT TYPE BOARS AND STRESS RESISTANT SOWS

F. Schmitten, A. Festerling, H. Jüngst and K.-H. Schepers

Institut für Tierzuchtwissenschaft der Universität Bonn
Endenicher Allee 15,
D-5300 Bonn,
Germany

ABSTRACT

Several crossbreeding experiments were conducted in which Pietrain boars were mated with stress resistant sows of a specially developed line of German Landrace selected for homozygosity in the halothane negative reaction (DL HN), or with German Landrace halothane positive sows (DL HP), or with stress resistant F_1 German Landrace x Edelschwein sows (DE ·x DL). Reproduction criteria for 559 litters from 396 sows indicated an improvement with the stress resistant purebred and crossbred sows, although the mean number of piglets born per litter was not significantly different between HN and HP sows.

The carcass composition of the progeny of DL (HN) sows was only slightly less favourable from that of the progeny of DL (HP) sows. In contrast all criteria of meat quality were superior to the progeny of DL (HN) sows. It was concluded that, despite the observed difference in leanness, the crossbred Pi x DL (HN) pigs combined superior meat quality with a sufficiently good carcass grade.

INTRODUCTION

Breeding strategies for decreasing the frequency of inferior meat quality in pigs are concentrated on crossbreeding systems. Apart from commercial hybrids in practical pig production the methods of simple crossings of meat type boars such as the Pietrain, with stress resistant (halothane negative) purebred sows or F_1-sows of DE x DL in a three-way crossing system are favoured.

Results of three-way crossbreeding, from several official tests comparing different breeds, indicate possibilities for improving meat quality combined with the desired high level of meatiness.

Introduction of simple crossing requires the development of purebred stress resistant lines based on selection with halothane. This task was undertaken in several German pig breeding associations. A halothane negative purebred line of German Landrace pigs (DL HN) has been developed at the experimental station Frankenforst with the intention of establishing a homozygous halothane negative herd. Some results of crossbreeding of Pietrain boars with halothane negative and halothane positive sows are presented here.

MATERIAL AND METHODS

The experiments comprised 559 matings of 7 Pietrain boars with 396 sows (299 HN, 64 HP, 33 F_1), including 89 litters from first mating sows and 470 litters from second and following matings. The halothane test was conducted in litters of 134 sows (122 HN, 12 HP). Carcass composition and meat quality were measured in the progeny of 105 sows (87 HN, 18 HP) and 4 Pietrain boars.

A least square means statistical analysis of reproduction criteria was carried out using a factorial design with respect to number of farrowing, boars and type of sows (HP, HN, F_1). The evaluation of the results of carcass composition and meat quality measurements by least square means analysis was based on the following statistical model:

$$Y_{ijkl} = \mu + E_i + HRS_j + B_{ik} + e_{ijkl}$$

E = Experiment
HRS = Halothane reaction of sows
B = Boar within experiment

RESULTS

In crossbred litters of 122 DL (HN) sows the average frequency of halothane negative piglets was 72 percent. In 55 litters none or at most one piglet showed a positive halothane reaction. These progeny tested sows were selected for integration in the special halothane negative line at the experimental station. In matings of Pietrain boars with DL (HP) sows an average frequency of 21.5 percent halothane negative pigs in the offspring was observed.

Preliminary results for reproductive performances are presented in Table 1. A trend for higher performance in the halothane negative sows was evident and also for the F_1 sows as expected, but the differences were not statistically significant with the exception of the weight at weaning on the 28th day. The weight of the litter and of the piglets at weaning was mainly influenced by the farrowing number of the sow. The LSQ analysis showed that the effect of halothane status on reproduction criteria was higher for the sows than for the boars.

The carcass composition and meat quality results are presented in

TABLE 1   Several reproduction criteria in crossbred litters from Pietrain (Pi) boars and German Landrace halothane positive (DL HP) sows, or halothane negative (DL HN) sows, or Edelschwein x German Landrace (DExDL) F_1 sows.   Data are presented as least square means

| | Matings | | |
	Pi x DL(HP)	Pi x DL(HN)	Pi x F_1(DExDL)
Number of sows	64	299	33
Number of litters	99	405	55
Piglets born per litter	9.72	10.01	10.13
Piglets reared per litter	8.28	8.61	8.25
Rearing losses (%)	15.21	14.19	18.73
Weight of piglet at 28 days (kg)	6.13	6.52	6.34
Weight of litter at 28 days (kg)	50.53	56.19	52.40

Tables 2 and 3. As expected, crossbred pigs of DL (HP) sows were significantly superior in those criteria expressing meatiness such as muscle area and conformation. The significant difference in carcass length in general characterises the more expressed meat type carcass of the progeny of DL(HP) sows. The measurements of backfat thickness and also the lean to fat ratio were not significantly different between the progeny of DL (HN) and DL (HP) sows. That means that, with respect to the high absolute level of meat yield in the carcasses of the crossbred pigs of DL (HN) sows, this crossbreeding system, independently from the advantages given by the halothane status of the sow, is a favourable method for producing pigs to satisfy the market demands. Furthermore the meat quality results (Table 3) demonstrate a significant superiority in all of the important traits for the crossbred pigs of DL (HN) sows.

It was concluded that the crossbreeding system of mating meat type boars with halothane negative sows is a favourable way to produce carcasses with a high lean meat content combined with the high level of meat quality demanded.

TABLE 2   Carcass composition of crossbred pigs from Pietrain (Pi) boars and German Landrace halothane positive (DL HP) and halothane negative (DL HN) sows. Least square means (LSM) and standard errors (SE) of LSM are shown

	Matings			
	Pi x DL(HP)		Pi x DL(HN)	
	n = 95		n = 472	
	LSM	SE	LSM	SE
Carcass length (cm)	93.30***	0.43	95.47	0.21
Muscle area (cm^2)	49.35***	0.71	46.69	0.35
Backfat area (cm^2)	17.92	0.52	18.16	0.26
Lean/fat ratio	0.37	0.01	0.39	0.01
Average backfat (cm)	2.38	0.04	2.41	0.02
Backfat, RM (cm)	1.51	0.07	1.50	0.06
Meatiness of belly, s.v.	4.65	0.20	4.55	0.10
Ham (%)	32.46*	0.15	32.08	0.07
Meat yield in carcass (%)	58.06**	0.30	57.18	0.15

* P $\leq$ 0.05;   ** P $\leq$ 0.01;   *** P $\leq$ 0.001

TABLE 3   Meat quality in crossbred pigs from Pietrain (Pi) boars and German Landrace halothane positive (DL HP) or halothane negative (DL HN) sows. Least square means (LSM) and standard errors (SE) of LSM are shown. Measurements of pH and electrical conductivity (EC) were made in M. longissimus dorsi (LD) and M. semimembranosus (SM) at 40 minutes and 24 hours post mortem (subscripts 40 and 24 respectively)

	Matings			
	Pi x DL(HP)		Pi x DL(HN)	
	n = 95[1]		n = 472[1]	
	LSM	SE	LSM	SE
pH_{40} (LD)	5.61	0.04	5.89***	0.02
pH_{40} (SM)	5.85	0.05	6.07***	0.02
EC_{40} (LD)	10.01	0.73	7.14***	0.29
EC_{40} (SM)	8.43	0.54	6.14***	0.21
Göfo value (LD)	50.47	1.19	58.76***	0.58
Elrepho value (LD)	24.22	0.44	21.34***	0.22
Transmission value (LD)	55.03	3.04	34.12***	1.50
Visual score[2] (LD)	4.42	0.29	6.23***	0.14
pH_{24} (LD)	5.51	0.01	5.47***	0.01
pH_{24} (SM)	5.63	0.01	5.54***	0.01
EC_{24} (LD)	11.12	0.33	9.81***	0.13
EC_{24} (SM)	12.31	0.41	11.62	0.17

See overleaf for footnote

[1]The number (n) of EC measurements was 46 and 402 for the Pi x DL (HP) and Pi x DL (HN) matings respectively.

[2]A higher score indicates better meat quality

***$P \leq 0.001$

EVALUATION OF SOME QUALITATIVE CHARACTERISTICS OF SEASONED HAM IN EIGHT PIG GENETIC TYPES

D. Matassino, A. Zullo, L. Ramunno, E. Cosentino

with the technical collaboration of A. Di Lucia

Istituto di Produzione animale
Facoltà di Agraria - Università degli Studi di Napoli
80055 - Portici, Italy

ABSTRACT

The study was performed on 141 seasoned hams (grouped in 4 weight classes: < 7, 7 → 8, 8 → 9 and > 9 kg) from 72 castrated males and 69 entire females of eight swine genetic types. The rheological, colour and chemical characteristics were evaluated on M. gluteobiceps, M. semitendinosus and M. semimembranosus of the 'pera' cut obtained by dissecting the ham according to the 'Modena' cutting system.

The results showed: (i) genetic type significantly affected many qualitative characteristics; (ii) sex did not affect rheology and colour; the castrated males had, on average, hams with more dry matter and less protein and ash; (iii) rheology and colour did not vary statistically with ham weight; (iv) the muscle used was important in determining all the variables considered; (v) genetic type interacted with sex for brightness, fat content and energy value and with ham weight for hardness, chewiness, brightness, dry matter, protein, fat and energy value.

INTRODUCTION

A previous study showed the importance of genetic type, sex, weight and age at slaughter in determining the main qualitative and economic properties of seasoned ham (Quadri et al., 1981). Subsequently the effect of genetic type, sex and ham weight category on panel score was confirmed (Matassino et al., 1985a,b).

The aim of this study was (a) to determine the significance of certain sources of variation on rheology colour and chemical characteristics, measured on the most representative muscle of seasoned ham; (b) to determine for each genetic type, at which weight the ham reached optimum quality; (c) to integrate the results already obtained to improve the evaluation of quality in seasoned ham.

MATERIAL AND METHODS

Seasoned hams (141, grouped in four weight classes: < 7, 7 → 8, 8 → 9 and > 9 kg) from 72 castrated males and 69 entire females of eight swine genetic types were used (Table 1). Details of rearing and feeding

TABLE 1 Number of pigs within each genetic type in relation to the sources of variation.

Genetic type	Symbol	Sex		Hams classified by weight (kg)				Total
		(♂♂)	♀♀	< 7 (x̄=6.5)	7→8 (x̄=7.5)	8→9 (x̄=8.6)	> 9 (x̄=9.6)	
Dutch Large White	DLW	9	9	1	7	5	5	18
Italian Large White	ILW	9	9	2	3	6	7	18
Landrace x Large White	LxLW	9	9	2	8	5	3	18
Belgian Landrace x (Spotted x Large White)	BLx(SPxLW)	9	6	2	4	3	6	15
Landrace x (Spotted x Large White)	Lx(SPxLW)	9	9	–	8	6	4	18
Camborough	Camborough	9	9	1	6	7	4	18
Hypor	Hypor	9	9	4	2	5	7	18
Suffolk	Suffolk	9	9	3	5	6	4	18
Total		72	69	15	43	43	40	141

were reported by Bergonzini et al. (1982) and the seasoning procedures by Matassino et al. (1985a,b).

The rheological, colour and chemical characteristics were determined as reported in Matassino et al. (1974, 1975, 1976a,b) on M. gluteobiceps (Gb), M. semitendinosus (St) and M. semimembranosus (Sm) from the 'pera' cut. Chemical composition and energy value were calculated on a dry matter basis. The energy value was calculated using the following coefficients: 4 kcal/g for the protein and 9 kcal/g for the fat (Fidanza et al., 1974).

An analysis of variance was carried out using the following factorial model where the factors were considered fixed and the effect of each factor was expressed as deviation from the overall mean μ:

$$\mathbf{y}_{ijklm} = \mu + \alpha_i + \beta_j + \gamma_k + \delta_l + (\alpha\beta)_{ij} + (\alpha\gamma)_{ik} + (\alpha\delta)_{il} +$$
$$+ (\beta\gamma)_{jk} + (\beta\delta)_{jl} + (\gamma\delta)_{kl} + \varepsilon_{ijklm} \tag{1}$$

where: $\mathbf{y}_{ijklm}$ = the value of the lth muscle of the mth ham belonging to i^{th} genetic type, j^{th} sex and k^{th} class of weight.

The means were estimated as reported in Matassino et al. (1985a,b) by using the following model:

$$\mathbf{y}_{ijklm} = (\alpha\beta\gamma\delta)_{ijkl} + \varepsilon_{ijklm} \tag{2}$$

and are mean values weighted for all other factors considered as single or interactive effects.

Differences between means were determined using Student's t test.

The percentage reflectance of the ham was measured at nine wavelengths in the visible spectrum between 426 nm and 684 nm, and a fourth degree polynomial curve was fitted to the data for each level of the factors considered.

RESULTS

Table 2 shows the statistical significance of the particular muscle analysed and of the interactions 'genetic type x sex' and 'genetic type x ham weight' in determining the quality and chemical composition of the meat.

The components of variance (Table 3) show that (a) in many cases 'muscle' absorbs most of the total variability, reaching the highest value for hardness (69%); (b) 'ham class weight' and 'sex' have some influence on sodium chloride (9%) and dry matter content (8%)

TABLE 2   An analysis of variance of the quality and chemical composition
of seasoned ham in relation to genetic type, sex and weight of ham.

Variable	muscle	F value[1] interaction genetic type x sex	genetic type x ham weight class
1. Rheological			
Hardness (kg)	362.8***	1.6	1.9*
Cohesiveness (TU)[2]	31.7**	1.4	1.1
Springiness (mm)	1.5	0.6	1.6
Adhesiveness (TU)[2]	37.2***	3.2**	1.1
Chewiness (TU)[2]	118.5***	1.5	1.7*
2. Colour			
Brightness (%)	95.8**	2.8**	1.3
λ complementary (mm)	3.0*	2.6*	1.0
Purity (%)	12.1***	1.0	1.0
3. Chemical			
Dry matter (DM) (%)	225.7***	1.1	2.4***
Protein (% on DM)	361.4***	1.5	2.7***
Fat (% on DM)	218.9***	3.0**	2.9***
Ash (% on DM)	11.8***	1.2	3.4***
NaCl (% on DM)	273.0***	1.9	1.4
Energetic value (kcal/g of DM)	112.3***	2.2*	2.2***

[1] * $P < 0.05$; ** $P < 0.01$; *** $P < 0.001$, the other interactions were not significant.

[2] TU = texturometric units.

TABLE 3   Percentage of variance attributable to the factors included in model 1.

Variable	genetic type, α	sex, β	ham weight class, γ	muscle, δ	$\alpha\beta$	$\alpha\gamma$	$\alpha\delta$	$\beta\gamma$	$\beta\delta$	$\gamma\delta$	random error, ε
1. Rheological											
Hardness	2.02	0.45	0.56	69.10	0.01	1.60	0.29	0.00	0.00	0.00	25.97
Cohesiveness	0.60	0.00	0.42	17.26	1.17	0.95	0.00	0.00	0.00	0.00	79.60
Springiness	0.69	0.00	0.00	0.32	0.41	4.87	0.00	0.59	0.00	0.00	93.12
Adhesiveness	0.00	2.56	2.26	18.41	5.60	0.57	0.60	0.96	0.00	1.73	63.71
Chewiness	1.99	0.32	0.11	42.85	1.47	3.53	0.00	0.00	0.00	0.00	49.73
2. Colour											
Brightness	3.63	0.00	0.27	37.02	4.76	1.70	0.00	0.35	0.00	0.00	52.27
λ complementary	0.00	0.56	0.46	1.31	2.92	0.00	0.00	8.48	0.00	0.57	85.78
Purity	0.31	0.04	0.16	7.00	0.00	0.84	0.00	0.95	0.00	1.00	89.70
3. Chemical											
Dry matter	0.80	7.90	2.51	52.19	0.32	2.47	0.86	1.86	1.12	0.00	29.70
Protein	1.15	1.01	1.34	66.91	1.05	3.11	0.83	0.69	1.36	0.01	22.54
Fat	3.00	0.14	0.00	55.84	3.33	4.36	1.20	0.92	0.44	0.00	30.77
Ash	4.86	3.13	0.00	6.09	1.09	13.36	0.00	0.00	0.84	0.13	70.50
NaCl	1.67	2.16	9.28	55.01	0.85	1.48	0.23	1.91	0.37	0.00	26.59
Energetic value	3.57	0.00	2.46	39.34	2.52	4.45	0.18	1.63	0.00	0.00	45.85

Source of variation — interaction

respectively; (c) 'genetic type' and most of the interactions absorb little of the total variability, except for the interactions (i) 'genetic type x sex' for adhesiveness and brightness (5%) and (ii) 'genetic type x ham weight class' for ash content (13%) and springiness (5%).

The effect of single factors was shown by the significance of the differences between the means calculated according to model 2. This procedure statistically evaluates differences between the estimated means of a factor and also examines whether interactions are present. The coefficient of variability (C V = $\sigma / \bar{x}$. 100) was less than 10% for all the variables.

Genetic type

Almost all the characteristics considered varied significantly in relation to genetic type. Tables 4 and 5 show that:

a. LxLW and Suffolk pigs provided seasoned hams that were on average more tender and required less energy for chewing (P < 0.05) than DLW and Camborough. Springiness was significantly higher in Camborough than in Lx(SPxLW).

b. hams from Suffolk and ILW were lighest while those from three-way crosses were darkest (P < 0.01).

c. regarding chemical composition:
   (i) Lx(SPxLW), DLW and ILW hams had the highest percentage of dry matter while Suffolk had the lowest (P < 0.01);
   (ii) protein content was significantly higher in BLx(SPxLW) and DLW compared to Suffolk hams;
   (iii) LxLW hams were fatter than those from the other genetic types (P < 0.05) with the exception of DLW and BLx(SPxLW);
   (iv) ash content was lowest in Lx(SPxLW) hams compared with all other types (P < 0.05) and highest in the Suffolk;
   (v) Suffolk hams compared with the other genetic types, had significantly more sodium chloride;
   (vi) as a consequence of the fat content, the energy value was highest in hams from Lx(SPxLW) and lowest in L x LW (P < 0.001).

The fourth degree polynomial functions (P < 0.001; Table 7) for ham reflectance data are shown in Figure 1 for each genetic type. Significant differences were observed between the functions, indicating that ham from ILW had a higher percentage reflectance than that from DLW (P < 0.001),

TABLE 4   Mean values for quality characteristics and chemical composition of seasoned ham within genetic type.

Variable	Genetic type							
	DLW	ILW	LxLW	BLx(SPxLW)	Lx(SPxLW)	Camborough	Hypor	Suffolk
1. Rheological								
Hardness (kg)	6.8	5.9	5.6	5.9	6.3	6.5	6.4	5.8
Cohesiveness (TU)[1]	0.622	0.606	0.596	0.619	0.610	0.607	0.610	0.611
Springiness (mm)	17.2	17.5	17.3	17.5	16.4	17.7	17.2	16.5
Adhesiveness (TU)[1]	74.8	75.6	70.9	72.6	66.9	80.1	79.1	74.0
Chewiness (TU)[1]	7,150	6,244	5,719	6,353	6,302	7,171	6,972	5,858
2. Colour								
Brightness (%)	19.1	20.4	19.6	18.4	18.2	19.9	19.4	20.5
λ complementary (nm)	505.9	505.9	505.3	505.8	505.7	504.7	506.0	504.9
Purity (%)	43.6	44.3	46.8	49.8	44.4	46.0	44.7	46.8
3. Chemical								
Dry matter (DM) (%)	38.0	38.0	37.4	37.5	38.1	37.9	37.7	37.0
Protein (% on DM)	69.45	68.52	68.79	69.94	68.57	68.55	69.18	67.85
Fat (% on DM)	5.13	6.34	5.05	5.47	6.93	6.03	6.25	6.43
Ash (% on DM)	2.55	2.42	2.50	2.43	2.14	2.53	2.46	2.66
NaCl (% on DM)	22.87	22.71	23.12	22.21	22.45	22.89	22.76	23.47
Energetic value (kcal/g of DM)	3.24	3.31	3.21	3.29	3.37	3.29	3.33	3.29

1 TU = Texturometric Units.

TABLE 5   Differences between genetic types in quality characteristics and chemical composition of seasoned hams.

Comparison		Variable[1]										
		rheological		colour		chemical						
		hardness	chewiness	brightness	purity	dry matter	protein	fat	ash	NaCl	energetic value
DLW	– ILW	0.9***	906	-1.3*	-0.7	0.0	0.93*	-1.21**	0.13	0.16	-0.07*
"	– LxLW	1.2****	1,431**	-0.5	-3.2	0.6	0.66	0.08	0.05	-0.25	0.03
"	– BLx(SPxLW)	0.9**	797	0.7	-6.2**	0.5	-0.49	-0.34	0.12	0.66	-0.05
"	– Lx(SPxLW)	0.5	849	0.9	-0.8	-0.1	0.88	-1.80***	0.41***	0.42	-0.13***
"	– Camborough	0.3	-21	-0.8	-2.4	0.1	0.90	-0.90*	0.02	-0.02	-0.05
"	– Hypor	0.4	178	-0.3	-1.1	0.3	0.27	-1.12**	0.09	0.11	-0.09**
"	– Suffolk	1.0***	1,292**	-1.4*	-3.2	1.0**	1.60***	-1.30***	-0.11	-0.60	-0.05
ILW	– LxLW	0.3	525	0.8	-2.5	0.6	-0.27	1.29**	-0.08	-0.41	0.10***
"	– BLx(SPxLW)	0.0	-109	2.0**	-5.5**	0.5	-1.42**	0.87*	-0.01	0.50	0.02
"	– Lx(SPxLW)	-0.4	-58	2.2***	-0.1	-0.1	-0.05	-0.59	0.28**	0.26	-0.06
"	– Camborough	-0.6*	-927	0.5	-1.7	0.1	-0.03	0.31	-0.11	-0.18	0.02
"	– Hypor	-0.5*	-728	1.0	-0.4	0.3	-0.66	0.09	-0.04	-0.05	-0.02
"	– Suffolk	0.1	386	-0.1	-2.5	1.0**	0.67	-0.09	-0.24*	-0.76*	0.02
LxLW	– BLx(SPxLW)	-0.3	-634	1.2	-3.0	-0.1	-1.15*	-0.42	0.07	0.91*	-0.08*
"	– Lx(SPxLW)	-0.7*	-583	1.4*	2.4	-0.7	0.22	-1.88***	0.36**	0.67	-0.16***
"	– Camborough	-0.9**	-1,452**	-0.3	0.8	-0.5	0.24	0.98*	-0.03	0.23	-0.08*
"	– Hypor	-0.8**	-1,253*	0.2	2.1	-0.3	-0.39	-1.20**	0.04	0.36	-0.12***
"	– Suffolk	-0.2	-139	-0.9	0.0	0.4	0.94	-1.38**	-0.16	-0.35	-0.08**
BLx(SPxLW)	– Lx(SPxLW)	-0.4	51	0.2	5.4*	-0.6	1.37**	-1.46***	0.29**	-0.24	-0.08*
"	– Camborough	-0.6	-818	-1.5*	3.8	-0.4	1.39**	-0.56	-0.10	-0.68	0.00
"	– Hypor	-0.5	-619	-1.0	5.1*	-0.2	0.76	-0.78	-0.03	-0.55	-0.04
"	– Suffolk	0.1	495	-2.1***	3.0	0.5	2.09***	-0.96*	-0.23*	-1.26***	0.00
Lx(SPxLW)	– Camborough	0.2	-869	-1.7**	-1.6	0.2	0.02	0.90*	-0.39***	-0.44	0.08**
"	– Hypor	-0.1	-670	-1.2	-0.3	0.4	-0.61	0.68	-0.32**	-0.31	0.04
"	– Suffolk	0.5	444	-2.3***	-2.4	1.1**	0.72	0.50	-0.52***	-1.02**	0.08*
Camborough	– Hypor	0.1	199	0.5	1.3	0.2	-0.63	-0.22	0.07	0.13	-0.04
"	– Suffolk	0.7*	1,313**	-0.6	-0.8	0.9*	0.70	-0.40	0.13	-0.58	0.00
Hypor	– Suffolk	0.6*	1,114*	-1.1	-2.1	0.7*	1.33**	-0.18	-0.20*	-0.71*	0.04

1   * = $P<0.05$; ** = $P<0.01$; *** = $P<0.001$.

TABLE 6 Mean values for quality characteristics and chemical composition of seasoned hams within sex, ham weight class and muscle[1]

Variable	Sex		Ham weight class (kg)				Muscle		
	(♂♂)	♀♀	< 7	7→8	8→9	> 9	Gb	St	Sm
1. Rheological									
Hardness (kg)	6.3	6.0	6.5a	5.9b	6.2ab	6.1ab	8.1A	4.0B	6.3C
Cohesiveness (TU)[2]	0.612	0.609	0.622	0.600	0.611	0.611	0.614A	0.640B	0.578C
Springiness (mm)	17.0	17.3	16.9	17.3	17.3	17.0	16.9	17.1	17.4
Adhesiveness (TU)[2]	78.3a	70.7b	85.0Aa	76.4ab	72.8bc	66.6Bc	56.9A	75.4B	90.7C
Chewiness (TU)[2]	6,542	6,394	6,986	6,183	6,550	6,275	8,544A	4,447B	6,404C
2. Colour									
Brightness (%)	19.3	19.6	18.8	19.8	19.6	19.4	21.1A	20.4A	16.8B
λ complementary (nm)	505.8	505.2	505.2	505.7	504.9	506.2	504.6a	506.1b	505.8ab
Purity (%)	45.1	46.7	45.7	47.1	45.1	45.9	44.6A	44.5A	48.7B
3. Chemical									
Dry matter (DM) (%)	38.4A	37.0B	38.4A	37.5B	37.8a	37.2Bb	35.8A	37.1B	40.2C
Protein (% on DM)	68.54a	69.17b	68.38A	68.59A	68.79a	69.53Bb	67.38A	66.10B	73.14C
Fat (% on DM)	6.03	5.83	5.69a	5.66A	6.39Bb	5.90AB	5.36A	8.51B	3.92C
Ash (% on DM)	2.41a	2.53b	2.39	2.51	2.46	2.51	2.63A	2.51A	2.28B
NaCl (% on DM)	23.16A	22.50B	24.19A	23.31B	22.47C	21.78D	24.84A	22.99B	20.62C
Energetic value (kcal/g of DM)	3.28	3.29	3.25AaB	3.25B	3.33	3.31AbC	3.18A	3.41B	3.28C

1 Mean values with different superscripts, within the factor, are significantly different at P < 0.05 (small letters) or P < 0.01 (capital letters).

2 TU = texturometric units.

TABLE 7   Significance of the polynomial function on reflectance data
within each level of the factors considered, and comparison between functions within factors.

Factor	R^2	Comparison between functions[1]
Genetic type		
DLW	0.827*	BbDEF
ILW	0.834*	AaC
LxLW	0.836*	AacE
BLx(SPxLW)	0.811*	BDF
Lx(SPxLW)	0.837*	AbcD
Camborough	0.861*	A
Hypor	0.808*	AbcF
Suffolk	0.827*	C
Sex		
(♂♂)	0.839*	
♀♀	0.822*	
Ham weight class		
< 7	0.826*	AB
7 —⊢ 8	0.842*	A
8 —⊢ 9	0.820*	AB
> 9	0.825*	B
Muscle		
Gb	0.876*	A
St	0.848*	B
Sm	0.847*	C

[1] Different letters between the levels of each factor   mean that the
functions are significantly different at $P < 0.05$ (small letters) or
$P < 0.01$ (capital letters).

* $P < 0.001$

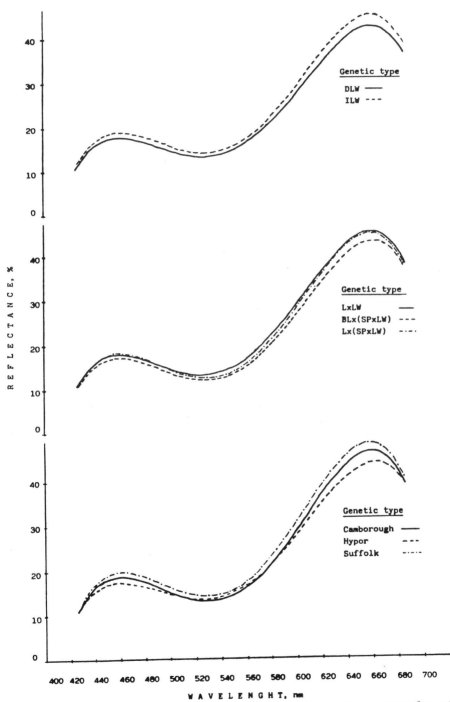

FIGURE 1   The effect of genetic type of pig on the reflectance value of seasoned ham. Pig breed types are listed in Table 1. The spectrophotometric curves are fourth degree polynomial functions.

Lx LW was higher than BLx(SPxLW), and Suffolk was lighter in colour than all other genetic types.

Sex

Rheological and colour characteristics were similar in males and females (Table 6), except adhesiveness which was higher in the ham of castrated males ($P < 0.05$). On the contrary sex was important in determining the chemical variables as the hams from entire females had less dry matter and NaCl ($P < 0.001$) and more protein and ash ($P < 0.05$). The polynomial functions of the spectrophotometric curve (Fig 2) did not differ significantly between the two sexes (Table 7).

Ham weight class

Weight significantly influenced the chemical composition of hams but not the rheological and colour characteristics, with the exception of the adhesiveness (Table 6). With increasing weight, the adhesiveness and sodium chloride percentage decreased ($P < 0.05$), protein and fat content and energy value increased ($P < 0.05$).

The functions representing the spectrophotometric curves of the four ham weight classes (Fig 2) were not significantly different, with the exception that class 7 $\longrightarrow$ 8 kg had a higher percentage reflectance especially in the red spectrum (600 to 684 nm) than class > 9 ($P < 0.001$; Table 7).

Muscle

Differences among muscles were significant for all the characteristics considered (Table 6). The St, compared to the other muscles, was on average: (i) more tender and required less chewing energy but had higher cohesiveness; (ii) was lighter in colour than the Sm; (iii) had less protein and more fat and energy content. The Gb had a higher value for hardness, chewiness, ash and sodium chloride but a lower value for dry matter.

The percentage reflectance varied ($P < 0.001$, Table 7) among the three muscles considered: the Gb was lighter than the St, which in turn was lighter than the Sm at all wavelengths (Figure 2).

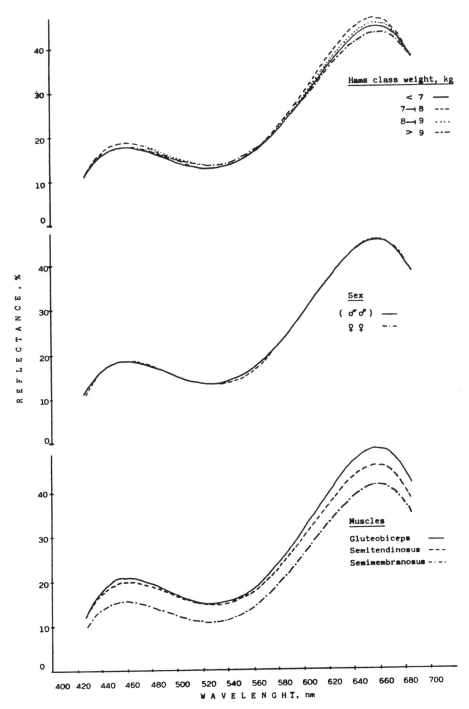

Figure 2 Polynomial functions of spectrophotometric curves in relation to ham weight class, sex and muscle

Interactions

The significance of the interaction 'genetic type x sex' for some characteristics (Tables 2 and 8) was due to the fact that hams from castrated males, compared to entire females, had higher adhesiveness in the DLW and ILW (P< 0.05) and lower in the Hypor (P < 0.05), were less bright in all genetic types with the exception of the three-way crosses and had higher fat content and consequently energy value in the BLx(SPxLW) and lower in the other types.

TABLE 8   Mean values[1] within genetic type and sex for quality characteristics and chemical composition of seasoned hams

Genetic type	Variable							
	adhesiveness, TU^2		brightness, %		fat, % on DM^3		energy value kcal/g of DM^3	
	$(\sigma\!\!\!/\sigma\!\!\!/)$	♀♀	$(\sigma\!\!\!/\sigma\!\!\!/)$	♀♀	$(\sigma\!\!\!/\sigma\!\!\!/)$	♀♀	$(\sigma\!\!\!/\sigma\!\!\!/)$	♀♀
DLW	86.6^A	59.2^B	19.0	19.1	5.59^a	4.51^b	3.24	3.24
ILW	86.1^a	65.1^b	20.3	20.5	6.20	6.47	3.29	3.33
LxLW	76.4	66.7	18.5	20.3	5.03	5.06	3.20	3.21
BLx(SPxLW)	80.7	64.5	19.3	17.5	6.52^A	4.43^B	3.35^A	3.23^B
Lx(SPxLW)	73.6	60.3	18.9	17.5	6.48	7.39	3.33	3.41
Camborough	80.9	79.5	18.7^a	20.9^b	5.96	6.09	3.29	3.28
Hypor	68.4^a	87.2^b	18.9	19.7	6.48	6.08	3.32	3.34
Suffolk	72.0	76.0	20.1	20.9	6.01	6.85	3.27	3.32

[1] Mean with different superscripts are significantly different at P<0.05 (small letters) or P<0.01 (capital letters).
[2] TU = Texturometric Units.
[3] DM = Dry Matter.

The interaction 'genetic type x ham weight class' showed that variation in quality with increasing ham weight was different for each genetic type. Data in Table 9 showed that:

(i)   hardness was least in hams weighing >9 kg in the LxLW and Suffolk 8 ⟶ 9 kg in the Camborough and 7 ⟶ 8 kg in the other types;

(ii) chewing energy was least in hams weighing 7 ⟶ 8 kg in the three-way crosses, 8 ⟶ 9 kg in the Camborough and Suffolk and >9 kg in the others;

(iii) brightness values were highest in hams weighing < 7 kg in the Hypor, at 7 ⟶ 8 kg in the DLW, LxLW, Ls(SPxLW) and Camborough and at

TABLE 9 Mean values[1] within genetic type and ham weight class for quality characteristics and chemical composition of seasoned hams.

1. Hardness, kg | 2. Springiness, mm

Genetic type	< 7	7→8	8→9	> 9	< 7	7→8	8→9	> 9
DLW	6.9^{ab}	6.4^{a}	6.6^{ab}	7.4^{b}	16.3^{AB}	18.7^{A}	18.3^{A}	14.9^{B}
ILW	6.3	5.6	6.1	5.6	17.2	18.0	17.7	17.2
LxLW	5.4	6.0	5.4	5.3	16.8	18.0	17.3	16.9
BLx(SPxLW)	6.3	5.6	5.9	5.7	16.3^{ab}	16.4^{a}	18.2^{ab}	19.0^{b}
Lx(SPxLW)	–	6.0^{a}	6.9^{b}	6.1^{ab}	–	17.0	15.8	16.3
Camborough	8.3^{Aa}	6.4^{b}	6.1^{B}	6.2^{b}	20.7	17.2	17.1	17.4
Hypor	6.8^{A}	4.9^{aB}	6.4^{b}	7.0^{A}	16.6	16.3	17.4	18.1
Suffolk	6.6^{a}	5.6^{ab}	5.5^{b}	5.4^{b}	16.2	16.6	16.9	16.4

3. Chewiness, TU ([2]) | 4. Dry matter (DM), %

Genetic type	< 7	7→8	8→9	> 9	< 7	7→8	8→9	> 9
DLW	6,873	7,716	7,091	6,781	40.4^{Aa}	37.8^{b}	38.0^{b}	37.1^{B}
ILW	6,637	5,931	6,553	5,894	40.0^{Aa}	37.7^{b}	38.4^{A}	36.9^{B}
LxLW	5,307	6,576	5,765	5,250	38.2^{a}	37.7^{ab}	36.5^{b}	36.8^{ab}
BLx(SPxLW)	6,374	5,929	6,456	6,651	37.4	37.4	37.7	37.5
Lx(SPxLW)	–	5,972	6,634	6,300	–	38.8	37.7	37.8
Camborough	$12,176^{A}$	$6,558^{B}$	$6,211^{B}$	$6,243^{B}$	40.7^{A}	37.4^{B}	37.6^{B}	37.1^{B}
Hypor	$6,902^{a}$	$4,245^{Ab}$	$7,501^{B}$	$7,878^{B}$	38.5^{a}	36.2^{b}	38.1^{a}	37.3^{ab}
Suffolk	6,995	5,440	5,796	5,201	36.5	36.6	37.5	37.3

5. Protein, % on DM | 6. Fat, % on DM

Genetic type	< 7	7→8	8→9	> 9	< 7	7→8	8→9	> 9
DLW	65.17^{Aa}	69.60^{B}	68.82^{ABb}	72.07^{C}	5.91^{AB}	3.96^{A}	5.93^{B}	5.10^{AB}
ILW	66.52^{A}	67.61^{a}	68.59^{ab}	69.91^{Bb}	6.87^{a}	7.00^{A}	7.06^{A}	5.01^{Bb}
LxLW	69.31	68.86	69.26	67.98	4.55	5.21	5.20	5.31
BLx(SPxLW)	71.32	69.91	68.86	69.65	3.48^{Aa}	5.61^{b}	6.44^{B}	6.36^{B}
Lx(SPxLW)	–	67.58^{a}	69.15^{b}	68.98^{ab}	–	6.92	6.61	7.28
Camborough	66.18^{a}	68.45^{ab}	69.02^{b}	69.36^{b}	6.79	5.59	6.21	5.93
Hypor	69.36	70.32	68.60	69.01	7.01^{A}	3.64^{B}	6.42^{A}	6.64^{A}
Suffolk	67.14^{a}	66.75^{A}	68.27^{ab}	69.25^{Bb}	6.46	7.03	6.63	5.61

7. Ash, % on DM | 8. Energetic value, kcal/g of DM

Genetic type	< 7	7→8	8→9	> 9	< 7	7→8	8→9	> 9
DLW	2.66^{a}	2.92^{A}	2.69^{A}	2.00^{Bb}	3.14^{a}	3.14^{A}	3.29^{B}	3.34^{Bb}
ILW	2.47	2.60	2.26	2.45	3.28^{AB}	3.33^{AB}	3.38^{A}	3.25^{B}
LxLW	2.33^{a}	2.42^{A}	2.90^{Bb}	2.55^{ab}	3.18	3.22	3.24	3.20
BLx(SPxLW)	2.25^{a}	2.36^{ab}	2.41^{ab}	2.71^{b}	3.17^{Aa}	3.30^{ab}	3.33^{b}	3.36^{B}
Lx(SPxLW)	–	2.13	2.11	2.18	–	3.33	3.36	3.41
Camborough	2.38^{ab}	2.62^{ab}	2.32^{a}	2.71^{b}	3.26	3.24	3.32	3.31
Hypor	2.04^{Aa}	2.63^{b}	2.76^{B}	2.50^{b}	3.40^{A}	3.14^{B}	3.32^{A}	3.36^{A}
Suffolk	2.76^{ab}	2.48^{a}	2.45^{A}	2.95^{Bb}	3.27	3.30	3.33	3.27

[1] Means with different superscripts are significantly different at P<0.05 (small letters) or P<0.01 (capital letters).

[2] TU = Texturometric Units.

$8 \longrightarrow 9$ kg in the other genetic types;

(iv)   dry matter was highest in hams weighing on average 6.5 kg in five genetic types, 7.5 kg in the Lx(SPxLW) and 8.5 kg in the BLx(SPxLW) and Suffolk;

(v)   protein content was highest in hams weighing <7 kg in the LxLW, $7 \longrightarrow 8$ kg in the DLW, BLx(SPxLW) and Hypor, $8 \longrightarrow 9$ kg in the Camborough and >9 kg in the DLW, Lx(SPxLW) and Suffolk;

(iv)   fat content was highest in hams weighing <7 kg hams from Camborough and Hypor, $7 \longrightarrow 8$ kg from Suffolk, $8 \longrightarrow 9$ kg from DLW, ILW and BLx(SPxLW) and>9 kg from LxLW and Lx(SPxLW);

(vii)   energy value was highest in the first weight class for the Hypor, in the third class for the ILW, LxLW and Suffolk and in the last class for the other types.

CONCLUSIONS

Significant effects of genetic type, sex and ham weight on the quality and chemical composition of seasoned hams were found. An empirical index was constructed by making the 'best' value of each characteristic equal to 100. Thus the lowest values of chewiness, adhesiveness, sodium chloride content and brightness and the highest values of dry matter, protein and fat content were scored 100 because these are the qualitative characteristics that mainly influence the palatability of a ham. The choice of the lowest value for brightness was justified by panel assessment (Matassino et al., 1985a,b).

Overall scores (Table 10) showed that hams from Lx(SPxLW) had the best characteristics with an index of 88, followed by the BLx(SPxLW), Suffolk and ILW (84.4), Hypor (83.6), LxLW (83.3), Camborough (82.4) and finally DLW (81.9). However, the order of merit was different within sex or within ham weight category. Ham from entire females, compared to that from castrated males, was better if it came from ILW, DLW, LxLW, Lx(SPxLW) and Suffolk, and worse if it came from BLx(SPxLW), Camborough and Hypor. The highest index value was reached at ham weights of $7 \longrightarrow 8$ kg in the ILW, at $8 \longrightarrow 9$ kg in the Camborough and Suffolk, and at >9 kg in the other types. Therefore, this interactive effect confirmed the results of Quadri et al. (1981) and Matassino et al. (1985a,b) that each genetic type provided ham with the best qualitative characteristics at a given weight, which was always above 7 kg.

TABLE 10 Quality scores for seasoned ham, using an empirical index in which the best value for each quality characteristic was set at 100. Results are overall scores for eight pig genetic types as well as scores within sex and weight categories

Genetic type	Sex		Hams weight class, kg				Total
	(♂♂)	♀♀	< 7	7→8	8→9	>9	
DLW	80.9	84.0	80.4	78.7	83.2	85.1	81.9
ILW	82.5	86.8	82.4	90.4	85.0	83.4	84.4
LxLW	82.9	83.7	81.9	81.7	85.4	85.9	83.3
BLx(SPxLW)	85.6	83.7	82.2	83.0	86.6	87.6	84.5
Lx(SPxLW)	86.4	89.9	--	85.7	86.8	92.6	88.0
Camborough	83.7	81.4	78.9	82.4	84.4	83.6	82.4
Hypor	84.8	83.1	82.9	83.7	85.3	84.2	83.6
Suffolk	83.8	85.1	83.3	84.6	85.0	85.7	84.4
Total	83.8	84.7	81.7	83.8	85.2	86.0	

REFERENCES

Bergonzini,E., Quadri, G., Cosentino, E., Zullo, A. and Pieraccini, L. 1982. Studio di alcuni tipi genetici di suino 'da salumificio' allevati con un piano alimentare 'medio'. I. Rilievi 'infra vitam' e alla mattazione. Prof. Anim., 1, n.s., 1-33.

Fidanza, F., Liguori, G. and Mancini, F. 1974. Lineamenti di nutrizione umana. Idelson, Napoli, XXIII-554.

Matassino, D., Cosentino, E. and Bordi, A. 1974. Quantizzazione di alcuni aspetti qualitativi della carne bovina con l'impiego della tecnica tessurometrica. Atti IX Simp. Int. Zootecnia, Milano.

Matassino, D., Cosentino, E., Bordi, A. & Colatruglio, P. 1975. Sulla composizione chimica di muscoli di carcassa di 10 tipi genetici bovini. Genet. Agr., 29, 23.

Matassino, D., Romita, A., Colatruglio, P. & Bordi, A. 1976a. Studio comparativo fra bufali e bovini sulla qualità della carne. I. Caratteristiche mioreologiche all'età di 20 settimane. Atti II Conv. Naz. ASPA, Bari, XIV-426.

Matassino, D., Romita, A., Girolami, A. & Cosentino, E. 1976b. Studio comparativo fra bufali e bovini sulla qualità della carne. II. Colore della carne all'età di 20 settimane. Atti. II Conv. Naz. ASPA, Bari, XIV-426.

Matassino, D., Grasso, F., Girolami, A. and Cosentino, E. 1985a. Eating quality of seasoned ham in eight pig genetic types. In: Evaluation and Control of Meat Quality in Pigs, EC Seminar, Dublin, Nov. 1985.

Matassino, D., Grasso, F., Girolami, A. & Cosentino, E. 1985b. Studio di
    alcuni tipi genetici di suino 'da salumificio' allevati con un piano
    alimentare 'medio'. IX. Prove di degustazione sul prosciutto
    stagionato Prod. Anim., 4, n.s., 1.
Quadri, G., Bergonzini, E., Zullo, A., Cosentino, E. & Matassino, D. 1981.
    Studio di alcuni tipi genetici di suino 'da salumificio' allevati
    con un piano alimentare 'medio'. IV. Aspetti quanti-qualitativi ed
    economici del prosciutto stagionato. Suinicoltura, 22, 5, 21-49.

EATING QUALITY OF SEASONED HAM IN EIGHT PIG GENETIC TYPES

D. Matassino, F. Grasso, A. Girolami and E. Cosentino
with the technical collaboration of M. Caiazzo

Istituto di Produzione animale
Facoltà di Agraria - Università degli Studi di Napoli
80055 Portici, Italy

ABSTRACT

The eating quality of seasoned hams from eight pig genetic types was evaluated using a sensory panel. The results showed that the factors examined (genetic type, sex, ham weight category and panellist) had significant effects on the organoleptic characteristics of the ham. The effects of sex and ham weight on meat quality varied between genetic types. It was concluded that the production standards for 'Modena' and 'Parma' hams should be re-examined in light of these results to ensure a quality seasoned ham from the genetic types of pig that are in use.

INTRODUCTION

The increasing demand for fresh pork in the past few years has influenced rearing techniques and choice of genetic types, in order to obtain pigs of 100 to 120 kg live weight, characterized by a low feed conversion index, a higher daily weight increase and a lower percentage of fat tissue. This trend has influenced the production cycle of heavy pigs suitable for the pork industry. In fact, these pigs, selected for fresh marketing, show some qualitative and quantitative improvements but the meat is less suitable for processing and preserving. These considerations concern the whole carcass but particularly the ham, since this cut represents nearly 40 per cent of the commercial value of the carcass and 25 per cent of its weight.

To obtain quality products, particularly hams, the pork industry asks for pigs of about 150 kg live weight; the age at slaughter is also a very important factor to achieve a ham that qualifies as 'very good'. The main qualitative and quantitative characteristics of seasoned ham (seasoning period, morphological aspects, etc.) are influenced, among other things, by genetic type, sex, weight and age at slaughter (Quadri et al. 1981).

Several other factors are important in determining the organoleptic characteristics of seasoned ham, as shown by taste panel procedures. For example, muscle biochemistry is an important factor in determining the flavour: a low pH, generally associated with PSE (pale, soft, exudative) meat, caused too rapid dehydration during ham seasoning (Kauffman et al., 1978) and consequently the meat was saltier (Goutefongea et al., 1978); on the contrary, DFD (dark, firm, dry) meat with a high pH is unsuitable for seasoning because it absorbed less salt (Goutefongea et al., 1978) and had a lower panel score for aroma (Lawrie 1983). Another important factor is mechanical massaging of hams at the beginning of the seasoning period. This technique (Weiss, 1973; Krause et al., 1978a, b; Gillett et al., 1982; Kemp and Fox, 1985) improved the qualitative characteristics of hams (colour, tenderness, flavour, juiciness, etc.), although Marriott et al. (1984) found no significant advantages.

Additives such as sodium nitrite, nitrate, ascorbate or potassium chloride produced marked changes in colour and taste (Watts and Lhemann, 1952; Brown et al., 1974; Price and Greene, 1978; Vollmar and Melton, 1981; Hand et al., 1982; Froehlich et al., 1983), also the growth of certain microorganisms during seasoning produced hams with a much appreciated aroma (Baldini et al., 1983; Lawrie, 1983).

The aim of this study was to show, by using a sensory panel, possible differences due to the sources of variation considered and to compare subjective evaluations to those obtained using objective methods (Matassino et al., 1985a, b, c; Zullo and Quadri, 1985).

MATERIALS AND METHODS

The trial used 140 hams taken from the right side of 72 castrated males (♂ ♂) and 68 entire females '♀ ♀' of eight genetic types. The hams were grouped into four weight classes (<7, 7→8, 8 → 9 and >9 kg) (Table 1) . The rearing technique and the feeding plan were reported by Bergonzini et al. (1982).

The hams were trimmed, salted using medium grain kitchen salt and put in a refrigerator cell at 1 to 3°C with 85 to 95% relative humidity (RH). After 4 to 5 days the residual salt was removed and a second salting was made,after a mechanical massage to re-establish the permeability of the muscle fibers. The total time was 18 to 24 days, depending upon ham weight and conditions in the refrigerator. Subsequently, the residual salt was removed and the hams were put in the 'resting' cell at 1 to 4°C and 70 to 80% RH for 35 days, to allow a gradual dehydration and a uniform penetration of the salt. After this second period, the hams were washed using warm water jets and, after dripping at room temperature, put in the 'dry' cell at 24 to 25°C and 85 to 90% RH (these values changed to 15 to 18°C and 80 to 85% RH after 60 to 100 hours). The hams were then put in the 'first seasoning' cell (18 to 20°C; 60 to 70% RH). Then, after smearing the part not covered by rind with a mixture of lard, wheat flour and pepper, the hams were placed in special rooms till the end of the seasoning period. This was determined for each ham on the basis of a subjective evaluation of the consistency, colour, presence of mould, etc.

At the end of the seasoning period, established by an expert using the 'whalebone needle' test, the hams were boned and dissected using the 'Modena' cutting system into 'pera', 'falsa pera' and 'gambo' (see Fig 1 in Santoro and Lo Fiego - this volume). A 10 cm thick sample, including M. gluteobiceps (Gb), M. semitendinosus (St) and M. semimembranosus (Sm), was collected from the middle of 'pera'. Slices were cut from the distal part of the sample using a slicing machine and judged by six panellists using the evaluation form reproduced in Table 2. The statistical analysis included an examination of the mean panel scores on colour, uniformity of colour and tenderness for the three separate muscles.

An analysis of variance was made according to the following factorial model where the factors are considered fixed and the effect of each factor is expressed as deviation from the overall mean μ:

$$y_{ijklm} = \mu + \alpha_i + \beta_j + \gamma_k + \delta_l + (\alpha\beta)_{ij} + (\alpha\gamma)_{ik} + (\alpha\delta)_{il} +$$
$$+ (\beta\gamma)_{jk} + (\beta\delta)_{jl} + (\gamma\delta)_{kl} + \varepsilon_{ijklm} \qquad (1)$$

where: Y_{ijklm} = the score given by the lth panellist on the mth ham belonging to the ith genetic type, j[th] sex and k[th] weight class. The other interactions were not included in model 1 because their explanation would be difficult; however, all types of interaction were taken into account for the 'estimated means' calculation, because every mean is a value weighted for all the other factors considered as single or inter-active effects (Matassino et al., 1984). For this purpose the model used for the means estimation was:

$$y_{ijklm} = (\alpha\beta\gamma\delta)_{ijkl} + \varepsilon_{ijklm} \qquad (2)$$

The differences between means were tested with Student's t-distribution.

RESULTS

An analysis of variance showed that the factor 'panellist' and the

TABLE 1   Number of pigs within each genetic type in relation to the sources of variation.

Genetic type	Symbol	Sex		Hams weight class, (kg)				Total
		(♂♂)	♀♀	<7 (x̄=6.5)	7—8 (x̄=7.5)	8—9 (x̄=8.6)	>9 (x̄=9.6)	
Dutch Large White	DLW	9	9	1	7	5	5	18
Italian Large White	ILW	9	9	2	3	6	7	18
Landrace x Large White	L x LW	9	9	2	8	5	3	18
Belgian Landrace x (Spotted x Large White)	BLx(SPxLW)	9	6	2	4	3	6	15
Landrace x (Spotted x Large White)	Lx(SPxLW)	9	9	–	8	6	4	18
Camborough	Camborough	9	9	1	6	7	4	18
Hypor	Hypor	9	9	4	2	5	7	18
Suffolk	Suffolk	9	8	3	5	6	3	17
Total		72	68	15	43	43	39	140

TABLE 2   Panel evaluation form for seasoned ham.

A RATIO OF FAT TO LEAN	B COLOUR OF MUSCLES[1]
	Gb St Sm
1 Much too fat ☐	1 Much too dark ☐ ☐ ☐
2 Moderately too fat ☐	2 Moderately too dark ☐ ☐ ☐
3 Ideal ☐	3 Ideal ☐ ☐ ☐
4 Moderately too lean ☐	4 Moderately too pale ☐ ☐ ☐
5 Very much too lean ☐	5 Much too pale ☐ ☐ ☐

C COLOUR OF EXTERNAL FAT	D UNIFORMITY OF MUSCLE[1] COLOUR
	Gb St Sm
1 Entirely satisfactory ☐	1 Not two toned ☐ ☐ ☐
2 Satisfactory ☐	2 Two toned ☐ ☐ ☐
3 Dislike ☐	3 Extremely two toned ☐ ☐ ☐
4 Dislike extremely ☐	

E AROMA	F SALTINESS OF MUSCLES[1]
	Gb St Sm
1 Very aromatic ☐	1 Not enough salt ☐ ☐ ☐
2 Aromatic ☐	2 Just enough salt ☐ ☐ ☐
3 Slightly aromatic ☐	3 Moderately salty ☐ ☐ ☐
	4 Excessively salty ☐ ☐ ☐

G TENDERNESS OF MUSCLES[1]	H TOTAL SALTINESS
Gb St Sm	
1 Very tender ☐ ☐ ☐	1 Not enough salt ☐
2 Tender ☐ ☐ ☐	2 Just enough salt ☐
3 Ideal ☐ ☐ ☐	3 Moderately salty ☐
4 Slightly tough ☐ ☐ ☐	4 Excessively salty ☐
5 Tough ☐ ☐ ☐	

I JUICINESS	J OVERALL ACCEPTABILITY
1 Very juicy ☐	1 Like extremely ☐
2 Juicy ☐	2 Like moderately ☐
3 Slightly juicy ☐	3 Like slightly ☐
4 Dry ☐	4 Dislike ☐

[1] Gb = M. gluteobiceps; St = M. semitendinosus; Sm = M. semimembranosus

interactions 'genetic type x sex' and 'genetic type x weight class' significantly affected the quality of seasoned ham (Table 3). By considering the components of variance (Table 4) it was deduced that: (a) 'panellist' absorbed most of the total variability for almost all the attributes, reaching the highest value for fat colour (40%); (b) the 'ham weight class' was important for the ratio of fat to lean (12%) and juiciness (7%); (c) 'genetic type', 'sex' and almost all the interactions absorbed little of the total variability, except for the interactions (i) 'genetic type x sex' for meat colour (4%) and (ii) 'genetic type x ham weight class' for the ratio of fat to lean (10%), colour uniformity and overall acceptability (8%) and juiciness (6%).

The effect of single factors was evaluated by the significance of the differences between the means calculated according to model 2. The coefficient of variability$(C.V. = \sigma / \bar{x})$was less than 10% for all variables.

TABLE 3   Analysis of variance of the organoleptic quality of seasoned ham in relation to genetic type, sex, ham weight and panellist.

Variable	F value[1]	
	interaction	
	genetic type x sex	genetic type x ham weight class
Ratio of fat lean	1.3	5.5***
Colour of muscles	4.1***	2.2***
Colour of external fat	3.3**	2.5***
Uniformity of colour of muscles	2.9**	3.5***
Aroma	1.5	1.1
Tenderness of muscles	0.7	2.1***
Total saltiness	0.4	2.4***
Juiciness	1.6	2.9***
Overall acceptability	2.5*	3.6***

[1]* = P < 0.05;   ** = P < 0.01;   *** = P < 0.001. The effects of 'panellist' and of the other interactions are not reported because the former is significant (P < 0.001) and the latter non significant for all the variables.

TABLE 4   Percentage of variance attributable to the factors included in model 1.

		ratio of fat to lean	colour of muscles	colour of external fat	uniformity of colour of muscles	aroma	tenderness of muscles	total saltiness	juiciness	overall acceptability
								Variable		
Genetic type,	α	4.70	5.05	1.03	2.07	0.0	3.74	1.07	0.67	2.13
Sex,	β	0.0	0.30	0.50	1.19	0.0	1.15	0.87	0.88	0.44
Ham weight class,	γ	12.43	0.33	0.94	0.14	0.59	3.90	1.31	7.36	2.74
Panellist,	δ	8.36	14.81	40.56	10.97	27.08	13.32	34.08	4.42	9.16
Interaction,	$\alpha\,\beta$	0.0	4.16	2.21	2.85	0.83	0.0	0.0	0.0	2.18
"	$\alpha\,\gamma$	10.15	4.78	2.93	8.35	0.52	3.04	2.72	6.45	7.79
"	$\alpha\,\delta$	0.0	0.74	0.0	0.0	0.0	0.0	1.46	0.0	1.58
"	$\beta\,\gamma$	2.18	0.37	0.27	0.57	0.42	0.0	0.69	0.42	0.0
"	$\beta\,\delta$	0.0	0.0	0.09	0.0	0.0	0.0	0.0	0.0	0.0
"	$\gamma\,\delta$	0.15	0.0	0.0	0.0	0.08	1.53	0.0	0.0	0.0
Random error,	ε	62.03	69.46	51.47	73.86	70.48	73.32	57.80	79.80	73.98

Genetic type

Differences between genetic types were significant for all quality characteristics. Tables 5 and 6 show that:

(i)   DLW hams had an 'ideal' ratio of fat to lean and were significantly different from the other types (except ILW and Suffolk) which were judged to be too fatty;

(ii) meat colour varied from 'ideal' to 'moderately too pale' for all genetic types; however DLW hams were lighter than others ($P < 0.05$);

(iii) concerning fat colour, no genetic type was 'entirely satisfactory': hams from LxLW and Camborough were the best and those from BLx (SPxLW) the worst ($P < 0.05$);

(iv) the hams from all genetic types were two toned in colour; this attribute was most evident in Suffolk hams ($P < 0.05$);

(v)   the aroma of the hams did not differ between genetic types;

(vi) the ham was 'moderately salty' in all genetic types and especially in Suffolk ($P < 0.05$);

(vii) hams from DLW and Hypor were significantly the least juicy;

(viii) concerning 'overall acceptability', none of the genetic types were judged 'excellent' but all the assessments varied significantly from 'like slightly' to 'like moderately'; however, the ILW hams scored best.

TABLE 5   Mean values within pig genetic type for organoleptic attributes of seasoned hams.

Variable[1]	Genetic type							
	DLW	ILW	LxLW	BLx(SPx xLW)	Lx(SPx xLW)	Cambo- rough	Hypor	Suffolk
Ratio of fat to lean	3.0	2.9	2.3	2.5	2.4	2.6	2.4	2.8
Colour of muscles	4.0	3.7	3.4	3.6	3.6	3.7	3.9	3.8
Colour of external fat	2.3	2.4	2.2	2.5	2.3	2.2	2.3	2.3
Uniformity of muscles' colour	2.1	2.2	2.1	2.3	2.1	2.3	2.2	2.4
Aroma	2.2	2.3	2.3	2.3	2.3	2.3	2.3	2.3
Tenderness of muscles	2.8	2.8	2.5	2.5	2.7	2.5	2.7	2.5
Total saltiness	2.6	2.4	2.4	2.6	2.5	2.5	2.4	2.7
Juiciness	2.4	2.3	2.2	2.2	2.2	2.2	2.4	2.3
Overall acceptability	2.9	2.6	2.7	2.9	2.8	2.8	2.7	3.0

[1] The hedonic scales are given in Table 2.

TABLE 6   Differences between genetic types in the organoleptic attributes of seasoned ham.

Comparison		ratio of fat to lean	colour of muscles	colour of external fat	uniformity of muscles' colour	aroma	tenderness of muscles	total saltiness	juiciness	overall acceptability
DLW	- ILW	0.1	0.3**	-0.1	-0.1	-0.1	0.0	0.2*	0.1	0.3**
"	- LxLW	0.7***	0.6***	0.1	0.0	-0.1	0.3***	0.2	0.2*	0.2
"	- BLx(SPxLW)	0.5***	0.4***	-0.2	-0.2	-0.1	0.3**	0.0	0.2**	0.0
"	- Lx(SPxLW)	0.6***	0.4***	0.0	-0.1	-0.1	0.1	0.1*	0.2**	0.1
"	- Camborough	0.4***	0.3***	0.1	-0.1	-0.1	0.3**	0.1*	0.2**	0.1
"	- Hypor	0.6***	0.1	0.0	-0.1	-0.1	0.1	0.2*	0.0	0.2*
"	- Suffolk	0.2	0.2*	0.0	-0.3***	-0.1	0.3**	-0.1	0.1	-0.1
ILW	- LxLW	0.6***	0.3**	0.2*	0.1	0.0	0.3***	0.0	0.1	-0.1
"	- BLx(SPxLW)	0.4*	0.1	-0.1	-0.1	0.0	0.3**	-0.2	0.1	-0.3
"	- Lx(SPxLW)	0.5***	0.1	0.1	0.1	0.0	0.1	-0.1	0.1	-0.2
"	- Camborough	0.3	0.0	0.2*	-0.1	0.0	0.3***	-0.1	0.1	-0.2*
"	- Hypor	0.5**	-0.2	0.1	0.0	0.0	0.1	0.0	-0.1	-0.1
"	- Suffolk	0.1	-0.1	0.1	-0.2*	0.0	0.3**	-0.3***	0.0	-0.4***
LxLW	- BLx(SPxLW)	-0.2	-0.2	-0.3*	-0.2*	0.0	0.0	-0.2	0.0	-0.2
"	- Lx(SPxLW)	-0.1	-0.2	-0.1	0.0	0.0	-0.2*	-0.1	0.0	-0.1
"	- Camborough	-0.3	-0.3	0.0	-0.2*	0.0	0.0	-0.1	0.0	-0.1
"	- Hypor	-0.1	-0.5***	-0.1	-0.1	0.0	-0.2*	0.0	-0.2*	0.0
"	- Suffolk	-0.5**	-0.4***	-0.1	-0.3***	0.0	0.0	-0.3**	-0.1	-0.3**
BLx(SPxLW)	- Lx(SPxLW)	0.1	0.0	0.2*	0.2*	0.0	-0.2	0.1	0.0	0.1
"	- Camborough	-0.1	-0.1	0.3*	0.0	0.0	0.0	0.1	0.0	0.1
"	- Hypor	0.1	-0.3**	0.2*	0.1	0.0	-0.2*	0.2	-0.2*	0.2
"	- Suffolk	-0.3	-0.2*	0.2*	-0.1	0.0	0.0	-0.1	-0.1	-0.1
Lx(SPxLW)	- Camborough	-0.2	-0.1	0.1	-0.2*	0.0	0.2*	0.0	0.0	0.0
"	- Hypor	0.0	-0.3**	0.0	-0.1	0.0	0.0	0.1	-0.2	0.1
"	- Suffolk	-0.4**	-0.2*	0.0	-0.3***	0.0	0.2	-0.2**	-0.1	-0.2*
Camborough	- Hypor	0.2	-0.2*	-0.1	0.1	0.0	-0.2*	0.1	-0.2*	0.1
"	- Suffolk	-0.2	-0.1	-0.1	-0.1	0.0	0.0	-0.2**	-0.1*	-0.2
Hypor	- Suffolk	-0.4*	0.1	0.0	-0.2*	0.0	0.2	-0.3**	0.1	-0.3**

1 * = P <0.05; ** = P <0.01; *** = P<0.001.

Sex

Differences were observed for some attributes; in particular, female hams were more 'two toned' in colour, more tender and less salty (Table 7).

Ham weight class

This factor had a great influence on panel assessments (Table 7). With increasing weight of hams:

(i)   the fat to lean ratio varied from 'ideal' in the first three classes of weight to 'moderately too fat' in the heaviest class ($P < 0.01$);

(ii) meat colour became lighter and more uniform while the colour of external fat tended to be disliked.

(iii) the aroma score rose, but tenderness varied from 'ideal' to 'tender';

(iv) total saltiness varied from 'moderately salty' in the first two weight classes to 'just enough salt' in the following two classes ($P < 0.05$);

TABLE 7   Mean values within sex and within ham weight class for the organoleptic properties of seasoned ham[1]

Variable	Sex		Ham weight class, (kg)			
	($\sigma\sigma$)	♀♀	< 7	7→8	8→9	> 9
Ratio of fat to lean	2.7	2.6	3.2^A	2.8^B	2.7^B	2.0^C
Colour of muscles	3.7	3.8	3.6^A	3.7^a	3.7^a	3.8^{Bb}
Colour of external fat	2.3	2.3	2.2	2.4	2.3	2.4
Uniformity of muscle colour	2.1^A	2.3^B	2.3^a	2.2^{ab}	2.1^b	2.2^{ab}
Aroma	2.3	2.3	2.4^{Aa}	2.3^b	2.3^B	2.2^B
Tenderness of muscles	2.7^a	2.6^b	2.8^{Aa}	2.7^A	2.6^b	2.5^{Bc}
Total saltiness	2.6^a	2.5^b	2.7^A	2.6^a	2.4^{Bb}	2.4^{Bb}
Juiciness	2.3	2.2	2.5^A	2.3^{aB}	2.2^B	2.2^{Bb}
Overall acceptability	2.8	2.8	3.1^A	2.8^B	2.7^B	2.8^B

[1] The hedonic scales are given in Table 2. Means with different superscripts, within the factor, are significantly different at P<0.05 (small letters) or P< 0.01 (capital letters).

(v) juiciness increased progressively, showing significant improvement between the first and the following classes;

(vi) 'overall acceptability' was significantly better at the three heavier ham weights compared with the lightest.

Panellist

There was significant variation between panel judges. This result does not invalidate the panel which, on the whole, provided useful information on the influence of the other sources of variation. Moreover, as reported by Tomassone and Flanzy (1977), the assessor, in this kind of test, behaves as an instrument with variable sensitivity of measurement. Nonetheless, the result emphasises the importance of finding instruments able to furnish meaningful objective assessments of the organoleptic characteristics of the food.

Interactions

The interaction 'genetic type x sex' affected only four of the nine attributes considered (Table 8). The castrated males, compared to entire females, showed:

(i) a lighter muscle colour in the DLW, LxLW and Camborough hams; the contrary in the other types;

(ii) fat colour less satisfactory in DLW, LxLW and Hypor hams; vice versa in the other types;

(iii) more uniformity of colour in all the genetic types with the exception of DLW;

(iv) lower 'overall acceptability' for hams from DLW, ILW and LxLW.

The interaction 'genetic type x ham weight class' was statistically significant for most organoleptic characteristics. Table 9 shows that:

(i) the fat to lean ratio nearest to 'ideal' was obtained at a ham weight < 7 kg from BLx(SPxLW), at > 9 kg from ILW and at 7 ⟶ 8 kg from the other genetic types;

(ii) the most appreciated colour of muscle occurred at the two extreme ham weight classes in ILW, at 7.5 kg in Lx(SPxLW) and at 7 to 9 kg in Hypor and Suffolk;

(iii) fat colour was more satisfactory in ILW hams weighing 9.5 kg on the average, in those from DLW and Hypor weighing 6.5 kg and at 8.5 kg in LxLW, Camborough and Suffolk;

TABLE 8   Mean organoleptic values for seasoned ham within pig genetic type and within sex[1].

Genetic type	colour of muscles		colour of external fat		uniformity of muscles' colour		overall acceptability	
	(♂♂)	♀♀	(♂♂)	♀♀	(♂♂)	♀♀	(♂♂)	♀♀
DLW	4.1	4.0	2.4	2.2	2.2^A	1.9^B	3.0	2.8
ILW	3.7	3.8	2.4	2.5	2.1	2.3	2.8^a	2.5^b
LxLW	3.5	3.4	2.6^A	2.0^B	2.1	2.1	2.9^a	2.5^b
BLx(SPxLW)	3.3^A	3.8^B	2.4	2.5	2.2	2.3	2.8	2.9
Lx(SPxLW)	3.6	3.7	2.3	2.3	2.0^a	2.2^b	2.7	2.8
Camborough	3.9^a	3.5^b	2.1	2.3	2.1^a	2.4^b	2.7	3.0
Hypor	3.9	3.9	2.4	2.2	2.1^a	2.3^b	2.6	2.8
Suffolk	3.5^A	4.1^B	2.3	2.3	2.2^a	2.5^b	3.0	3.0

[1] For the meaning of hedonic scales see Table 2. Means with different superscripts are significantly different at $P < 0.05$ (small letters) or $P < 0.01$ (capital letters).

TABLE 9   Mean values within the genetic type and within ham weight class[1].

Genetic type	<7	7—8	8—9	>9	<7	7—8	8—9	>9
	1. Ratio of fat to lean				**2. Colour of muscles**			
DLW	4.2^{Aa}	3.1^B	3.2^{ABb}	2.2^C	4.3	4.0	4.0	4.0
ILW	3.4^{Aa}	2.3^B	2.7^b	3.1^A	3.6	3.7	3.9	3.6
LxLW	2.4^a	2.9^A	2.5^A	1.5^{Bb}	2.9^{Aa}	3.5^b	3.6^B	3.8^B
BLx(SPxLW)	3.1^A	2.6^A	2.6^A	1.7^B	3.4^{ab}	3.6^{ab}	3.4^a	3.8^b
Lx(SPxLW)	–	2.5	2.4	2.1	–	3.5	3.7	3.6
Camborough	3.2^A	3.0^A	2.7^A	1.9^B	2.7^A	3.8^B	3.6^{aB}	4.1^{Bb}
Hypor	2.7^a	2.8^a	2.3^{ab}	2.1^b	4.0	3.8	3.8	3.9
Suffolk	4.0^A	2.9^B	2.7^B	1.5^C	3.9	3.7	3.7	3.9
	3. Colour of external fat				**4. Uniformity of muscles colour**			
DLW	1.8^a	2.3^{ab}	2.5^b	2.4^b	2.7^{Aa}	2.2^{AbC}	2.0^{BC}	1.8^B
ILW	2.4^{AB}	2.7^A	2.5^{AB}	2.3^B	1.9^a	2.1^{ab}	2.3^b	2.3^b
LxLW	2.2^{AB}	2.2^{AB}	2.0^A	2.5^B	2.2	2.0	2.2	2.0
BLx(SPxLW)	2.2^a	2.4^{ab}	2.6^{ab}	2.7^b	2.1^{ab}	2.4^{ab}	2.1^a	2.4^b
Lx(SPxLW)	–	2.5^{Aa}	2.2^b	2.1^B	–	2.2^a	2.0^b	2.1^{ab}
Camborough	2.5	2.2	2.2	2.2	2.3	2.3	2.2	2.2
Hypor	2.2	2.3	2.3	2.3	2.3^a	1.8^{Ab}	2.2^a	2.3^B
Suffolk	2.3^{AB}	2.4^A	2.0^B	2.5^A	2.5	2.3	2.3	2.4

— TABLE 9 continued

Genetic	ham weight class, (kg)							
type	< 7	7→8	8→9	> 9	< 7	7→8	8→9	> 9
	5. Aroma				**6. Tenderness of muscles**			
DLW	2.3	2.2	2.3	2.1	3.3^A	2.8^a	2.9^a	2.5^{Bb}
ILW	2.5^a	2.1^b	2.3^{ab}	2.2^{ab}	3.1^a	2.9^{ab}	2.7^b	2.8^{ab}
LxLW	2.4	2.3	2.3	2.2	2.6^a	2.7^A	2.6^A	2.1^{Bb}
BLx(SPxLW)	2.5^a	2.3^{ab}	2.1^b	2.2^{ab}	2.9^A	2.4^{AB}	2.4^{AB}	2.3^B
Lx(SPxLW)	-	2.3	2.2	2.4	-	2.9^A	2.7^{AB}	2.5^B
Camborough	2.5	2.3	2.3	2.1	2.5^{ab}	2.7^A	2.6^a	2.2^{Bb}
Hypor	2.3	2.5	2.3	2.2	2.8	2.6	2.7	2.8
Suffolk	2.5	2.2	2.3	2.3	2.7	2.4	2.4	2.7
	7. Total saltiness				**8. Juiciness**			
DLW	3.2^{Aa}	2.6^b	2.8^A	2.3^B	3.3^A	2.3^{aB}	2.3^B	2.1^{Bb}
ILW	2.5	2.6	2.3	2.3	2.4	2.3	2.3	2.2
LxLW	2.6	2.5	2.4	2.3	2.3	2.3	2.1	2.0
BLx(SPxLW)	2.7	2.5	2.5	2.6	2.7^A	2.1^B	1.9^B	2.1^B
Lx(SPxLW)	-	2.7^{Aa}	2.3^B	2.4^b	-	2.5^{Aa}	2.1^B	2.1^b
Camborough	2.7^{ab}	2.6^a	2.5^{ab}	2.2^b	2.0	2.3	2.2	2.1
Hypor	2.6	2.4	2.3	2.4	2.6^A	2.1^B	2.4^{AB}	2.2^B
Suffolk	2.7^{AB}	2.7^{AB}	2.5^A	3.0^B	2.5^a	2.2^{ab}	2.2^b	2.5^a
	9. Overall acceptability							
DLW	3.7^{Aa}	2.7^B	3.0^b	2.7^B				
ILW	2.8^a	2.7^{ab}	2.9^A	2.3^{Bb}				
LxLW	2.7	2.7	2.5	2.8				
BLx(SPxLW)	3.2^A	2.9^a	2.3^{Bb}	3.0^A				
Lx(SPxLW)	-	3.1^{Aa}	2.5^B	2.7^b				
Camborough	3.0	2.9	2.9	3.7				
Hypor	3.2^A	2.5^B	2.5^B	2.5^B				
Suffolk	3.1^{ab}	2.8^a	2.8^a	3.2^b				

[1] For the meaning of hedonic scales see Table 2. Means with different superscripts are significantly different at P<0.05 (small letters) or P<0.01 (capital letters).

(iv)   tenderness was 'ideal' if obtained at the extreme classes of weight in Hypor and Suffolk, at 8 ⟶ 9 kg in DLW and at 7 ⟶ 8 kg in other types;

(v)   the 7 ⟶ 8 kg hams from BLx(SPxLW), those >9 kg from DLW, ILW, LxLW and Camborough, and those weighing 8 ⟶ 9 kg from the other genetic types were judged 'just enough salt';

(vi)   overall, the most appreciated ham was the one weighing 7 ⟶ 8 kg in DLW, Hypor and Suffolk, 8 ⟶ 9 kg in Hypor, Suffolk, LxLW, Lx(SPxLW) and BLx(SPxLW), >9 kg in DLW, ILW, Camborough and Suffolk.

CONCLUSIONS

The factors examined in this paper had statistically significant effects on the qualitative characteristics of seasoned ham. A meat quality score was derived taking into account the effect of interactions previously discussed, by constructing an 'empirical index' in which a score of 100 represented the 'best' or 'ideal' judgements on the hedonic scales. Among the eight genetic types, the ILW was best with an index value of 71.4, followed by the LxLW (70.2), Lx(SPxLW) (69.6) and Suffolk (67.3), (Table 10).

Regarding the index values within sex, it was again confirmed that ILW was superior for the castrated males while for the females the best genetic type was LxLW. Moreover, DLW, LxLW and Camborough females had an index higher than that of males from the other genetic types.

Overall, the index was higher in the two intermediate classes of ham weight and lower in the extreme ones. However, each genetic type performed best at different live weights and, consequently, at different ham weights: the DLW and ILW were best at ham weights >9 kg, the three-way crosses and the two commercial hybrids Camborough and Suffolk at 8 ⟶ 9 kg and finally the LxLW and Hypor at 7 ⟶ 8 kg.

An interesting conclusion from the commercial point of view, is that none of the genetic types studied provided a seasoned ham below 7 kg which received a high index value either on a single attribute or on the overall acceptability score. This result suggests that producers of seasoned hams should re-examine the standards of 'Modena' and 'Parma' hams in order to obtain a qualitatively superior product.

Finally, the superiority of DLW and ILW pigs could be explained by

TABLE 10 Ham quality index values within and independently from the factors considered.

Genetic type	Sex		Ham weight class, (kg)				Total
	(♂♂)	♀♀	< 7	7→8	8→9	> 9	
DLW	67.5	69.9	54.0	69.7	67.8	70.3	69.5
ILW	72.0	71.5	69.3	69.1	69.2	73.9	71.4
LxLW	66.7	72.3	69.6	71.9	71.7	66.0	70.2
BLx(SPxLW)	70.1	67.2	67.7	68.8	72.0	63.7	67.7
Lx(SPxLW)	69.5	69.2	–	67.3	72.5	69.0	69.6
Camborough	69.0	70.4	69.3	69.6	70.2	67.5	69.0
Hypor	69.0	67.6	66.1	71.4	68.8	68.9	67.9
Suffolk	68.8	65.4	61.9	68.8	69.5	59.2	67.3
Total	69.3	68.8	67.5	69.2	70.7	67.4	

the genetic improvement performed to meet the needs of the pork industry for the production of hams with organoleptic characteristics that are particularly appreciated by consumers.

REFERENCES

Baldini, P., Campanini, M., Pezzani, G. e Palmia, F. 1983. Possibilità di ridurre la quantità di cloruro di sodio nei prodotti stagionati valutando opportunamente le varie tecniche di preparazione. Proc. 29th European Meeting of Meat Research Workers, Salsomaggiore (Parma), Italy.

Bergonzini, E., Quadri, G., Cosentino, E., Zullo, A. e Pieraccini, L. 1982. Studio di alcuni tipi genetici di suino 'da salumificio' allevati con un piano alimentare 'medio'. I. Rilievi 'infra vitam' e alla mattazione. Prod. Anim., 1, n.s., 1-33.

Brown, C.L., Hedrick, H.B. and Bailey, M.E. 1974. Characteristics of cured ham as influenced by levels of sodium nitrite and sodium ascorbate. J. Food Sci., 39, 977-979.

Froehlich, D.H., Gullet, E.A. and Usborne, W.R. 1983. Effect of nitrite

and salt on the color, flavor and overall acceptability of ham.
J. Food Sci., 48, 152-154.

Goutefongea, R., Girard, J.P. et Jacque, B. 1978. Caracteristiques de la
viande de porc de transformation. Journées de la Rech. Porcine en
France, 10, 235-248.

Gillett, T.A., Cassidy, R.D. and Simon, S. 1982. Ham massaging. Effect of
massaging cycle, environmental temperature and pump level on yield,
bind and colour of intermittently massaged hams. J.Food Sci., 47,
1083-1088.

Hand, L.W., Terrell, R.N. and Smith, G.C. 1982. Effects of complete or partial replacement of sodium chloride on processing and sensory properties of hams. J. Food Sci., 47, 1776-1778.

Kauffman, R.G., Wachholz, D., Henderson, D. and Lochner, J.Y. 1978.
Shrinkage of PSE, normal and DFD hams during transit and processing.
J. Anim. Sci., 46, 1236-1240.

Kemp, J.D. and Fox, J.D. 1985. Effect of needle tenderization on salt
absorption, yields, composition and palatability of dry-cured hams
produced from packer style and skinned green hams. J. Food Sci., 50,
295-299.

Krause, R.J., Plimpton, R.F., Ockerman, H.W. and Cahill, V.R. 1978a.
Influence of tumbling and sodium tripolyphosphate on salt and nitrite
distribution in porcine muscle. J. Food Sci., 43, 190-192.

Krause, R.J., Ockerman, H.W., Krol, B., Moerman, P.C. and Plimpton JR.,R.F.
1978b. Influence of tumbling time, trim and sodium tripolyphosphate
on quality and yield of cured hams. J. Food Sci., 43, 853-855.

Lawrie, R.A. 1983. Scienza della carne. Ed.Italiana a cura di R. Chizzoli-
ni. Edagricole, Bologna, XVI-348.

Marriott, N.G., Kelly, R.F., Shaffer, C.K., Graham, P.P. and Boling, J.W.
1984. The effect of blade tenderization on the cure and sensory
characteristics of dry-cured hams. Proc. 30th European Meeting of Meat
Research Workers. 264-266.

Matassino, D., Cosentino, E., Freschi, P. e Grasso, F. 1984. Studio comparativo fra bufali e bovini alimentati con fieno e mangime concentrato
composto. XIV. Variazione delle caratteristiche colorimetriche della
carne dagli 8,5 ai 15 mesi di età. Prod. Anim., 3, n.s., 81.

Matassino, D., Zullo, A., Ramunno, L. and Cosentino, E. 1985a. Evaluation
of some qualitative characteristics of seasoned ham in eight pig genetic types. In: Evaluation and Control of Meat Quality in Pigs,
EC Seminar, Dublin, Nov. 1985.

Matassino, D., Grasso, F. e Cosentino, E. 1985b. Studio di alcuni tipi genetici di suino 'da salumificio' allevati con un piano alimentare 'medio'. XI. Caratteristiche reologiche del prosciutto stagionato. Prod.
Anim., 4, n.s. (in c.d.s.).

Matassino, D., Ramunno, L. e Bordi, A. 1985c. Studio di alcuni tipi geneti-
ci di suino 'da salumificio' allevati con un piano alimentare 'medio'.
XII. Caratteristiche colorimetriche del prosciutto stagionato. Prod.
Anim., 4, n.s. (in c.d.s.).

Price, L.G. and Greene, B.E. 1978. Factors affecting panelists perception of cured meat flavor. J. Food Sci., 43, 319-322.

Quadri, G., Bergonzini, E., Zullo, A., Cosentino, E. e Matassino, D. 1981. Studio di alcuni tipi genetici di suino 'da salumificio' allevati con un piano alimentare 'medio'. IV. Aspetti quanti-qualitativi ed economici del prosciutto stagionato. Suinicoltura, 22, 5, 21-49.

Tomassone, R. et Flanzy, C. 1977. Presentation synthétique de diverses méthodes d'analyse de données fournies par un jury de dégustateurs. Ann. Technol. Agric., 26, 373-418.

Vollmar, E.K. and Melton, C.C. 1981. Selected quality factors and sensory attributes of cured hams as influenced by different phosphate blends. J. Food Sci., 46, 317-320.

Watts, B.M. and Lhemann, B.T. 1952. The effect of ascorbic acid on the oxidation of hemoglobin and the formation of nitric oxide hemoglobin. Food Res., 17, 100-104.

Weiss, J.M. 1973. Ham tumbling and massaging. Western Meat Industry, 14, 23-28.

Zullo, A. e Quadri, G. 1985. Studio di alcuni tipi genetici di suino 'da salumificio' allevati con un piano alimentare 'medio'. XIII. Composizione chimica di muscoli del prosciutto stagionato. Prod. Anim., 4, n.s. (in c.d.s.).

THE QUALITY IN PARMA HAMS AS RELATED TO HALOTHANE REACTIVITY

P. Santoro, D.P. Lo Fiego

Università de Bologna
Istituto di Allevamenti Zootecnici
42100 Reggio Emilia, Italy

ABSTRACT

In the present study data are reported concerning hams from halothane negative and halothane positive crossbred pigs, slaughtered at about 130 kg live weight. Seasoning losses were higher in hams from halothane positive pigs, but the difference was not statistically significant. Also separation into boneless cuts and laboratory analysis did not show significant differences. Meat processing techniques probably prevented the detection of meat quality differences between the seasoned products from halothane negative and positive pigs. Moreover, even though halothane positive pigs produce poorer quality fresh meat, it is not easy to ascertain to what extent the tendency towards PSE may cause problems during processing and seasoning.

INTRODUCTION

These data are part of a more extensive research programme aimed at evaluating carcass and meat characteristics in heavy pigs intended for the processing industries.

Previously Santoro (1982) Santoro et al. (1982) Camporesi et al (1982) and Santoro (1983) studied carcass and meat characteristics in pigs of ca. 130 kg live weight which were submitted to a halothane test between nine and thirteen weeks of age. Halothane positive pigs produced carcasses with a greater proportion of lean cuts and a lower percentage of adipose tissue. The meat from halothane positive animals, however, showed a tendency to PSE at least for measurements in M. longissimus dorsi.

The data reported here refer to measurements taken on the ham throughout seasoning, at separation into boneless cuts and subsequently during laboratory analysis. Separation data are given since in Italy the ham is sometimes dissected at the end of the seasoning period and merchandised in boneless pieces.

MATERIALS AND METHODS

A total of 35 raw hams from the right sides of carcasses produced by crossbreeding Belgian Landrace (BL) and Large White (LW) - BLO'x(BL x LW)♀ -, were studied. The animals, classed as halothane negative

(9 castrated males and 9 females) and halothane positive (9 castrated males and 8 females), were slaughtered at ca. 130 kg live weight.

The data concerning the characteristics of carcass and meat quality in the pigs on trial have already been reported in the work mentioned above. During the entire seasoning period the hams were kept under control. They were weighed after 8 months and at the end of seasoning (12 months). Subsequently the seasoned hams, Parma type, were divided into three boneless cuts: 'falsa pera', 'pera' and gambo' (Fig 1), as is frequently done in practice (Quadri et al., 1981). On the freshly cut surface of the 'falsa pera', meat colour was determined by the L, a, b values of the Hunter system using the Hunterlab D25D2M colour meter (Santoro et al., 1978; Santoro, 1980; Santoro, 1984) and pH by means of a portable pH meter. Subsequently a sample of lean meat was taken for chemical analysis. Other samples, obtained from the centre of the 'falsa pera' surface, were used to determine myoglobin concentration (Wierbicki et al., 1955), sodium chloride content (A.O.A.C., 1975) and water activity (A_w) (Lerici et al., 1981).

RESULTS AND DISCUSSION

Data taken on fresh hams and during seasoning are presented in Table 1. No substantial differences were found between hams from pigs classed as halothane negative or halothane positive. The higher trimming losses recorded in the hams from halothane negative pigs (P< 0.05) can be ascribed to a larger quantity of covering fat in the pigs from this group. In fact, previous results from carcass cutting indicated that the same non reactors had a higher percentage of fat and a significantly lower percentage of lean (Santoro, 1982; Santoro, 1983).

Although pH_1 values were lower in hams from halothane positive pigs (P < 0.01), they did not give substantially worse results during seasoning, at least with regard to the few data reported in Table 1, except for a greater, but not significant, seasoning loss.

On separation of the seasoned ham into boneless retail cuts, no statistically significant differences were found between halothane negative and positive pigs (Table 2). The hams from reactors had a higher percentage of the most valuable cuts ('pera' and 'falsa pera') and hence a lower percentage of the 'gambo', although the differences were not significant at P < 0.05.

Chemical composition and meat quality characteristics were not

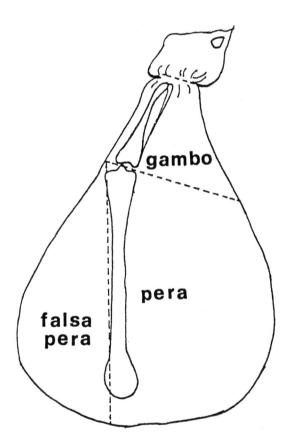

Fig. 1   Method of separation of the seasoned ham into boneless
retail cuts.

TABLE 1   Measurements taken on fresh hams and during seasoning for 12 months.   Mean values and standard deviations are presented.

		Halothane negative		Halothane positive		Significance[1]
		Mean	s d	Mean	s d	
pH_1 (45 min)		6.47	0.17	5.97	0.21	**
pH_2 (24 h)		5.86	0.11	5.85	0.15	n s
Hot weight	kg	12.39	0.77	12.29	0.93	n s
Refrigeration losses (24 h)	%	0.92	0.12	0.94	0.26	n s
Refrigerated and trimmed ham weight	kg	10.04	0.68	10.10	0.77	n s
Trimming losses	%	18.22	1.68	17.07	1.25	*
Seasoning losses (8 months)	%	25.17	1.70	25.98	1.72	n s
Seasoning losses (12 months)	%	28.07	1.97	29.30	1.96	n s

[1] *P < 0.05;   **P < 0.01;   n s   not significant

TABLE 2   Percentage yield of boneless retail cuts from ham seasoned for twelve months. The results are mean values and standard deviations[1]

		Halothane negative		Halothane positive	
		Mean	s d	Mean	s d
Seasoned ham (bone-in)	kg	7.22	0.47	7.14	0.60
"Falsa pera"	%	18.65	1.70	18.98	1.65
"Pera"	%	48.38	1.72	49.00	2.52
"Gambo"	%	18.16	1.20	17.49	1.53
Losses (from the bone-in product)	%	14.82	1.00	14.53	1.58

[1] Mean values were not significantly different

TABLE 3 Chemical composition and meat quality of lean meat from hams seasoned for twelve months. Samples and measurements were taken from the centre of 'falsa pera' surface. Mean values and standard deviations are presented[1]

		Halothane negative		Halothane positive	
		Mean	s d	Mean	s d
Moisture	%	57.78	2.18	58.38	1.39
Protein (N x 5.9)	%	24.83	1.06	25.26	0.90
Ether extract	%	4.46	1.47	3.72	1.37
Ash	%	10.08	1.18	10.05	1.03
Unaccounted matter	%	2.85	0.39	2.59	0.50
pH		5.97	0.16	5.96	0.13
Colour { L		31.21	1.47	30.32	1.21
a		10.59	1.30	10.31	0.57
b		6.04	1.05	5.56	0.44
Myoglobin	mg/g	1.03	0.25	1.18	0.48
Sodium chloride	%	9.56	1.01	9.90	0.86
Water activity (A_w)		0.86	0.02	0.85	0.01

[1]Mean values were not significantly different.

significantly different between the two groups of pigs (Table 3). The hams from halothane positive pigs appear to have less intramuscular fat content. This may suggest a less delicate flavour of the seasoned meat obtained from these animals. However, this difference is too far from the minimum level of statistical significance to draw definite conclusions. The results suggest that processing techniques may prevent the detection of differences between the meat from negatively and positively reacting animals.

CONCLUSIONS

Although fresh meat from pigs with a positive halothane test was poorer in quality (Santoro et al., 1982), with a tendency towards PSE, the present data from seasoning, separation and laboratory analysis did not show substantial differences between hams from negative and positive animals. Hams from positive pigs showed higher, but not statistically significant seasoning losses. Laboratory analysis did not reveal any difference in quality between the two groups of animals.

It is generally assumed that halothane positive pigs produce carcasses containing more lean meat but of poorer quality. However, it is not clear to what extent a tendency to PSE may be considered a serious defect in fresh meat and to what extent it can cause problems in the processing industry and hence in seasoned products.

ACKNOWLEDGEMENT

This study was supported by a grant of the Consiglio Nazionale delle Ricerche, Italy.

REFERENCES
A.O.A.C. 1975. Official methods of analysis of the A.O.A.C., Association of Official Analytical Chemists, Washington, DC 20044, p. 417.
Camporesi, A., Santoro, P. and Lambertini, L. 1982. Reattività all'alotano e dati di stagionatura dei prosciutti. Atti Soc. Ital. Sci. Vet., 36, 442-443.
Lerici, C.R., Piva, M., Pinnavaia, G. and Aimi, M. 1981. Ricerche su alcune proprietà dei formulati alimentari a contenuto intermedio di umidità. -Nota 1. Attività dell'acqua (A$_w$) e composizione chimica. Riv. Soc. Ital. Sci. Alimentazione, 10, (5), 287-294.
Quadri, G., Bergonzini, E., Zullo, A., Cosentino, E. and Matassino, D. 1981. Studio di alcuni tipi genetici di suino da salumificio allevati con un piano alimentare medio. -IV. Aspetti quanti-qualitativi ed economici del prosciutto stagionato. Suinicoltura, 22 (5), 21-49.

436

Santoro, P., Rizzi, L. and Mantovani, C. 1978. Colore e contenuto di mioglobina in carni di tacchini macellati a diverse età. Zoot. Nutr. Anim., 4, 147–155.

Santoro, P. 1980. Acquisizioni ed indagini su frequenti alterazioni della carne suina. Riv. Zoot. Vet., (2), 112–122.

Santoro, P. 1982. Aspetti genetici della qualità della carne nel suino pesante. Atti del Progetto Finalizzato del Consiglio Nazionale delle Ricerche 'Difesa delle risorse genetiche delle popolazioni animali', Obiettivi e Risultati, 383–399.

Santoro, P., Lambertini, L. and Camporesi, A. 1982. Reattività all'alotano e parametri relativi alla qualità della carne suina. Atti Soc. Ital. Sci. Vet., 36, 439–441.

Santoro, P. 1983. Qualità delle carcasse suine e reazione al test alotano. Suinicoltura, 24, (7), 57–60.

Santoro, P. 1984. L, a, b colour values as related to meat quality in pigs. Proc. Scientific Meeting 'Biophysical PSE-muscle Analysis' (Ed. H. Pfützner) (Technical University, Vienna), pp 259–269.

Wierbicki, E., Cahill, V.R., Kunkle, L.E., Klosterman, E.W. and Deatherage, F.E. 1955. Effect of castration on biochemistry and quality of beef. J. Agric. Food Chem., 3, 244–249.

RESEARCH ON STRESS SUSCEPTIBILITY IN BELGIAN PIG POPULATIONS;
INFLUENCE ON ZOOTECHNICAL PARAMETERS AND THE PRACTICAL USE
OF BLOOD MARKER SYSTEMS

Ph. Lampo, A Van Zeveren, Y. Bouquet
Laboratory of Animal Genetics and Breeding
Faculty of Veterinary Medicine
State University of Ghent
Heidestraat, 19 - B-9220 Merelbeke , (Belgium)

ABSTRACT

Penetrance of halothane genes was shown to be incomplete (91 percent) in Belgian Landrace pigs. Determination of polymorphic variants of blood marker systems, combined with halothane testing, allowed more accurate diagnosis of halothane sensitivity. Depending on the position of each particular marker in the halothane linkage group an accuracy up to 98.5 percent in halothane phenotyping can be reached. In addition halothane genotyping can also be performed.

The influence of stress susceptibility on carcass, meat quality and reproductive traits was studied by correlating the activities of creatine kinase and lactate dehydrogenase with breeding and fattening results and by comparing the results obtained in halothane positive and halothane negative Belgian Landrace pigs. Little effect was found in daily weight gain and food conversion. Stress susceptible pigs yielded on average two percent more lean but meat quality, determined by temperature and pH at 1 h after slaughter and by the Göfo value, was significantly worse. In farrowings subsequent to the first halothane negative sows provided more piglets, this amounted to 0.4 extra piglets in the second and one extra piglet in the following litters at birth, with an additional gain of 0.2 piglets at weaning in all litters. A cross breeding scheme is proposed in order to obtain good reproductive and meat quality characteristics combined with a good carcass.

INTRODUCTION

Stress susceptibility is high in Belgian pig populations, both in the Landrace and in the Pietrain breeds (Lampo, 1978; Hanset et al., 1983). Up to now the diagnosis was done using serum enzyme tests, based on the studies by Bickhardt et al. (1969, 1977), or by using the method of halothane anaesthesia as described by Eikelenboom and Minkema (1974).

Details of the results of the Belgian studies have been published (Lampo, 1977, 1981). Both techniques distinguish between phenotypically sensitive and resistant animals but not between homozygously and heterozygously resistant ones.

The halothane genes are linked to some blood marker systems on the porcine chromosome number 15 (Andresen, 1971, 1981; Rasmusen, 1981, Imlah, 1982; Juneja et al., 1983; Van Zeveren et al., 1985). These marker systems include blood group and related systems (H and S systems) and polymorphic systems of soluble blood substances (PHI, PGD, Po2 systems). All marker systems show polymorphism in Belgian pig populations (Van Zeveren et al., 1983).

This paper considers a practical application for blood marker determination in addition to halothane testing. In particular its usefulness in performing more accurate halothane phenotyping in Belgian Landrace pigs and therefore in producing homozygously halothane resistant breeding animals will be discussed.

Correlations between halothane sensitivity and growth, fattening, and meat quality characteristics are summarised by Webb (1981). The general trend found is that stress susceptible pigs have a higher meat percentage but a lower meat quality, while daily growth and food conversion are relatively independent of the halothane gene. Research on the effect of the halothane gene on reproduction revealed that stress resistant sows are more prolific (Webb and Jordan, 1978; Schneider et al., 1980; Lampo et al., 1985).

MATERIALS AND METHODS

To study the blood marker system we used animals kept at the Experimental Station on the Faculty of Veterinary Medicine of the State University of Ghent on litters from sows and boars kept on seven commercial farms in West Flanders. Part of that material was also used for the study of halothane sensitivity and meat quality. Research on fattening and carcass traits was performed on piglets sent to the pig testing stations of East and West Flanders. The blood serum enzymes creatine kinase (CK), lactate dehydrogenase (LDH) and the isoenzyme LDH_5 were determined on ten-week-old pigs using methods described earlier (Lampo, 1977).

The halothane test was performed on pigs at eight weeks or older by administering five percent halothane (Fluothane I C I) in an oxygen flow of 3 litres per minute for a maximum period of five minutes.

Polymorphism in the H and S marker systems was detected serologically by means of monospecific antisera (hemolysis and indirect Coombs test). PHI and PGD phenotypes were determined by one-dimensional electrophoresis of erythrocyte hemolysates, while Po2 variants were revealed by two-dimensional electrophoresis of serum.

In addition, twelve other blood group systems and six other electrophoretic systems were included in a parentage control programme in order to check the pedigree data provided.

Activities of CK and LDH were expressed logarithmically, while the arcsin transformation of the percentage of LDH_5 activity was used. Statistical parameters were calculated using the standard methods including analysis of variance and analysis of covariance.

RESULTS AND DISCUSSION

Blood marker determination

In an extensive family analysis of the recombination phenomenon between the different marker systems, including the halothane locus, an estimation of the relative distance between the corresponding genes was performed. These distances, calculated from 2 point crosses and expressed in centimorgans, are presented in Table 1.

As a result of this and additional data (in preparation) gathered mostly from 3 and 4 point crosses, which are even more informative in establishing the relative position of the marker genes in a genetically more uniform animal material, the gene order S - PHI - HAL - H - Po2 - PGD was established.

When selecting for homozygous halothane resistant breeding animals and for the early prediction of an individual's halothane sensitivity, the recombination frequencies between each blood marker and the halothane locus are especially important. Indeed, using blood marker information only, the latter recombination frequencies equal the degree of predictive inaccuracy in practical breeding work against stress susceptibility.

In addition, a particular problem concerns the assignment of the halothane locus into the linkage group. As can be deduced from Table 1, the so-called recombination frequencies (in brackets) between the halothane locus and the other markers reach about 10 to 12 percent, which should mean that, contrary to earlier presumptions, the halothane locus should not fit into the chromosomal region S - PGD. This is certainly due to a systematic effect, namely the incomplete penetrance of the halothane genes, since animals showing incomplete penetrance cannot be distinguished from animals really showing recombination. This leade to an overestimate of the real recombination frequencies. In practice this

TABLE 1   Recombination rates between different markers in the
halothane linkage group in different 2 point crosses

Systems	Parental	Recombinant	Total	Rec. Rate + Deviation (cm)	
S - PHI	166	6	172	3.49	1.40
S - H	525	19	544	3.49	.79
S - Po2	647	13	660	1.97	.54
S - PGD	624	38	662	5.74	.90
PHI - H	896	26	922	2.82	.55
PHI - Po2	1197	36	1233	2.92	.48
PHI - PGD	1063	69	1132	6.10	.71
H - Po2	1742	36	1778	2.02	.33
H - PGD	1444	98	1542	6.36	.62
Po2 - PGD	1696	98	1794	5.46	.54
S - HAL	641	(51)	692	(7.37)	(.99)
PHI - HAL	1487	(170)	1657	(10.26)	(.75)
H - HAL	1671	(180)	1851	(9.72)	(.69)
Po2 - HAL	2191	(240)	2431	(9.87)	(.60)
PGD - HAL	2126	(307)	2433	(12.62)	(.67)

means that the observed recombination rate, marker locus - halothane
locus, is not the real one, but the sum of the real one on one hand and
the inaccuracy due to incomplete penetrance of the halothane gene on the
other hand. Considering the PHI - HAL - H chromosomal region, the
distance PHI-H (= 2.82) should equal the sum of the distances PHI-HAL
and HAL-H (10.26 + 9.72 = 19.98), which is not the case. Since the
halothane locus was involved twice, the difference 19.98 - 2.82 = 17.16
represents twice the loss of penetrance of the halothane locus. Thus
in Belgian Landrace pigs the accuracy of the halothane diagnosis after
halothane testing is only 100 - (17.16/2) = about 91 percent. By also
including determination of blood markers the accuracy is improved to the
level of the marker-HAL recombination rates. Of course, the closer to
the halothane locus the markers used are, the higher the accuracy to be
expected. For example combining halothane testing and PHI-H
determination will improve the accuracy to at least (100 - 2.82) = about
97 percent and more probably to (100-2.82/2) = about 98.5 percent when
supposing the halothane locus to be in the middle of the PHI-H region.
It may be concluded that a better diagnosis of halothane sensitivity in
pigs is obtained by combining halothane testing and blood marker
determination. Moreover, using accumulated haplotype information over

subsequent generations, homozygous and heterozygous halothane negative animals can be distinguished, since the former possess two detectable halothane negative parental haplotypes, while the latter possess only one of such haplotypes, originating from one of the halothane negative parents.

Stress susceptibility and zootechnical parameters

The effect of stress susceptibility was studied in two ways: firstly by calculating the correlation between the amount of CK, LDH and LDH_5 and different growth, carcass and meat characteristics and secondly by comparing absolute values for growth and carcass measurements in halothane positive and negative animals. The results are summarised in Tables 2 and 3.

TABLE 2   Phenotypic correlations between CK, LDH and LDH_5 and some fattening, carcass and meat quality parameters (n = 879, 22 Belgian Landrace sires)

	Log CK	Log LDH	Arc. sin $\sqrt{LDH_5 \%}$
Daily weight gain	0.011[1]	−0.030	0.012
Food conversion	−0.067	−0.079	−0.068
Loin percentage	0.252	0.182	0.254
Ham percentage	0.168	0.064	0.170
Shoulder percentage	0.206	0.116	0.207
Meat percentage	0.261	0.150	0.264
Meat/fat ratio	0.194	0.110	0.197
Backfat thickness	−0.216	−0.201	−0.101
Temperature[2]	0.08	0.09	0.08
pH	−0.31	−0.25	−0.16
Göfo reflectance value	−0.14	−0.31	−0.14

[1]The S D of the correlation coefficient equals 0.032 for all r-values

[2]The measurements were made in M. longissimus dorsi at one hour postmortem

From Table 2 it can be seen that daily weight gain and food conversion are only weakly correlated with the activity of the serum enzymes studied. For daily weight gain this is confirmed by the comparison between halothane positive and halothane negative animals where no difference was found (Table 3). However that comparison shows a somewhat better food conversion in halothane positive animals (Table 3).

These results agree with mean differences found by Webb (1981) of 2 g for daily weight gain and 0.06 units for food conversion. Stress susceptible pigs gave a better yield of butchers joints in both sows and hogs. Concerning meat quality measurements the halothane negative pigs were better than the positive one (Table 3). The differences observed for carcass and for meat quality traits fall within the ranges given by Webb (1981).

TABLE 3   Comparison of production, carcass and meat quality measurements in halothane positive and negative Belgian Landrace pigs

Hogs	Hal + n=418	Hal − n=126	d
Daily weight gain (g)	682 ± 103.6	682 ± 93.8	N S
Food conversion	3.39 ± 0.480	3.54 ± 0.541	* *
Loin percentage	22.6 ± 1.26	21.6 ± 1.41	* *
Ham percentage	23.4 ± 1.16	22.9 ± 1.42	* *
Shoulder percentage	14.0 ± 0.84	13.5 ± 0.76	* *
Backfat thickness (cm)	2.86 ± 0.44	3.12 ± 0.53	*
Meat/fat ratio	6.24 ± 1.22	5.38 ± 1.10	* *
Temperature ($^{\circ}$C)	41.46 ± 0.78	40.88 ± 0.86	* *
pH[2]	5.80 ± 0.30	6.06 ± 0.23	* *
Göfo reflectance value[3]	62.6 ± 7.01	65.5 ± 4.32	*

Sows	Hal + n=431	Hal − n=138	d
Daily weight gain (g)	636 ± 93.7	635 ± 85.0	N S
Food conversion	3.54 ± 0.587	3.62 ± 0.52	N S
Loin percentage	23.5 ± 1.29	22.9 ± 1.44	* *
Ham percentage	24.1 ± 1.37	23.4 ± 1.33	* *
Shoulder percentage	14.1 ± 0.80	13.8 ± 0.77	* *
Backfat thickness (cm)	2.55 ± 0.50	2.81 ± 0.52	* *
Meat/fat ratio	7.61 ± 2.09	6.37 ± 1.70	* *
Temperature ($^{\circ}$C)	41.36 ± 0.28	40.79 ± 0.88	* *
pH[2]	5.82 ± 0.28	6.10 ± 0.22	* *
Göfo reflectance value[3]	62.1 ± 6.16	66.2 ± 4.55	* *

* $P < 0.05$;   ** $P < 0.01$

[1] Kg feed intake/kg liveweight gain

[2] Measured in M. longissimus dorsi at one hour post mortem

[3] Mean Göfo value for M. gluteus medius and M. longissimus dorsi

The effect of the halothane gene on some reproduction traits is treated in detail in a separate publication (Lampo et al., 1985) and only the conclusions are presented here. There was no difference in litter size at birth between HP and HN sows in the first farrowing. For the second litter HN sows gave 0.4 piglets more (9.6 vs 9.2) than HP ones, while for the following farrowings the difference was one piglet (10.3 in HN vs 9.3 in HP). At weaning there was an additional gain of 0.2 piglets in all litters.

CONCLUSIONS

In addition to mass selection against stress susceptibility, based on both halothane testing and variation of gene frequencies in blood marker loci, more refined family analysis over subsequent generations can be performed. By this means a diagnosis of sensitivity to halothane in pigs can be reduced. Indeed problems of incomplete penetrance in halothane expression are occurring. Thus, penetrance reached only 0.91 in the present experiments with Belgian Landrace pigs.

By means of blood marker determination, piglets showing incomplete penetrance are detectable. Depending on the chromosomal position and on the polymorphism of the blood marker loci involved, an accuracy in halothane diagnosis up to 98.5 percent can be obtained. Using haplotype information from halothane negative parents it is also possible to distinguish between homozygously and heterozygously halothane resistant animals.

A selection programme for a more resistant Belgian Landrace pig will result in a lower percentage of lean. Although meat quality and reproduction should be better, we cannot advise such a selection scheme be undertaken under present circumstances, because Belgian exports are based on the carcass yield more than on other characteristics. Nevertheless, Belgian Landrace breeders want better reproduction results. A proposed solution is the foundation of a new breed, called the 'HN'. The foundation stock consists of halothane negative Belgian Landrace pigs selected for better growth and reproduction. Sows of the HN breed are being crossed with heavily muscled (mostly HP) Belgian Landrace boars to obtain heterozygote HN piglets and advantages in the numbers born and weaned and in muscularity, while still retaining good meat quality.

444

ACKNOWLEDGEMENT

This research was supported by the I W O N L Foundation.

REFERENCES

Andresen, E. 1971. Linear sequence of the autosomal loci PHI, H and 6-PGD in pigs. Anim. Blood Grps biochem. Genet., 2, 119-120.

Andresen, E. 1981. Evidence for a five-locus linkage group involving direct and associate interactions with the A-O blood group locus in pigs. In: E. Brummerstedt (Ed), Papers dedicated to Prof. J. Moustgaard, Royal Danish Agric. Soc., Copehagen, 208-212.

Bickhardt, K. 1969. Ein enzymatisches Verfahren zur Erkennung von Muskelschäden beim lebenden Schwein. Dtsch. Tieräzl. Wschr. 76, p. 601 and 699.

Bickhardt, K., Flock, D.K., Richter, L. 1977. Creatine-Kinase test as a selection criterion to estimate stress resistance and meat quality in pigs. Vet. Sci. communs. 1, 225-233.

Eikelenboom, G. and Minkema, D. 1974. Prediction of pale, soft, exudative muscle with a non-lethal test for the halothane-induced porcine malignant hyperthermia syndrome. J. Vet. Sci. 91, 421-426.

Hanset, R., Leroy, P., Michaux, C., Kintaba, K.N. 1983. The Hal locus in the Belgian Pietrain breed. Z. Tierz. u. Züchgsbiol. 100, 123-133.

Imlah, P. 1982. Linkage studies between the halothane (Hal), phosphohexose isomerase (PHI) and the S(A-O) and H red blood cell loci of Pietrain/Hampshire and Landrace pigs. Anim. Blood Grps. biochem. Genet., 13, 245-262.

Juneja, R.K., Gahne, B., Edfors-Lilja, I., Andresen, E. 1983. Genetic variation at a pig serum protein locus, Po-2 and its assignment to the Phi, Hal, S, H, Pgd linkage group. Anim. Blood Grps. biochem. Genet., 14, 27-36.

Lampo, Ph. 1977. La créatine phosphokinase et la déhydrogenase lactique comme indicateurs du stress chez le porc. Rev. Agr. 30, 382-402.

Lampo, Ph. 1978. Quantification of the Halothane-anaesthesia test for stress susceptibility diagnosis in pigs. Proceeding 5th I.P.V.S. Congress Zagreb (Yugoslavia), publ. K.A. 28.

Lampo, Ph. 1981. La sensibilité au stress chez le Landrace Belge. La relation entre le test d'anaesthésie à l'Halothane, les caracthéristiques d'engraissement et de la carcasse. Rev. Agr. 34, 213-220.

Lampo, Ph., Nauwynck, W., Bouquet, Y., Van Zeveren, A. 1985. Effect of stress susceptibility on some reproductive traits in Belgian Landrace Pigs. Livest. Prod. Sci. (In press).

Rasmusen, B.A., 1981. Linkage of genes for PHI, halothane sensitivity, A-O inhibition, H red blood cell antigens and 6-PGD variants in pigs. Anim. Blood Grps. biochem. Genet., 12, 207-209.

Schneider, A., Schwörer, B., Blum, J. 1980. Beziehung der Halothan genotyps zu den Produktions- und Reproduktionsmerkmalen der Schweizerischen Landrasse. 31st Ann. Meeting E.A.A.P. Mnchen, Germany, G.P. 3.9.

Van Zeveren, A., Bouquet, Y., Hojny, J., Van de Weghe, A., Varewyck, H. 1983. Genetic variability expressed by marker-systems in 2 Belgian pig populations. Livestock Prod. Sci., 10, 373-386.

Van Zeveren, A., Van de Weghe, A., Bouquet, Y., Varewyck, H. 1985. The position of the epistatic S locus in the halothane linkage group in pigs. Anim. Blood Grps. biochem. Genet., 16, (in press).

Webb, A.J. 1981. The halothane sensitivity test. Proceedings Symposium Porcine stress and meat quality. Refness gods, Jeloy, Norway, 1980, 105-124.

Webb, A.J. and Jordan, C.H.C. 1978. Halothane sensitivity as a field test for stress susceptibility in the pig. Anim. Prod. 26, 157-168.

GLYCOLYTIC POTENTIAL, HALOTHANE SENSITIVITY AND MEAT QUALITY
IN VARIOUS PIG BREEDS

G. Monin[1] and P. Sellier[2]

1 - Station de Recherches sur la Viande - I.N.R.A. - Theix - 63122 Ceyrat
2 - Station de Génétique Quantitative et Appliquée - I.N.R.A - 78350 Jouy
   en Josas - France

ABSTRACT

Three experiments were designed to clarify the relationships between halothane sensitivity, ultimate pH (mainly depending on muscle glycolytic potential) and meat quality in the five porcine breeds extensively used for crossbreeding in France, i.e. Large White, Pietrain, French Landrace, Belgian Landrace and Hampshire. The low technological quality of meat from Hampshires was explained by the low ultimate pH characterizing this breed. Conversely, in Landrace and especially in the Belgian Landrace, the high value of meat ultimate pH prevented to some extent the unfavourable influence of halothane susceptibility on meat quality. It is concluded that there is a need to further study the factors governing muscle glycolytic potential at slaughter, especially those controlling resting glycolytic potential.

INTRODUCTION

Cooked cured ham (Paris ham) is one of the most important modes of pig carcass utilization in France, since more than 80 % of the hams are treated in this way. So the technological quality of pig meat is generally appreciated as the Paris ham processing ability. This ability largely depends on the post mortem evolution of muscle pH, i.e. the rate and the total extent of pH drop. There is some evidence that the latter characteristic plays the major role (e.g. Jacquet et al., 1984) and for this reason ultimate pH -and not pH at 45 or 60 minutes post mortem- is taken into account in the 3-trait Meat Quality Index (MQI) incorporated since 1981 in the aggregate genotype used in France for the genetic improvement of growth and carcass traits (see Tibau i Font and Ollivier, 1984). Analysis of records collected in central progeny-test stations during recent years, as well as results of earlier breed comparisons (Dumont, 1974 ; Goutefongea et al., 1977) indicate that Belgian Landrace pigs give on average a meat of satisfying technological quality, in spite of the well-known high incidence of halothane susceptibility in this breed (e.g. Ollivier et al., 1978). Conversely, it has been repeatedly found in the French "commercial product evaluation" programme that pigs from crossbreeding schemes using the Hampshire breed as a component of the sire line show a markedly inferior meat quality (Runavot and Sellier, 1984), which was also noted by Sellier (1982) on the basis of earlier comparisons between sire lines carried out at the Institut National de lu Recherche Agronomique. However, the Hampshire breed is likely to be free of the halothane sensitivity gene.

A series of experiments were designed to clarify the relationships between halothane sensitivity, ultimate pH (mainly depending on muscle glycolytic potential) and meat quality in the five porcine breeds extensively used for crossbreeding in France, i.e. Large White, Pietrain, French Landrace, Belgian Landrace and Hampshire. Some of the results presented here (experiments 1 and 2) were reported previously (Sellier et al., 1984 ; Monin and Sellier, 1985).

HAMPSHIRE BREED

The comparison including purebred Hampshire pigs is referred to as experiment 1.

Material and methods

One hundred and twenty nine pigs (47 Large Whites, 20 halothane negative (HN) Pietrains, 27 halotane positive (HP) Pietrains and 35 Hampshires) were fattened from 25-30 kg liveweight in an experimental farm of our Institute. All Large Whites and Hampshires were halothane negative. The animals were killed in six weekly series of 16 to 25 pigs, after a transportation lasting about half an hour, at around 103 kg liveweight.

One hour after slaughter, a sample of M. longissimus dorsi (LD) was removed from 56 pigs in 3 series, using a punch, to measure pH (0.005 M iodoacetate technique, referred to as pH_1) and glycolytic potential as the sum of the concentrations of glycogen, glucose-6-phosphate, glucose and lactic acid (Monin et al., 1981). On the day after, the right-hand sides of the 129 pigs were cut and the following measurements were taken, among others :

- pH (pH_2) in M. longissimus dorsi (LD), M. adductor femoris (AF), M. biceps femoris (BF) ;
- reflectance on BF and M. gluteus superficialis (GS), using a Retrolux reflectometer ;
- water holding capacity (WHC, assessed by the paper imbibition technique) on BF muscle ;
- fibre optic probe value (Fop, measured using TBL equipment) on LD, BF and

AF muscles.

A MQI index was calculated according to Jacquet et al. (1984). The hams were processed into Paris ham by using a brine containing 150 g of NaCl and 0.9 g of $NaNO_2$ per kg and by cooking to a core temperature of 68°C. Technological yield was calculated as 100 (weight of cooked ham/weight of trimmed deboned ham).

Results

Results are shown in Table 1.

The glycolytic potential in LD muscle was much higher in Hampshires than in the other genetic types (P < 0.001). The two types of Pietrains showed a lower pH_1 than both Large Whites and Hampshires, especially HP Pietrains which differed at the P < 0.001 level from the three other types. No difference was observed between Large Whites and Hampshires in this respect.

The pH_2 varied largely between breeds whatever the muscle ; as a general rule, the Large whites had the highest pH and the Hampshires the lowest ones. The effect of halothane sensitivity on pH_2 was on average of low magnitude, as judged by the differences between HP and HN Pietrains.

Halothane-negative Pietrains had the most coloured BF muscle, whereas HP Pietrains showed the palest one. The water holding capacity was the highest in Large Whites and the lowest in HP Pietrains, being intermediate in Hampshires and HN Pietrains. Fibre optic probe values were strongly affected by the genetic type : Hampshires and HP Pietrains generally showed the highest values, whereas HN Pietrains and Large Whites had rather similar values .

Meat Quality Index was the highest in Large Whites and the lowest in Hampshires and HP Pietrains, which did not differ in this respect. The technological yield was found to be the highest in HN Pietrains, which were significantly better than both Hampshires and HN Pietrains, with Large Whites being intermediate.

The high glycolytic potential found here in Hampshire pigs agrees with previous observations by Sayre et al. (1963) of a very high content of glycogen in the LD muscle of this breed. The results on the meat quality of Hampshire pigs confirm numerous reports of European or American researchers.

Finally, it is possible to divide the animals of experiment 1 in two groups according to meat quality : on one hand the Large Whites and the HN

TABLE 1 Glycolytic potential and meat quality characteristics of Large White, HN and HP Pietrain and Hampshire pigs (experiment 1 ; from Monin and Sellier, 1985)

Means with the same superscript are not significantly different at the 5 % level

Trait	rsd(1)	Large White	HN Pietrain	HP Pietrain	Hampshire
No. of pigs		10	7	9	14
Glycolytic potential LD(2)	34	136^a	159^a	161^a	229^b
No. of pigs		15	11	13	17
pH_1 LD	0.28	6.39^a	6.08^b	5.56^c	6.36^a
No. of pigs		47	20	27	35
pH_2 LD	0.08	5.53^a	5.45^b	5.41^c	5.40^c
AF	0.16	5.75^a	5.65^a	5.74^a	5.45^b
BF	0.10	5.61^a	5.49^b	5.54^b	5.36^c
Reflectance BF	35	353^b	314^a	375^{bc}	376^c
GS	50	342^a	330^a	417^c	389^b
WHC BF	5.2	18.1^a	11.8^b	6.2^c	13.3^b
Fop value LD	11	32^a	33^a	49^c	43^b
AF	9	34^a	35^a	37^a	50^b
BF	10	44^{ab}	41^a	52^c	51^{bc}
MQI	1.5	87.5^a	86.2^b	85.2^c	84.7^c
Technological yield %	3.3	82.7^{ab}	84.7^a	81.5^b	81.2^b

(1) Residual standard deviation from an analysis model including the following effects : genetic type, sex, slaughter date, genetic type x sex and genetic type x date interactions, and linear regression on liveweight at slaugther

(2) In terms of potential lactate formation, in μmol.(g fresh tissue)$^{-1}$

Pietrains, on the other hand the Hampshires and the HP Pietrains. The latter group shows deficiencies in various traits : pale colour in HP Pietrains and Hampshires, reduced water holding capacity in HP Pietrains and low technological yield of Paris ham processing in Hampshires. However, there is some evidence that the mechanisms of the defects are different in HP Pietrains and Hampshires. The low meat quality of HP Pietrains undoubtedly results from the fast post mortem pH drop in muscles, whereas that of Hampshires appears to be a consequence of the low ultimate pH. In fact, the high correlations between ultimate pH on one hand, fresh meat quality traits or technological yield of Paris ham processing on the other hand are well-established (e.g. Charpentier et al., 1971 ; Jacquet and Ollivier, 1971 ; Jacquet et al., 1984). The low ultimate pH observed in meat from Hampshires is probably related to the high muscle glycolytic potential characterizing this breed.

BELGIAN LANDRACE BREED

Two experiments involving Belgian Landrace pigs were carried out.

Experiment 2

Material and methods

Measurements were made in one progeny-test station on 1053 gilts from the Large White, French Landrace, Pietrain and Belgian Landrace breeds. All the pigs were halothane-tested at a weight of 25-30 kg.

The pigs were killed at liveweights of about 100 kg for Large Whites and Landrace and of about 90 kg for Pietrains, in a commercial abattoir. The transportation lasted around one hour and a half, and the pigs were slaughtered within the 30 minutes following arrival. The day after slaughter, the right-hand sides were cut and some meat quality measurements were made according to the techniques described above.

Results

Results are reported in Table 2.

Both Landrace types presented the highest pH_2 and Pietrains the lowest one. Among HN pigs, water holding capacity and reflectance were similar in every breed, except the case of water holding capacity in French Landrace, which was slightly lower than in the other breeds. MQI was similar in Large Whites and both Landraces, but was significantly lower in Pietrains ; this agrees with results of comparisons by Jacquet and Ollivier (1971 ; Pietrain vs. Large White) and Goutefongea et al. (1977 ; Pietrain vs. Belgian and French Landrace) who found that Pietrains constantly have the lowest technological yields.

Halothane sensitivity did not influence pH_2. Water holding capacity was lower and reflectance higher in HP pigs than in HN pigs whatever the breed. However, the difference was more marked in Pietrains than in both

TABLE 2   Meat quality characteristics in Large White and in HN and HP French Landrace, Belgian Landrace and Pietrain pigs (from Sellier et al., 1984)

Means with the same superscript are not significantly different at the 5 % level
ns non significant   & P< 0.10   * P< 0.05   ** P< 0.01   *** P< 0.001

Trait	rsd(1)	Large White HN	French Landrace HN	French Landrace HP	Belgian Landrace HN	Belgian Landrace HP	Pietrain HN	Pietrain HP	HP − HN(2)
No. of pigs		365	203	41	67	330	7	40	
pH$_2$ AF	0.26	5.99^{ab}	6.04^{bc}	6.09^{c}	6.08^{c}	6.06^{c}	5.81^{a}	5.91^{a}	
HP − HN		—	0.05ns		− 0.02ns		0.10ns		0.05ns
WHC BF	5.2	14.6^{a}	12.6^{bcd}	10.4^{e}	13.7^{ab}	12.3^{cd}	15.3^{abc}	10.5^{de}	
HP − HN		—	− 2.2*		− 1.4*		− 4.8*		− 2.8***
Reflectance GS	89	554^{a}	561^{ab}	588^{b}	545^{a}	575^{b}	597^{ab}	688^{c}	
HP − HN		—	27&		30*		91*		49***
MQI	2.6	86.5^{a}	86.3^{a}	86.0^{a}	86.9^{a}	86.2^{a}	85.1^{ab}	84.1^{b}	
HP − HN		—	− 0.3ns		− 0.7&		− 1.0ns		− 0.7ns

(1) Analysis model including the following effects : genetic type, control group, slaughter date and linear regression on liveweight at slaughter
(2) Difference between HP and HN pigs pooled over French Landrace, Belgian Landrace and Pietrain breeds

Landraces. Halothane sensitivity influences meat quality by accelerating post mortem pH fall and so inducing muscle protein denaturation. Such an effect requires that pH reaches a rather low value (Charpentier, 1969), and it can be assumed that this effect is even more diminished if the pH keeps a high value, as previously pointed out by Monin et al. (1981). It appears that in Landrace pigs and especially in the Belgian Landrace, the high value of ultimate pH prevents to some extent the unfavourable influence of halothane sensitivity on colour and water holding capacity.

A further experiment was set up to establish if the differences in ultimate pH - and consequently related meat characteristics - between Pietrains, Large Whites and Belgian Landrace are related to differences in glycolytic potential at slaughter.

Experiment 3
Material and methods

The experiment used 145 gilts reared in 3 progeny-test stations, i.e. 45 Large Whites, 9 HN and 33 HP Pietrains, 16 HN and 42 HP Belgian Landrace. All the pigs were tested for halothane reaction at the weight of 25-30 kg. They were killed at a liveweight of about 100 kg for the Large Whites, 97 kg for the Belgian Landrace and 90 kg for the Pietrains, in six slaughter series at 3 commercial abattoirs. Transport and slaughter conditions greatly varied between abattoirs, but in every case the pigs were slaughtered in the hour following arrival. All genetic types, except HP Pietrains in two series, were represented in each of the 6 series.

One hour on average after slaughter, a sample of M. semimembranosus (SM) was taken using a punch from the right-hand side of 101 pigs (15 to 20 pigs in each series). pH (pH_1) was measured after homogenization of muscle in 0.005 M iodoacetate ; a part of the sample was frozen in liquid nitrogen for subsequent estimation of glycolytic potential.

On the day after slaughter, the right-hand sides of the 145 pigs were cut and meat quality measurements were made according to the techniques previously described. The ham was processed into Paris ham and the technological yield was determined.

Results

Results are shown in Tables 3 and 4.

As in experiment 1, halothane sensitivity did not affect glycolytic otential. There was a trend to a higher glycolytic potential in Pietrains,

TABLE 3 Glycolytic potential and pH in muscles of Large White and HN and HP Pietrain and Belgian Landrace pigs (experiment 3)

Means with the same superscript are not significantly different at the 5 % level
ns non significant   & P<0.10   * P<0.05   ** P<0.01   *** P<0.001

Trait	rsd(1)	Large White HN	Pietrain HN	Pietrain HP	Belgian Landrace HN	Belgian Landrace HP	HP - HN(2)
No. of pigs		28	7	23	15	28	
Glycolytic potential SM	24	99^a	114^a	109^a	93^a	102^a	
HP - HN		-		- 5ns		9ns	2ns
pH_1 SM	0.34	6.18^c	5.97^{abc}	5.85^a	6.00^{bc}	5.77^a	
HP - HN		-		- 0.12ns		- 0.23*	- 0.17&
No. of pigs		45	9	33	16	42	
pH_2 LD	0.17	5.55^{ab}	5.47^a	5.50^a	5.54^{ab}	5.60^b	
HP - HN		-		0.03ns		0.06ns	0.04ns
pH_2 AF	0.30	5.97^{bc}	5.79^{ab}	5.82^a	6.09^c	6.03^c	
HP - HN		-		0.03ns		- 0.06ns	- 0.01ns
pH_2 SM	0.25	5.80^a	5.74^a	5.76^a	5.84^a	5.87^a	
HP - HN		-		0.02ns		0.03ns	0.02ns

(1) Analysis model including the following effects : genetic type, control station, slaughter date and linear regression on liveweight at slaughter
(2) Difference between HP and HN pigs pooled over Belgian Landrace and Pietrain breeds

TABLE 4   Meat quality characteristics in Large White and HN and HP Pietrain and Belgian Landrace pigs (experiment 3)

Means with the same superscript are not significantly different at the 5 % level
* $P < 0.05$ ** $P < 0.01$ *** $P < 0.001$
& $P < 0.10$ ns non significant

Trait	rsd(1)	Large White HN	Pietrain HN	Pietrain HP	Belgian Landrace HN	Belgian Landrace HP	hP - HN(2)
No. of pigs		45	9	33	16	42	
Reflectance BF	75	253^a	289^{ab}	327^b	258^a	284^a	
HP - HN		-		38ns		26ns	31&
Reflectance GS	95	323^a	388^{ab}	432^b	375^{ab}	365^a	
HP - HN		-		44ns		- 10ns	17ns
WHC BF	5.1	12.6^c	7.7^{ab}	7.9^a	11.5^{bc}	12.2^c	
HP - HN		-		0.2ns		0.7ns	0.5ns
Fop LD	16	31^a	46^b	59^c	40^{ab}	49^b	
HP - HN		-		13*		9&	11*
Fop AF	14	28^a	46^{ab}	46^b	28^a	30^a	
HP - HN		-		0ns		2ns	1ns
Fop BF	14	34^a	40^{ab}	50^c	36^{ab}	42^b	
HP - HN		-		10*		6ns	8*
MQI	3.0	86.0^c	83.2^{ab}	83.1^a	85.7^c	85.2^{bc}	
HP - HN		-		0.1ns		- 0.5ns	- 0.3ns
Technological yield (%)	3.7	86.1^c	81.1^a	81.5^a	83.6^{ab}	84.4^b	
HP - HN		-		0.4ns		0.8ns	0.6ns

(1), (2) : see Table 3

which approached significance (P < 0.10) when the 3 breeds are compared irrespective of halothane sensitivity. pH fall was faster in HP pigs than in HN pigs, but to a lesser extent than in experiment 1. There was also a general tendency to find the highest pH_2 values in HN or HP Belgian Landrace and the lowest ones in HN or HP Pietrains. As halothane sensitivity did not affect pH_2, the 3 breeds can be compared without taking account of halothane sensitivity : Belgian Landrace gave significantly higher pH_2 than Pietrains whatever the muscle, whereas the difference between Large Whites and Pietrains was significant at the 5 % level only in AF muscle. Colour and water holding capacity were generally similar in Belgian Landrace and Large White pigs. Pietrain pigs, especially the HP ones, showed less favourable values. Fibre optic probe values were higher in LD and BF muscles from Pietrains and HP Belgian Landrace than in those from Large Whites ; in these muscles, the differences between HP and HN pigs were more marked than in AF muscle, which presented a higher pH_2. When ignoring halothane sensitivity which did not significantly affect this trait, the technological yield of Paris ham processing was the highest in Large Whites (86.1 ± 0.6), intermediate in Belgian Landrace (84.2 ± 0.5) and the lowest in Pietrains (81.4 ± 0.6).

The results from the comparison confirmed and extended those of experiment 2 : Belgian Landrace pigs, irrespective of halothane sensitivity, showed meat quality traits close to those observed in Large Whites, except for fibre optic probe values. Pietrain pigs associated a lower meat quality with a tendency toward a lower pH_2 and a higher glycolytic potential at slaughter.

CONCLUSION

Jacquet et al. (1984) reported that, in the slaughter conditions prevailing in France, ultimate pH plays a greater role than rate of post mortem pH drop in determining pork technological quality. The results of the present study agree with this observation. Ultimate pH mainly depends on muscle glycolytic potential at slaughter, whereas the rate of pH fall is primarily influenced by halothane sensitivity. The genetic basis of the latter and the implications for meat quality are now rather well-known, and this source of deficient meat quality is now coming under better control with the use of the halothane-test in testing stations or breeding herds, and with appropriate use of crossbreeding. The glycolytic potential at slaughter is determined by both resting glycolytic potential and glucidic

catabolism during the preslaughter period. The latter has attracted more
attention from research workers, principally as influenced by the nature
and the conditions of the preslaughter treatment (for a review, see
Barton-Gade, 1985) ; such factors as breed x treatment interactions, which
could be very important in the light of the work of Dantzer and Mormède
(1978), has not received as much consideration. There is also a need to
take into consideration the factors governing the glycolytic potential of
the resting muscle, in order to get as many ways as possible of controlling
glycolytic potential at slaughter and in this way controlling the ultimate
quality of the meat.

ACKNOWLEDGEMENTS
    The authors are grateful to all persons who have participated in
collecting data of the 3 experiments reported here. Special thanks are due
to Dominique Laborde (C.E.M.A.G.R.E.F.), A. Talmant and P. Vernin
(I.N.R.A., Theix)), J. Gruand (I.N.R.A., Rouillé), B. Jacquet
(C.T.S.C.C.V., Jouy en Josas), D. Brault (I.N.R.A., Le Rheu), Y. Houix
(I.T.P., Le Transloy) and M. Renault (I.T.P., Le Deschaux).

REFERENCES

Barton-Gade, P., 1985. Developments in the preslaughter treatment of
    slaughter animals.31st Europ. Meet. Meat Res. Workers, 1.1, 1-6.
Charpentier, J., 1969. Influence de la température et du pH sur quelques
    caractéristiques physico-chimiques des protéines sarcoplasmiques du
    muscle de porc. Conséquences technologiques. Ann. Biol. anim. Bioch.
    Biophys., 9, 101-110.
Charpentier, J., Monin, G. and Ollivier, L., 1971. Correlations between
    carcass characteristics and meat quality in Large White pigs. 2nd Int.
    Symp. on Condition and Meat Quality of Pigs, Pudoc, Wageningen, 255-
    260.
Dantzer, R. and Mormède, P., 1978. Behavioural and pituitary adrenal
    characteristics of pigs differing by their susceptibility to the
    malignant hyperthermia syndrome induced by halothane anaesthesia. Ann.
    Rech. Vét., 9, 559-567.
Dumont, B.L., 1974. Propriétés sensorielles et qualités technologiques de
    la viande de trois races (Landrace Belge, Landrace Français et
    Piétrain. Journées Recherche Porcine en France, 6, 233-239.
Goutefongea, R., Jacquet, B. and Sellier, P., 1977. Influences génétiques
    et non-génétiques sur l'aptitude du jambon à la transformation. 23rd
    Europ. Meet. Meat Res. Workers, A.11.
Jacquet, B. and Ollivier, L., 1971. Résultats d'une expérience de
    croisement Piétrain x Large White. 2. Aptitude du jambon à la
    transformation en jambon de Paris. Journées Recherche Porcine en
    France, 3, 23-33.
Jacquet, B., Sellier, P., Runavot, J.P., Brault, D., Houix, Y. Perrocheau,
    C., Gogué, J. and Boulard, J., 1984. Prediction of the technological
    yield of Paris ham processing by using measurements at the abattoir.
    Scientific Meeting on Biophysical PSE-muscle analysis, Ed. H.
    Pfützner, Technical University of Vienna, 143-153.

458

Monin, G. and Sellier, P., 1985. Pork of low technological quality with a normal rate of muscle pH fall in the immediate post mortem period : the case of the Hampshire breed. Meat Sci., 13, 49-63.

Monin, G., Sellier, P., Ollivier, L., Goutefongea, R. and Girard, J.P., 1981. Carcass characteristics and meat quality of halothane-negative and halothane-positive pigs. Meat Sci., 5, 413-423.

Ollivier, L., Sellier, P. and Monin, G., 1978. Fréquence du syndrome d'hyperthermie maligne dans des populations porcines françaises ; relations avec le développement musculaire. Ann. Génét. Sél. anim., 10, 191-208.

Runavot, J.P. and Sellier, P., 1984. Bilan des dix tests de contrôle des produits terminaux réalisés en France de 1970 à 1983. Journées Recherche Porcine en France, 16, 425-438.

Sayre, R.N., Briskey, E.J. and Hoekstra, W.G., 1963. Comparison of muscle characteristics and post mortem glycolysis in three breeds of swine. J. Anim. Sci., 22, 1012-1020.

Sellier, P., 1982. Le choix de la lignée mâle du croisement terminal chez le Porc. Journées Recherche Porcine en France, 14, 159-182.

Sellier, R., Monin, G., Houix, Y. and Dando, P., 1984. Qualité de la viande de quatre races porcines : relation avec la sensibilité à l'halothane et l'activité créatine phosphokinase plasmatique. Journées Recherche Porcine en France, 16, 65-74.

Tibau i Font, J. and Ollivier, L., 1984. La sélection en station chez le Porc. Bull. Tech. Dépt. Génét. Anim. INRA, 37, 69-75.

MEAT QUALITY IN THE DUROC BREED

P McGloughlin, P V Tarrant & P Allen

The Agricultural Institute
Grange/Dunsinea Research Centre
Dunsinea, Castleknock
Dublin 15, Ireland

ABSTRACT

Meat quality traits are presented for Duroc sired crossbred pigs in comparison with Landrace and Large White crossbreds. This work is part of an overall evaluation of the Duroc as a terminal sire breed. Objective and subjective measurements indicated that meat quality was good in both crossbred groups. The Duroc crosses had significantly paler M. longissimus dorsi, M. gluteus medius and M. semimembranosus, while the Landrace x Large White crosses had a tendency towards DFD, indicated by significantly higher ultimate pH values in M. longissimus dorsi at the 3rd/4th rib and in the ham muscles. Differences between the breed groups for carcass composition were determined and Duroc crosses had significantly less fat than Landrace x Large White crosses. Phenotypic correlations between meat quality and carcass composition indicated that leaner pigs had paler meat.

INTRODUCTION

The American Duroc breed is currently being evaluated as a terminal

sire in comparison with the Irish Landrace and Large White breeds

(McGloughlin et al. 1986). This paper presents results on meat quality

and its relationship with carcass composition.

MATERIALS AND METHODS

The evaluation of the Duroc breed as a terminal sire was carried out

in a herd where all sows are the products of a continuous reciprocal back-

crossing programme using the Irish Landrace (L) and Large White (W) breeds;

all sows are therefore either (2/3 L.1/3W) or (1/3L.2/3 W). Eleven Duroc

sires were used to produce three-way cross progeny (Duroc X) and were

compared to reciprocal backcross progeny of five Landrace and six Large

White boars (Landrace.Large White X). The sample of Duroc sires included

seven from North American A.I. studs (frozen semen) and four from Irish

breeders (live boars); the Landrace and Large White sires were performance

tested at a national testing station and had average indices of 129 points

(1.2 standard deviations above their contemporary breed average). A

total of 184 litters were produced on this trial. At weaning one male

pig from each litter (the middle weight male pig in the litter) was

removed and individually fed to slaughter weight. Remaining pigs were

allocated to a group feeding regime, and are not included in this present-

ation. From 28 kg to approximately 86 kg, the sample of boar pigs were

penned in groups of 8 and individually fed, with access to feeders for
one hour periods 13 times per week. On reaching slaughter weight (86 $\pm$
4 kg), pigs were transported in small groups (2 to 13 pigs)   190 km
to the experimental slaughter unit at Dunsinea; the journey lasted
approximately 3 hours. Transport groups were always mixed, i.e. composed
of pigs from more than one pen. On arrival, the length of time in
lairage before slaughter was recorded (84 $\pm$ 30 min.). For slaughter, pigs
were herded into a small pen and stunned while standing, using a double
handed tongs and a low voltage (90 v) power supply (50 Hz). The following
meat quality measurements were recorded. With the exception of pH, all
measurements were made on the left side of the carcass only.

pH_1 pH measured in M. longissimus dorsi (LD) at the level of the
last rib, 45 min. after slaughter, averaged for readings on both
sides of the split carcass.

FOP_1 light scattering coefficient measured using the fibre optic
probe (TBL Fibres, Leeds, UK), inserted in the LD at the level
of the last rib at 1 h after slaughter - the average reading
across the muscle was recorded.

FOP_u light scattering coefficient at about 24 h after slaughter,
measured as for FOP_1, but on the cut surfaces of the LD at the
shoulder (3rd/4th rib, mean of 3 readings) loin (last rib, mean
of 4 readings) and in M. gluteus medius (GM) and M. semimembranosus (SM) in the ham.

pH_u ultimate pH measured at about 24 h after slaughter, in the LD
muscle at the shoulder and loin sites, and in the GM and SM
muscles of the ham.

Panel
Score   Pork quality subjectively assessed by a panel of three or more
experienced judges and expressed on a scale from 1 for extreme
DFD to 5 for extreme PSE (2 = mild DFD, 3 = normal, 4 = mild
PSE). Panellists scored the exposed surfaces of the LD muscle
and the GM muscle approximately 30 min. after cutting on the
day after slaughter.

Drip
loss   drip loss determined on a 2.5 cm thick loin chop cut at the
level of the last rib at about 24 h after slaughter. The fat
and adjoining muscles were trimmed off without removing the
connective tissue on the LD. The sample was suspended within
a closed bag and the weight of drip loss in 48 h at 4°C was

recorded and expressed as a percentage of the weight of the
LD muscle in the chop

pH ratio pH_l/pH_u where both measurements were taken at the level of the
last rib on the left side of the carcass; a low ratio was
indicative of poor meat quality.

Backfat was measured on the left side of the split hot carcass, using
an optical probe at the 3rd last rib, the last rib and the 3rd lumbar
vertebra, 6.5 cm from the mid line; average backfat was the mean of the
three readings. "Eye muscle" area was measured on the cut surface of
the LD muscle at the last rib. The left side of a proportion of the
carcasses was fully dissected into lean, fat, bone and skin. Total fat
(subcutaneous and intramuscular) and lean were expressed as percentages
of the side. Intramuscular fat, in a sample of LD taken at the last rib,
was extracted by the Soxhlet method using petroleum ether as a solvent.
A total of 173 pigs were slaughtered of which 64 were dissected (32 Duroc
X and 32 Landrace.Large WhiteX). In the statistical analysis of the data
the effects of breed of sire (Duroc, Landrace, Large White), sire within
breed, month of slaughter and length of holding period before slaughter
were included. A mean value for the Landrace and Large White-sired pigs
was calculated by combining the separate estimates for the two breeds
giving both equal weight. Phenotypic correlations between meat quality
and carcass traits were based on residual variances.

RESULTS AND DISCUSSION

The meat quality measurements of the Durox X and other Landrace.Large
White X pigs are given in Table 1. In the LD muscle at the mid-loin,
FO P, and FOP_u values were significantly higher in the Duroc X pigs,
indicating paler meat. There were no significant differences in pH
measurements, drip loss or panel score. At the shoulder site of the LD
muscle and in the two ham muscles there were significant differences between
the breed types for three measurements; Duroc X pigs had higher FOP_u values
and panel scores, and lower pH_u values, all of which are indicative of
paler meat. The absolute size of the differences were small, and overall

the sample was relatively free of pork quality defects, as judged by the
panel. Only 1% of pork loins were judged to be mild PSE in both breeds
and 4% to be mild DFD in the Landrace.Large White X (Table 2).

The percentages of FOP_u values that were outside the range for normal
pork are shown in Table 3. There were low incidences of values indicative

TABLE 1   Meat quality of Duroc cross and Landrace.Large White cross pigs

Meat Quality	Duroc X	Landrace.Large White X	$S \bar{x}$	
LD shoulder[1]				
FOP_u	38.9	33.5	1.8	**
pH_u	5.65	5.76	.03	**
Panel Score[2]	3.17	2.99	.05	**
LD loin[1]				
pH_l	6.26	6.28	.03	NS
FOP_l	11.4	10.5	.2	**
FOP_u	31.0	27.3	1.1	**
pH_u	5.57	5.61	.02	NS
pH_u ratio	1.123	1.120	.008	NS
Panel score[2]	3.07	3.01	.03	NS
Drip loss (percentage)	2.64	2.54	.24	NS
GM[1]				
FOP_u	30.1	27.3	1.0	*
pH_u	5.70	5.82	.03	***
Panel score[2]	3.08	2.93	.04	***
SM[1]				
FOP_u	33.7	28.5	1.2	***
pH_u	5.70	5.79	.03	*
Number of pigs	83	90		
Number of sires	11	11		

[1]LD shoulder: M. longissimus dorsi at the 3rd/4th rib; LD loin: M. longissimus dorsi at the last rib; GM: M. gluteus medius; SM: M. semimembranosus.

[2]1 = extreme DFD, 2 = mild DFD, 3 = normal, 4 = mild PSE, 5 = extreme PSE

TABLE 2: Classification of carcasses for meat quality at three muscle sites according to panel score (percentages)

Panel Score	Duroc X			Landrace.Large White X		
	Shoulder[1]	Loin[1]	Ham (GM)[1]	Shoulder	Loin	Ham (GM)[1]
< 1.5 (Extreme DFD)	0	0	0	0	0	0
1.5-2.4 (mild DFD)	0	0	0	4	4	8
2.5-3.5 (normal)	88	99	93	94	95	89
3.6-4.5 (mild PSE)	12	1	7	2	1	3
> 4.5 (Extreme PSE)	0	0	0	0	0	0

[1]See Table 1, footnote 1

TABLE 3: Classification of carcasses at four muscle sites according to fibre optic probe value at 24 h after slaughter (FOP_u) (percentages)

FOP_u value	Duroc X				Landrace.Large White X			
	Shoulder[1]	Loin[1]	Ham[1]		Shoulder	Loin	Ham	
			GM	SM			GM	SM
< 20 (DFD)	1	2	7	2	9	19	23	14
21-49 (normal)	86	97	93	92	83	81	75	86
> 50 (PSE)	13	1	0	6	8	0	2	0

[1]See Table 1, footnote 1

of PSE in both breed groups. Landrace.Large White X pigs had more FOP_u values indicative of DFD.

Percentages of pH_1 values below 6.0 were 12% for both the Duroc X and Landrace.Large White X pigs. This figure is intermediate to the values reported for purebred Irish Landrace and Large White over the past decade (McGloughlin & McLoughlin, 1975; Tarrant et al., 1979; Somers et al., 1986). The percentages of pH_u values above 6.0 at the shoulder, loin, and two ham sites (GM and SM) were 7, 1, 11 and 12 for the Duroc X and 18, 3, 23 and 16 for the Landrace.Large White X pigs respectively. These figures are generally higher than those reported for purebred Landrace and Large White pigs (Tarrant et al., 1979; Somers et al., 1986) and possibly reflect differences in preslaughter treatments used in the present experiment and for test station pigs.

The results indicate that the pork quality of the Duroc X pigs is
similar to the mean of the Landrace and Large White breeds, with a tendency
to be slightly paler, and to have a lower predisposition to produce DFD
meat in the ham and shoulder cuts.   The finding that meat quality in the
Duroc breed is relatively good is in agreement with observations from
several other European countries where the breed has been evaluated in
comparison with local and imported breeds (Brascamp et al., 1979;
Sellier, 1982, Sutherland, Webb and King, 1984).

There were differences in the degree of heterozygosity of the two
breed types. The Duroc X pigs were derived from a three-way cross, while
the Landrace.Large White X pigs were derived from a two-breed reciprocal
backcross expected to exhibit 2/3 of the heterosis found in F_1 hybrids.
However since heterotic effects on meat quality traits are expected to
be negligible (Johnson, 1981; Sellier, 1982) this factor is unlikely to
be the cause of any bias in the results.

The effect of the length of the holding period following transport
and before slaughter on meat quality was examined. The holding period
varied from 20 to 154 min. There was a significant effect on the pH_u
in the GM and SM muscles in the ham, and on the panel score in the ham;
longer holding periods resulted in higher pH_u and lower panel scores.

Phenotypic correlations between meat quality and carcass traits are
given in Table 4.   The differences between the breed types for carcass
composition traits expressed as Duroc X - Landrace.Large WhiteX, were
-1.8, -1.0, -.8, + .5 and + .9 for backfat (mm), eye muscle area (cm^2),
fat (%), lean (%) and intramuscular fat (%) respectively. Only backfat
and intramuscular fat were significantly different between the breed
groups. These correlations are based on relatively small numbers of pigs,
yet all the significant correlations indicated that pigs which were
leaner or had less fat (with the exception of intramuscular fat) had
paler pork. This finding is in agreement with correlations reported for
purebred Landrace and Large White breeds based on indicators of leanness
(average backfat, eye muscle area, ham conformation score) by Somers et
al. (1986). Unfavourable relationships between FOP_1 values and carcass
composition were also found in a sample of 50 Irish commercial crossbred
pigs which were fully dissected (Allen & Tarrant, unpublished results).
Low but significant phenotypic correlations in the same unfavourable
direction were also reported for British pigs (Evans et al., 1978).
Swedish pigs (Lundström, 1975) and Dutch pigs (Walstra et al., 1971).

TABLE 4   Correlations between meat quality and carcass traits in crossbred pigs

	Average backfat	Eye muscle area	Fat percentage [Subcutaneous + intermuscular]	Lean percentage	Intramuscular fat percentage
LD shoulder[1]					
FOP$_u$	-.18*	-.03	-.01	.06	.10
pH$_u$	.05	.10	.09	-.13	-.10
Panel	-.11	.05	.05	.05	.16
LD loin[1]					
pH$_1$	.10	-.10	.23	-.30*	-.01
FOP$_1$	-.15*	.05	-.08	.14	.22
FOP$_u$	-.19*	.13	-.10	.17	.06
pH$_u$	.15	.00	.21	-.36**	-.26*
pH$_u$ratio	.07	.08	.03	.09	.23
Panel	-.14	.02	-.23	.33**	.14
Drip Loss	.00	-.09	.11	-.11	.15
GM[1]					
FOP$_u$	-.14	.04	-.24*	.31*	.16
pH$_u$	.08	-.03	.28*	-.39**	-.11
Panel	-.11	.03	-.16	.31*	.25*
SM[1]					
FOP$_u$	-.11	-.09	-.12	.09	-.08
pH$_u$	.08	.02	.19	-.25*	-.09
No. of pigs	173	173	64	64	64

[1]LD shoulder: M. longissimus dorsi at the 3rd/4th rib;
LD loin: M. longissimus dorsi at the last rib;   GM: M. gluteus medius;
SM: M. semimembranosus

Intramuscular fat content was higher in the Duroc X (2.91%) than
in the Landrace.Large White X pigs (2.04%), in agreement with Barton-Gade
(1981). The higher content of intramuscular fat is unlikely to fully
explain the paler meat in the Duroc X pigs. While the phenotypic
correlations between intramuscular fat in the LD and both FOP and panel
scores (at the shoulder and loin sites) were positive (Table 4) they were
low and non-significant and unlikely to explain much of the breed
difference in FOP and panel scores

CONCLUSIONS

The present results indicate that meat quality in the Duroc breed
is similar to the mean for Irish Landrace and Large White pigs. Therefore
a crossbreeding strategy including the Duroc breed is not expected to

result in a deterioration of meat quality in the Irish pig population.

ACKNOWLEDGEMENTS

We wish to acknowledge our collaborators on this project:
T Hanrahan, P B Lynch and S Arkins (Pig Husbandry Department, Moorepark
Research Centre) and also the technical assistance of P Ward, M Murray
and J Dalton (Meat Research Department).

REFERENCES

Allen, P., and Tarrant, P.V. Unpublished results.

Barton-Gade, P.A. 1981. The measurement of meat quality in pigs post
    mortem. In Proc. "Porcine stress and meat quality - causes and
    possible solutions to the problems" (Ed. T. Froystein, E. Slinde and
    N. Standal). (Agricultural Food Research Society, As, Norway).
    pp 205-218.

Brascamp, E.W., Cop, W.A.G. and Buiting, G.A.J. 1979. Evaluation of six
    lines of pigs for crossing. 1. Reproduction and fattening in pure
    breeding. Z. Tierzüchtg. Züchtgsbiol., 96, 106-169.

Evans, D.G., Kempster, A.J. & Steane, D.E. 1978. Meat quality in British
    crossbred pigs. Livest. Prod. Sci., 5, 265-275.

Johnson, R.K. 1981. Crossbreeding in swine: experimental results. J.
    Anim. Sci., 52, 906-923.

Lundström, K. 1975. Genetic parameters estimated on data from the Swedish
    pig progeny testing with special emphasis on meat colour. Swedish J.
    agric. Res., 5, 209-221.

McGloughlin, P and McLoughlin, J.V. 1975. The heritability of pH_1 in
    longissimus dorsi muscle in Landrace and Large White pigs.
    Livest. Prod. Sci., 2, 271-279.

McGloughlin, P., Arkins, S., Lynch, P.B., Hanrahan, T.J., Allen, P and
    P.V. Tarrant. 1986. An evaluation of the Duroc breed. In Proc.
    Pig Health Society 14th Annual Symposium (Ed. P. McGloughlin,
    P.J. O'Connor, M. Martin). (Pig Health Society, c/o The Agricultural
    Institute, Dunsinea, Castleknock, Dublin 15).

Sellier, P. 1982. Le choix de la lignée male du croisement terminal chez
    le porc. Journées Rech. Porcine en France, 14, 159-182.

Somers, C., McGloughlin, P. and Tarrant, P.V., 1986. Pork quality in
    Irish purebred pigs. Ir. J. agric. Res (submitted).

Sutherland, R.A., Webb, A.J. and King, J.W.B. 1984. Evaluation of over-
    seas pig breeds using imported semen. 1. Growth and carcass perform-
    ance. J. Agric. Sci., Camb., 103, 561-570.

Tarrant, P.V., Gallway, W.J., and McGloughlin, P. 1979. Carcass pH values
    in Irish Landrace and Large White pigs. Ir. J. agric. Res., 18,
    167-172.

Walstra, P., Minkema, D., Sybesma, W. and van de Pas, J.G.C. 1971. Genetic
    aspects of meat quality and stress resistance in experiments with
    various breeds and breed crossing. Proc 22nd Annual meeting of the
    EAAP, Versailles.

DISCUSSION - Session IV

K Lundström   The incidence of halothane positive reaction in European countries, presented by Dr Webb in Table 2, is misleading as far as Sweden is concerned. We have no calculations of the average halothane gene frequency in Sweden and there are no average figures available for our elite herds, and herds that have used the halothane test for some time and have got rid of the gene have stopped testing and are not included in the figures.

A J Webb   Of course there is the problem that the incidences of positive reaction quoted in 1978 and 1984 will be sampled in different ways. I was not aware that low incidence herds were excluded from the 1984 incidence in Sweden.

In spite of the dangers of the comparison, it is perhaps useful to try to get some 'feel' for the change in frequency in European strains over the last few years in order to make a decision on strategy. We can take comfort from the observation that at least there have been no serious increases.

J J A Heffron   Since the halothane test may be inadequate in the screening for MH, have you considered the use of suxamethonium plus halothane for detecting those pigs resistant to halothane alone in order to detect the true incidence of MH?

A J Webb   We have investigated the use of succinylcholine with halothane in detecting the heterozygote. We can show a significant difference between Nn and NN in mean response. However, the overlap of the distributions was large, so that the accuracy of genotyping individual pigs would be limited.

J J A Heffron   In your work relating halothane sensitivity and anaerobic Ca^{2+} efflux from mitochondria, no overlap of the Ca^{2+} efflux values was observed despite the fact that halothane does not detect all pigs susceptible to MH. I would expect an increase in the Ca^{2+} efflux from mitochondria of animals that only react to succinylcholine plus halothane. Can you explain the lack of overlap?

A J Webb   I am reluctant to comment in the absence of Dr K S Cheah. However, it may be helpful to point out that the mitochondrial measurements were taken post mortem, and correlated with the presence/absence of PSE and halothane reaction after the event (MH does not come into our definition). One of the objectives of this work is to develop a method of genotyping which might eventually be used in the live animal. Hence the lack of overlap in Ca^{2+} efflux values is very promising. I'm afraid I do not know enough detail of the work to answer your question.

J J A Heffron   In fact, one would not expect to find an overlap of the Ca^{2+} efflux values if the material was post mortem, due to the denaturing conditions present in the early stages of PSE development. I accept that the test has been used in a practical context, but unfortunately Dr Cheah is claiming that this work will also establish the molecular aetiology of the syndrome, and it is that with which I disagree.

K Lundström   I have a comment to Dr Vestergaard. We have had the same negative genetic trend for meat colour in Sweden too. I suppose we have a common difficulty when there is a mixture of ordinary genetic variance (additive genetic variance) and a major gene (the halothane gene) so even if you update the genetic parameters, it is very difficult to have the right weightings in the index. As the heterozygotes are leaner and will be about normal in meat quality, they will always be selected. So meat quality will not work very well in the selection index.

T Vestergaard   The KK value is not an all-or-none trait, it is a continuous trait with a normal distribution. At the beginning of the seventies it had two peaks, but after some years of selection it was quite normally distributed. That is the background to the use of our index. I think the problem arose from the change in the stunning system from electric to carbon dioxide gas. In that period we used the wrong genetic parameters in our index and that is the reason, I think, why we had the negative trend. We have recalculated these parameters now and have put more economic weight on the KK value to maintain the present level of meat quality.

P A Barton-Gade   I would like to add a further comment to T Vestergaard's answer. Dr Lundström states that the meat quality in heterozygotes is more or less normal. We do not find this, at least for Landrace. We get quite a significant proportion of poor meat quality in our heterozygotes. Heterozygotes lie between the two homozygotes, not completely intermediate but slightly closer to the homozygous negative. Therefore, this should not be a confounding factor for the pig breeding programme.

C P Ahern   I am not convinced that the halothane gene exists. It has recently appeared in the Large White - where did the gene come from in this breed? The halothane reaction in the pig behaves like a single gene trait, but it is an assumption to say that it is a single gene.

A J Webb   The 'gene' can be regarded at the level of a statistical model which allows the consequences of breeding strategies to be predicted in terms of changes in frequencies of PSE and stress-susceptibility. Mapping of the gene to a known chromosomal linkage group provides strong evidence for its physical existence. The 'gene' may well be a piece of chromosome containing separate units of DNA with different effects.

A L Archibald   The segregation of sensitivity to halothane, and its linkage to discrete genetic markers provides convincing evidence that a major contribution to porcine stress is under the control of a limited region of chromosomal DNA. Molecular genetic analysis of this chromosomal region should help elucidate over the next three to five years the structure of this region - be it one locus or several closely linked loci.

J J A Heffron   I should like to point out that the two low molecular weight proteins reported to exist in human MH muscle in 1984 (see Dr Archibald's paper) have now been found in normal and in MH human muscle by another group of researchers (Fletcher, J.E. and Rosenberg, H. (1985), Anesthesiology, 62, 849-850) and cannot therefore be used as markers. These findings illustrate the difficulty of finding markers for MH!

G Eikelenboom   I have a question for the geneticists. If we assume a pig population in which the percentage reactors for halothane is decreased through selection to such an extent that it is not economically feasible to continue to conduct the halothane test, how should we proceed? Under such conditions can we recommend a restricted selection based on the meat quality of sibs or progeny?

A J Webb   Well, if you do nothing at all the frequency of reactors will go back up again and if you continue halothane testing it will be a very long time indeed, if ever, until you eliminate it. Any selection to maintain the frequency of the halothane gene at a very low level will be wasteful in loss of selection differential on performance traits and in incurring testing costs in every future generation. A case therefore exists for eliminating the gene altogether. The cost-benefit of present methods of elimination, for example test mating and blood typing, should be examined rather than waiting for cheaper genotyping techniques in the future. I suspect that using a meat quality index, in particular using sib testing to maintain the low frequency, is extremely inefficient.

T Vestergaard   I will try to answer that question by showing a table in which I have calculated the expectation of genetic progress in two examples, one without economic weight on meat quality and one with a weight expected to maintain meat quality at the present level. The parameters are taken from my paper (see Table 2, period 3). In example 1, where we put no economic weight on meat quality, the economic gain is distributed over the other three traits and adds up to one hundred per cent. As can be seen, the cost of maintaining the existing level of meat quality (example 2) is a reduction of twenty-one per cent of the genetic gain in the other three traits (average daily gain, feed conversion ratio and percentage meat in the carcass).

TABLE   Expectation of genetic progress in one generation of selection. (Information for a sow was based on one full sib, three half sibs and one progeny group, i = 1)

	Example 1			Example 2		
	V^1	G^2	Value %	V	G	Value %
Average daily gain	0.22	10.8	24	0.22	9.2	20
Feed conversion ratio	-120	-0.038	46	-120	0.030	36
Meat percentage	7	0.42	30	7	0.33	23
Meat quality index	0	-0.167	-	21	0.00	-
			100			79

[1] Economic weight in Danish kroner per unit

[2] Genetic change

P Sellier   Concerning the 'Hampshire effect' on meat quality Dr Monin and myself have sometimes the impression that this effect is a sort of 'French Hampshire syndrome! Many people indeed have claimed that Hampshire pigs give meat of good quality. May I suggest that research workers from countries keeping Hampshire pigs try to confirm or refute the

French results, particularly the lower ultimate pH values and the higher glycolytic potential we observed in this breed. The 'Hampshire effect' may also explain some of the residual genetic variation, for example in meat colour, remaining after the effects of the halothane locus have been accounted for.

H J Swatland   I have a general comment on the nature of PSE. When Dr E H Briskey invented the designation 'PSE' for pork of low pH it was quite beneficial for the scientific development of work on the topic. However, there is increasing evidence that paleness, softness and exudation are partly independent variables. Thus, pork may sometimes be extremely exudative with a normal colouration, and so on. As new and more sensitive techniques are developed for measuring PSE, the independence of P, S and E may become more obvious.

R G Kauffman   Since PSE is a variable condition, with several possible combinations of colour, firmness and wateriness, how can the trait(s) be interpreted to be assured that genetic selection is effective? Furthermore if there is a strong environmental component in the PSE phenomenon, how is this second problem dealt with in the selection process?

A J Webb   There are a number of possibilities. There are strategies for selecting on discontinuous variables. The problem is that where there is an index of performance on which breeders are already selecting, an all-or-nothing trait is extremely difficult to incorporate into that index. An all-or-none trait such as PSE can be transformed onto an underlying normal distribution of liability. In principle this can be included in a selection index of performance traits. However, it may be easier to include continuously distributed traits such as pH or meat colour.

The best approach will be to try to understand the nature of the underlying genetic liability. For example, controlling the frequency of the halothane gene may be the most efficient method of reducing the incidence of PSE.

D Lister   I am sensing a change in the attitude to the use of the halothane gene. A few years ago, the attitude was very distinctly that the halothane gene was something to be manipulated to the benefit of the pig industry. Am I right to say that this is no longer the case and the attitude is that we should reduce its incidence?

A J Webb   It is now recognised that the advantage of the halothane gene in lean content without the disadvantage of PSS may be exploited by producing a heterozygous (Nn) slaughter generation. There is a strong case for eliminating the gene from 'maternal' lines. It remains to be seen whether the maintenance of halothane positive (nn) terminal sire lines will present an animal welfare issue.

We therefore have various options for exploiting the gene, but we do not have enough information on the economic advantage of the heterzygote to make a decision.

P Mormède   The use of the halothane gene in breeding programmes can have different consequences for animal welfare depending on the frequency that we want to maintain the gene at. Keeping the gene at a high level is probably dangerous from the point of view of welfare because these animals are obviously less resistant to acute stress. In this context I wonder whether the economic evaluations of the advantages of keeping this gene at a high level have always taken into account the differences of management quality between testing stations and ordinary breeding farms, where adverse consequences on productivity would be more readily seen.

If homozygotes are kept only as a small percentage of the herd, these animals may be considered as sensitive indicators of management problems. They would be a kind of safeguard against deteriorations of the quality of management practice and may even stimulate improvements in sensitive periods of the pigs life, such as the immediate period before slaughter and slaughter itself. All the animals would benefit from such results.

The third possibility is the complete elimination of the gene from the populations. It would be naive to think that this would be a 'cure for stress'. Firstly because the environmental pressure is still there even if its consequence is not as obvious as in stress-sensitive animals, and secondly because halothane susceptibility does not cover the whole field of porcine stress syndrome and stress sensitivity in general. I tried to demonstrate this in my paper.

P V Tarrant   Dr Mormede has defended the practice of producing halothane reactors, particularly for use in research programmes, and I think indeed that scientists should have the opportunity to have experimental animals of this kind, held under the appropriate conditions to ensure their welfare. However, at this meeting we are concerned about the welfare of farm animals, that is, animals which are held for commercial purposes, as distinct from laboratory animals. The regular commercial production of halothane reactors, for example halothane positive sire lines with distinct production advantages, does pose a serious welfare problem because these animals have a defective physiology which makes them liable to very high mortality rates. Obviously a mortality rate only reflects the animals that succumb. We must also consider the welfare of animals that manifest the stress syndrome but do not actually die.

On the issue of PSS we are faced with the classical dilemma, namely the opposing viewpoints of production economics and welfare. We are not asked to decide on the issue, but we are asked to provide the best available evidence so that objective decisions can be reached by the legislators. I believe there is not enough evidence to oppose and condemn the practice of producing halothane positive breeding stock for commercial purposes, but it is questionable from the welfare point of view.

A J Kempster Where would you draw the line on the issue of welfare and the halothane gene because there are many areas of genetic selection for commercial performance where the individual animal might be disadvantaged. Consider for example the locomotive ability of double-muscled cattle and the survivability of triplets in sheep in comparison with singles or twins.

P V Tarrant   You have raised a number of possibilities in animal production which could lead to welfare issues. Concerning the halothane gene there is a practical solution to the problem. Stress susceptible pigs must be produced under the appropriate environmental conditions to ensure their welfare, as Dr Mormède has suggested. The individual issues ought to be considered on their special merits. In general we can say that if we breed, or produce by genetic engineering, farm animals that are incapable of normal survival and performance, then we must either alter the environment to alleviate the problem or stop producing such animals.

B W Moss   I feel that in the halothane positive pig we have an animal whose stress response we cannot change by environmental influences. However, in the heterozygote we have a pig whose lean potential we might employ in production and whose stress response we might control by environmental manipulation. We need to consider the welfare of the pig from birth to slaughter. It is fairly well established that early stress experiences may be beneficial during later exposure to stressors. Could we then control the stress response of the heterozygous pig by early exposure to stressors? The total stress a pig experiences can be given by the product of stressor x response during the life of the pig, and it is this total stress that should be considered.

P Mormède   From my previous comment it is clear that I draw no absolute line between what is good and what is bad for animal welfare, so that each particular case merits a specific evaluation both from the welfare and economic points of view. Furthermore my personal opinion is that dissociation between welfare and economical objectives is frequently artificial because economic evaluations are made in 'protected' conditions such as testing stations. In ordinary breeding farms welfare problems may be correlated with poor productivity.

SESSION V

Chairman: R.G. Kauffman

REPORT ON SESSION I

Aetiology of the Porcine Stress Syndrome

K O Honikel

We know that genetic deficiencies and environmental stress are responsible for PSS. When the syndrome occurs in pigs before or at death it causes PSE meat due to the coincidence of fast anaerobic glycolysis (fast fall in pH) and prevailing high temperatures. Muscle proteins are denatured and membrane lesions occur. About the sequence of events numerous details are known but the connecting links between the events are missing. Extra-cellular signals induce the PSS, leading to an accelerated energy turnover in muscle cells, terminating after slaughter in PSE meat. The related malignant hyperthermia (MH) syndrome is induced in animals tending towards PSS by anaesthetics like halothane.

The papers presented in this session deal with these two syndromes. There are common symptoms in MH and PSS, but both syndromes also differ. Muscle rigidity is a common, but not a universal feature of MH, which, however, does not occur in PSS. PSE meat, the final state of muscles from animals with PSS, shows no muscle contracture either. Also, as Dr Lister pointed out, α-blockade prevents MH, whereas β-blockade reduces the incidence of PSE meat in stress susceptible pigs. In MH, Ca^{2+} ions are widely accepted as one of the triggering substances of the syndrome. Prof Heffron reviewed the literature and concluded that Ca^{2+} ion release may not be the primary cause of the syndrome, but may be indicative of a widespread membrane lesion. Cell membranes as well as the membranes of the sarcoplasmic reticulum and mitochondria could be affected. Membrane deficiencies may be located in the associated proteins, enzymes and membrane lipids.

In PSS there is a rapid utilization of energy in the muscle resulting in heat production. The rapid turnover of energy rich compounds is caused by the hydrolysis of ATP. ATP breakdown is the central reaction in PSE muscle, controlling anaerobic glycolysis. There are a number of possibilities for rapid ATP turnover in PSS muscle.

1. Increased expenditure of energy to maintain ionic gradients across leaky membranes.
2. ATP may be used up by futile cycling, for example the opposing reactions catalysed by phosphofructokinase and hexose diphosphatase[1].
3. The energy status of the cell and the relative concentrations of its allosteric modulators ATP, ADP and AMP are expressed by the 'energy charge of Atkinson'. Malfunctions of the associated enzymes (adenylate kinase, AMP deaminase and ATPases) may trigger an increased rate of glycolysis through activation of phosphorylase and phospho-fructokinase.
4. An early switch from aerobic to anaerobic metabolism would result in ATP production per glucose unit dropping to one twelfth of the aerobic rate.
5. Increased ATP turnover may result from contraction of myofibres induced by an increase in Ca^{2+} concentration. Dr Kozak-Reiss reported that muscle biopsy samples of halothane positive pigs lose contract-ility rather early at a high ATP concentration. This is in agreement

with my own observation in Session IV that PSE muscles show no
contracted sarcomeres.

Concluding from the above five possible causes of increased ATP turnover
the following three events may occur, induced by signals from outside and/
or by genetic defects.

(a) An altered membrane permeability.
(b) Altered enzyme activities or altered control mechanisms of glycolytic
     or adenine nucleotide converting enzymes.
(c) Mitochondrion malfunction with early and solely anaerobic ATP
     production.

One or several of these events may occur in PSE prone muscles. The
sequence of reactions in the muscle cell may be caused by the nervous
system and/or stressors and induced in the cell by hormones.

With regard to the action of hormones, Dr Moss found differences in thyroid
function between stress resistant and stress susceptible pigs with a
possibly higher degradation rate of thyroxine in stress susceptible pigs
and an inability of these animals to secrete sufficient hormone during
stress. Prof Ludvigsen favoured the view that there is a borderline change
in the hormonal balance in favour of anabolic hormone systems in animals
with more lean meat. STH may be primarily involved, stimulating protein
synthesis and deposition and antagonising the activity of TSH and ACTH,
which are important hormones in the adaptive capacity to strain situations.

With regard to the nervous system in MH susceptible pigs, Prof McLoughlin
examined the concentration of noradrenaline, dopamine and their non-O-
methylated metabolites. Differences in the activities of enzymes
metabolising these substances could not be found in various organs. He
observed, however, differences in the steady state concentrations of two
of the metabolites, 3,4-dihydroxymandelic acid (DHMA), and
dihydroxyphenylglycol (DHPG). Both were higher in MH susceptible pigs.

Dr Jorgensen found that the sensitivity of pigs to halothane depended on
various factors, which may be an additional reason why the reliability of
the halothane test in predicting PSE meat in the live animal is limited.
Low potassium supply in MH susceptible pigs caused a modified reaction to
halothane.

Dr Mormède viewed PSS as evidence of an excessive environmental pressure
on a sensitive animal, and pleaded for a thorough search for stressful
environmental factors not only to improve meat quality but also the
welfare of the animals.

In summary, stressful environmental factors cause hormonal signals, which
induce in cells a sequence of metabolic reactions that terminate in the
formation of PSE meat. How these signals are transferred to the cell and
how the increased ATP turnover in the cell starts is so far not clear.
There are a lot of possible mechanisms for ATP breakdown and events in
PSE prone muscles post mortem. As we are concerned ultimately with meat
quality, in the rapporteur's opinion these problems should receive priority
in research. If these problems within the cell are solved it may become
obvious how the hormonal signals are transferred into the cell and what
kinds of events and extracellular signals are important.

[1] Prof Heffron pointed out that calculations have shown that futile cycling can only account for an insignificant fraction, a maximum of ten per cent, of heat production in MH. Contractile activity was the most likely cause of ATP breakdown, with a small contribution from ion transport ATPases.

REPORT ON SESSION II

The Evaluation of Meat Quality

G Eikelenboom

Here we are dealing almost exclusively with intrinsic quality aspects. These aspects, which include traits such as waterbinding, colour, texture and taste, influence consumer appreciation of meat and its suitability for the manufacturing of processed products. Therefore, the improvement of these factors, including also the reduction of their variability, is of direct importance to our meat industry.

It is almost twenty years ago since Dr Briskey introduced the term PSE (pale, soft and exudative) and DFD (dark, firm and dry) muscle. Although these terms are commonly used nowadays, we should also realize their limitations. Firstly, the criteria for calling a carcass or muscle PSE, normal or DFD can be fairly arbitrary, while we know that it is a continuous scale. Secondly, colour, waterbinding and texture are frequently interrelated, although this relationship may vary considerably from one study to another. Moreover, the effect of genotype, ante- or post mortem handling may have differential effects on these quality traits. For instance, the so-called 'Hampshire-type meat' has poor waterbinding characteristics. As our French colleagues have shown, this is not associated with a rapid pH fall and protein denaturation as in PSE, but instead with a very low ultimate pH.

Meat quality may be measured for a number of purposes. These are: research, for example to determine the effect of a certain treatment; breeding and selection; and finally commercial practice, including slaughterline measurements and identification of produce for different destinations such as fresh versus processed meat outlets. We have seen examples of these applications during this session.

Dr Kallweit made clear to us that there is a large variability among EC member countries in methods for evaluation and assessment of meat quality, both for research purposes and in the selection of breeding stock. Among the sources of variation were the exact time of measurement and the techniques and muscles used. Few of the measurements were comparable and this may conflict with the interpretation of data by others. It was the general opinion during the discussion that these measurements need standardisation. Earlier work by a specific working group instituted by the Commission, has resulted in recommended procedures for the assessment of beef quality (Boccard et al., 1981). This was later followed by ring sample testing of beef by research institutes in various countries, for eating quality (Dransfield et al., 1982) and chemical composition (Dransfield et al., 1983). We should apply and adapt these procedures to pork muscle and eventually improve and extend these methods. As a result of this meeting, we already have a good start. The standardised method suggested by Dr Honikel for the assessment of drip and cooking loss should be considered for application in our research programmes. I remind you of Dr Honikel's request to contact him concerning your personal experiences and comments on this proposed method, so that it can if necessary be improved. Furthermore, Dr Swatland has offered to initiate a

reference bank, where methods currently available or to be developed, will
be described, registered, assembled and made available to others. He
would welcome descriptions of methods and apparatus, and will be happy to
supply information about this initiative.

In attempting to internationally standardise and develop our methods, we
should recognise that some methods may be used or adapted in certain
countries to evaluate quality for a specific destination, as in the case of
a typical locally produced processed product. Both Italian papers in this
session are examples of that type of research. Moreover, under commercial
conditions scientific methods or techniques may be subject to certain
constraints with respect to their practical application. Indeed, as one
participant stated, it might be difficult to identify carcasses the day
after slaughter and to sample them. However, as Dr Moss illustrated, a pH
measurement alone is frequently insufficient to predict ultimate meat
quality.

Alternative methods for the assessment of meat quality in the slaughterline
are becoming available. The development of fibre optics has opened new
perspectives in this field. The new instrumental grading equipment such as
the Fat O Meater and the Hennessy Grading Probe which will soon be used
throughout the EEC allows operators not only to determine carcass
composition, but also meat quality, by measuring light reflectance or
scatter in muscle. There are also fibre optic probes, either commercially
available or under development, which are operated independently from the
compositional grading equipment. Such equipment may be used in the
slaughterhouse, at the wholesale level or at central packaging units of
retail chains. Can and will these methods ultimately be used for payment
on meat quality to the producer, as with body composition determined by
instrumental classification? If so, how do we differentiate between
undesired quality traits caused by breeding and management, and those
resulting from inadequate techniques at the meat factory. Are the
measurements of light reflectance on the slaughterline detecting, in
particular, halothane positive pigs with fast glycolyzing muscle, as
preliminary research evidence in Denmark and Holland suggest? These
measurements can also be used for sorting carcasses for the various
destinations, including both fresh and processed meat. Evidence has been
presented at this seminar by Dr Lundström and Dr Barton-Gade that additional measurements at later moments may be required to correctly determine
meat quality.

New or modified instruments have also been developed in recent years for
the measurement of (complex) electrical conductivity. This technique is
based on changes in resistivity and capacitance of cell membranes in PSE
muscle which is associated with an inherently fast post mortem metabolism.
The two German papers on this technique showed quite promising results. It
would be interesting to test these new instruments in other countries with
different pig populations and industrial conditions. Undoubtedly both the
measurement of light scatter and electrical conductivity will be further
improved and refined, and their predictive power for determining ultimate
meat quality will increase. Eventually they may even be combined in on-
line measurements.

As Dr Swatland pointed out, there are potent new techniques coming up for
directly or indirectly monitoring the dynamic post mortem processes which
may lead to an abnormal pork quality. These techniques will not only

require further research, but also a further development to a practical application in the industry. At present we are clearly facing new changes and challenges in meat quality evaluation.

REFERENCES

Boccard R., Touraille C, Buchter, L., Casteels, E., Cosentino, E., Dransfield, E., MacDougall, D.B., Rhodes, D.N., Hood, D.E., Joseph, R.L., Schön, I., and Tinbergen , B.J., 1981. Procedures for measuring meat quality characteristics in beef production experiments. Report of a Working Group in the Commission of the European Communities (CEC) Beef Production Research Programme. Livest. Prod. Sci., 8, 385-397.

Dransfield, E. , Rhodes, D.N., Nute, G.R., Roberts, T.A., Boccard, R., Touraille, C., Buchter, L., Hood, D.E., Joseph, R.L., Schön, I., Casteels, M., Cosentino, E. and Tinbergen, B.J. 1982. Eating quality of European beef assessed at five research institutes. Meat Science, 6, 163-184.

Dransfield, E., Casey, J.C., Boccard, R., Touraille, C., Buchter, L., Hood , D.E., Joseph, R.L., Schön, I., Casteels, M. Cosentino, E. and Tinbergen, B.J., 1983. Comparison of chemical composition of meat determined at eight laboratories. Meat Science, 8, 79-92.

REPORT ON SESSION III

Control of Meat Quality in Transport, Lairage, Slaughter and Chilling

G Monin

Session III contained three very good papers followed by a rich discussion. These papers deal with the control of meat quality in transport, lairage, slaughter and chilling, i.e. essential points in the pork production chain.

Reviewing the effect of time and conditions of transport and lairage on pig meat quality, Dr Warriss pointed out that the factors involved are numerous, and that the results from   most of the studies may hardly, if at all, be compared. This is due to the tremendous variation in experimental conditions, measurement techniques, classification criteria or types of animals under study. Anyway, two rules of great practical interest can be derived from the authors conclusions: firstly, improvements of transport conditions will often lead to improvements of meat quality, as well as being desirable from the point of view of animal welfare. Secondly, time in lairage of more than about 6 hours is generally not favourable to meat quality; in many cases, 2 hours or less is sufficient.

This paper leads again to two questions which are central to research on meat quality. Firstly the necessity, and the possibility, of establishing standardised methods for use in the European Community. This question has been addressed by Dr Eikelenboom in his report on Session II. Another question of primary interest is the definition of stress susceptibility. The concept is used to classify animals as stress susceptible or stress resistant animals. In my opinion, it is confusing to use for classification a criterion which is not clearly defined: if one uses a term whose sense is not clear, one cannot know exactly what the term covers. In fact, stress susceptibility is generally confused with halothane susceptibility, which is not acceptable. To give an example, let us look at the results obtained in France on Corsican pigs:

Meat quality characteristics of purebred Large White and Corsican x Large White crossbred pigs (from Goutefongea et al., 1983)

	LWxLW	CxLW	Difference	Pen	Pasture	Difference
No.of animals	57	62		49	70	
pH_2	5.99	6.20	−0.21**	6.09	6.11	−0.02ns
MQI	8.16	9.25	−1.09ns	8.59	8.83	−0.24ns

LW: Large White   C: Corsican   MQI: Meat Quality Index   ** P < 0.01

Corsican pigs live freely in the forests of Corsica. They experience a very hard life: practically no food from their owners, cold, sometimes snow, occasionally running before dogs or fighting them, and so on.

They can hardly be considered as stress susceptible according to the usual terminology, and moreover, there are no reports of sudden deaths in the breed. However, even when born and grown in pens on experimental farms and slaughtered under commercial conditions, Corsican crossbred pigs give a lot of DFD meat as compared to Large White pigs. So they can be said to be more sensitive to preslaughter stress than the Large White. This illustrates the need for a precise definition of stress susceptibility before using it for classification of animals. In the present state of knowledge let us speak about halothane susceptibility, DFD proneness and so on, rather than of stress susceptibility.

Dr Gregory gave a very well documented report on stunning. Pigs are stunned by means of electric current or carbon dioxide, and there has been for years a debate about the respective advantages of each technique. A meeting was especially devoted to this subject three years ago in the Netherlands, (Eikelenboom, 1983). No clear conclusion was drawn regarding what technique is preferred. At the moment, electrical stunning is preferred in most European countries, but Danish pigmeat processors continue to prefer carbon dioxide. This difference could simply express the fact that the reaction of the market to such defects as 'blood splash' varies from one country to another; also the other attributes of quality may not be the same for pork processors in different countries, nor the price that these processors are ready to pay for quality. This subject is open to further research.

Dr Honikel examined the influence of different rates of meat chilling on the quality attributes of fast glycolyzing pork muscles. He used hot-boned pork, a process which is very little used at the moment in EC countries, but which undoubtedly has a great future and will undergo substantial development. One of the main interests in this technique is to preserve the waterholding and emulsifying capacities of pork by breaking the low pH/high temperature combination and so prevent to some extent the development of the PSE condition. The danger comes from cold-shortening, inducing drip loss and probably toughness. Dr Honikel established in a very astute way the conditions which must be followed to obtain the best results. At the same time, he showed that the technique was limited in preventing the PSE condition. Hot boning and rapid chilling do not allow the complete elimination of PSE meat. Moreover, DFD meat is unsuitable for vacuum packing and so may cause problems in the development of hot boning. The application of the technique will require methods for assessing meat quality on the slaughterline, particularly for detecting DFD meat, and probably will increase the need for reducing variability in meat quality. The best ways to achieve this again rely on better control of slaughter conditions and on the improvement of muscle quality in the live animal. So, the development of a more efficient post mortem technology could in turn oblige us to progress in the intra vitam treatment of animals.

REFERENCES

Eikelenboom, G. 1983 (Editor). Stunning of Animals for Slaughter. Current Topics in Veterinary Medicine and Animal Science, Martinus Nijhoff Publishers, The Hague.

Goutefongea, R., Girard, J.P., Labadie, J.L., Renerre, M. and Touraille, E. 1983. Utilisation d'aliments grossiers pour la production de porc lourds. Interactions entre type génétique, sexe et mode de conduite. Journées Recherche Porcine France, 15, 193-200.

REPORT ON SESSION IV

Control of Meat Quality in Breeding and Selection

P McGloughlin

The discovery of a major gene with large economic effects is a rare occurrence in animal production. However, such a discovery was made with the finding of the halothane gene, which is associated with the porcine stress syndrome and has large effects on meat quality and other traits. This discovery has stimulated much research in the past ten years into the mode of inheritance of the gene, its effect on performance, and its possible exploitation in pig improvement schemes. In the fourth session of this seminar, the majority of the fifteen presentations reflected the overwhelming importance of the halothane gene in the genetic control of meat quality in pig breeding.

The current state of knowledge on the halothane gene and the role of the halothane test in improving meat quality was comprehensively reviewed by Dr Webb. Research has confirmed that the genetic response to the halothane test is largely controlled by a single locus, but there is evidence of both genetic and maternal effects on the expression of the gene. Furthermore, it appears that the halothane gene may not be fully recessive for its effect on halothane reaction, or on meat quality. Dr Webb identified immediate priorities for research: (1) to provide a quick cheap method for genotyping at the halothane locus, and (2) to determine the relative performance of the heterozygote. Furthermore he listed as longer term priorities: (3) the separation of the beneficial and deleterious effects of the halothane gene, (4) the investigation of the possible mitochondrial inheritance of the porcine stress syndrome, and (5) the evaluation of eating quality of pork in the presence and absence of the halothane gene.

The accuracy of the present range of methods for identifying the heterogygote tend to be positively correlated with cost. The most accurate, test mating, may be prohibitively costly in most practical situations. The use of blood markers in the linkage group which includes the halothane locus, to aid selection for the halothane genotype, was reviewed by Dr Andersen. This approach, while dependent on linkage disequilibrium, can be very successful in changing the frequency of the halothane gene. However its complete elimination would require absolute determination of the halothane genotype. Dr Andresen reiterated the need for research to find an accurate diagnostic test for the halothane genotype.

The economic benefit of exploiting the halothane gene in practical breeding schemes depends on the magnitude of the favourable and unfavourable effects of the gene on performance traits in both the homozygote and heterozygote. There is evidence that this varies between breeds and populations. However there appears to be a significant advantage to be gained from producing a heterozygous slaughter generation, through the development and crossing of a halothane positive sire line and halothane negative dam line. There is however considerable variation between countries in the degree to which this approach has been pursued.

Complete elimination of the halothane gene in the dam line, as it has
deleterious effects on reproductive rate, appears to be an agreed goal;
however achievement of this goal in most populations where the gene is
present is likely to have to await the arrival of an accurate test for
identifying the heterozygote. Furthermore in situations where payment is
not based on lean content, the benefits of exploiting the halothane gene
will not be fully realised.

The probable answer to the search for a method of identifying the
heterozygote was outlined in a 'preview of the future' by Dr Archibald.
By applying molecular biology techniques to study the porcine stress
syndrome, it is hoped to provide an alternative means of determining
halothane genotypes within the next five years. Furthermore the underlying genetic lesion may also be identified. This work could also have
implications for the understanding of the malignant hyperthermia syndrome
in man. Genetic engineering techniques could also allow other exciting
prospects to become a reality; the development of a genetically stable
heterozygous line of pigs by the introduction of the halothane positive
gene into an egg which was homozygous for the halothane negative gene.
The realisation of the longer term objectives (numbers 3 and 4) listed by
Dr Webb could be possible through genetic engineering skills. This area
of research deserves full support.

In his review of the use of cross-breeding to improve meat quality,
Dr Sellier suggested that the classical view that there is no heterosis
for meat quality traits, should be revised when considering specific traits
in specific crosses. He gave two examples: firstly, the fast rate of
pH fall (low pH_1) appears to be inherited as a recessive, or partially
recessive, trait in Pietrain crosses, where the halothane gene probably
plays a major role; the observed heterosis was favourable. Secondly,
there is evidence that the large extent of the fall of pH (pH_u) found in
Hampshire crosses may be inherited as a dominant or partially dominant
trait, possibly controlled by a major dominant gene; the observed
heterosis was unfavourable. Further research is needed to substantiate
these tentative interpretations. In another presentation Dr Monin and
Sellier reported that the Hampshire breed has a high glycolytic potential,
resulting in a low pH_u and a low technological yield. Dr Sellier urged
researchers who had Hampshire pigs available to them to further investigate
this 'Hampshire effect'; furthermore he suggested that a study of the
glycolytic potential in breeds where the frequency of the halothane gene
was low or zero, might begin to explain the residual variation remaining
in meat quality after accounting for the effect of the halothane gene.

In the breeding programmes in several countries, meat quality is included
in the aggregate genotype of the selection index. Dr Vestergaard drew
attention to the adverse effect on the efficiency of the selection index
that occurs if genetic parameters for meat quality are poorly estimated.
In populations where the halothane gene is at a low frequency, or absent,
genetic improvement in meat quality will rely on selection for additive
variance. It is also possible that selection for additive variance in
meat quality in populations where the halothane positive gene is fixed,
could result in some genetic improvement.

In order to exploit the halothane gene, it is necessary to maintain
homozygous halothane positive terminal sire lines. The welfare aspects of
this strategy of selecting animals considered to be physiologically

extreme, or defective, were discussed. It is apparent that this was seen to be a complex subject, deserving further serious consideration, but that the debate should not necessarily be confined to pigs alone.

SUMMARY

P V Tarrant

The European Community is the second largest producer of pigmeat in the world, slaughtering 129 million pigs in 1983. Pig production in the EC is an efficient industry. Further growth will depend heavily on increasing pigmeats' share of EC consumer demand for meat products. Consequently prices and consumer habits are becoming increasingly relevant. Against this background, the control of meat quality is a matter of some urgency, hence the necessity for this seminar.

There is concern that the eating quality of pork may be declining. Organoleptic problems including dryness, hardness and inferior flavour have been reported. Furthermore the classical stress-related meat quality defects of PSE and DFD are still of major concern. On another front, there is a growing interest among the public in pig welfare during production, transport and slaughter.

It is well known that there is an unfavourable correlation between meat quality and meat quantity in pigs. This is manifested in higher mortality rates (PSS) and lower meat quality (PSE) in otherwise superior stock. The pathogenesis of PSS is well documented. Affected animals are susceptible to a cardiorespiratory complex induced by relatively common stressors and rigor mortis develops rapidly after death, inducing the PSE condition in the musculature.

The aetiology of PSS is not well understood. Investigators have used the closely related malignant hyperthermia (MH) syndrome as an experimental model. Swine that are prone to PSS are likely to develop MH upon exposure to halothane. Sessions I and IV concluded that the primary defect in MH is still unidentified. However, a basic lesion may exist in the membranes of the sarcolemma and/or the sarcoplasmic reticulum, preventing normal intracellular calcium homeostasis in the contracting myofibre. Such a membrane lesion may be responsible for PSS. Furthermore the autonomic nervous system is considered to be an essential part of the triggering mechanism for PSS.

The relative merits of exploiting or eliminating the halothane gene, and the importance of the residual genetic variation in meat quality after accounting for the effect of the halothane gene, was explored in Session IV. The economic argument for retaining the gene is that there appears to be a significant advantage to be gained from producing a heterozygous slaughter generation from a halothane positive sire line and a halothane negative dam line. The halothane test will continue to be used to control the gene or to select against it. Selection based on halothane tests combined with blood typing may be more efficient than selection based on halothane testing alone, but has not yet been fully evaluated. The complete removal of the gene would require a quick, cheap method for exact diagnosis of heterozygosity. At present this can only be achieved by costly test mating. The application of DNA technology may provide an alternative means of determining halothane genotypes within the next five years.

There is considerable genetic variation in meat quality apart from that controlled by the halothane gene, for example the 'Hampshire effect' investigated in France. Selection on meat quality criteria in breeding programmes in several countries is being improved. However, there is now a diminishing availability of information from slaughtered progeny, because that procedure is becoming prohibitively costly. Future selection may be based on the measurement of meat quality in the live animal. Finding suitable methods to do this is an important objective for immediate research and development.

It is evident that the handling, transport and slaughter of pigs is associated with welfare problems. General principals for the correct performance of these operations were presented in Session III. Lairage should not extend beyond six hours other than in exceptional situations. The lairage and resting requirements in long distance and international transport are not understood and research is required on this topic. With respect to meat quality, very stress susceptible pigs may not respond favourably to improved or even 'ideal' preslaughter handling, because the trauma of slaughter itself is sufficient to initiate development of the PSE condition. Accelerated processing of pork carcasses, including hot deboning and fast chilling, may alleviate the PSE condition but cannot eliminate it.

Although there is not yet full agreement on what is the best stunning method, there is a strong arguement in favour of high voltage head to back electrical stunning. This method may satisfy welfare requirements by inducing an immediate cardiac arrest in the animal. However, much more practical experience of the method is needed to establish whether it aggravates bone breakages and blood splash. In addition, legislative changes are necessary in some EC member states before this method can be used.

Meat quality is measured in the member states for the purposes of meat research, selection in pig breeding and commercial quality control. Session II discussed the large variation in methodology between laboratories in the different countries. Recognising that the application of various methods is frequently subject to theoretical and practical constraints, there is a need for better international standardisation. Initiatives were taken to obtain more uniformity in methods of determining waterbinding capacity.

The development of robust optical and electrical probes for use on intact carcasses opens the way for the routine determination of meat quality on the slaughterline or at a later time post mortem. Depending on the particular purpose of these measurements, their ability to predict meat quality may need further improvement. This may be achieved by improving the separate methods, or by combining them, or by supplementing them with other measurements at the same time or at a later period post mortem. Emphasis should be laid on their practical implementation in the meat industry, preferably with a high degree of automation. The availability of such methods is essential to give the industry greater control of meat quality.

Two aspects of pig welfare were considered, the general issue of pig welfare during transport, lairage and slaughter and the specific issue of the welfare of pigs that are prone to PSS. With regard to the general issue, the guidelines presented for the reduction of economic losses in carcass and meat quality are also valid when applied to welfare. These are the avoidance of overloading and temperature stress during transport, moderate duration of transport and proper methods and equipment for handling pigs. Tight supervision is essential at all points during transport. Short times in lairage, avoidance of social regrouping, and effective stunning and exsanguination are the essential welfare requirements for pigs at slaughter.

Regarding the specific issue of the welfare of PSS prone pigs, any increase in pig morbidity or mortality as a consequence of PSS is unacceptable on welfare grounds. Thus, measures aimed at reducing the incidence of PSS, for example by eliminating the halothane gene, are seen as beneficial to welfare. However, there is no evidence, either physiological or behavioural, of distress in PSS prone swine when the syndrome is not evoked. Consequently, it may be possible to produce stress susceptible stock humanely where the environment has been designed to be compatible with the stress-prone nature of the animals. This could be readily achieved for experimental animals and may also be attainable for stock held for commercial cross-breeding.

The evaluation and control of meat quality is a very active area of research and development. The following topics, inter alia, were recommended at this seminar for immediate research attention:

(a) transport, rest and lairage requirements of pigs in long distance and international transportation;
(b) methods of stunning, including the investigation of carcass and meat quality aspects of high voltage electrical stunning;
(c) harmonisation of methods and standards used for the evaluation of pork quality in member states;
(d) development of methods for the measurement of meat quality in the live animal, in the presence and absence of the halothane gene;
(e) development of a fast, cheap method for the exact diagnosis of the heterozygote at the halothane locus.

LIST OF PARTICIPANTS

BELGIUM

Mr J Connell
Commission of the European Communities
Rue de la Loi 200
1049 Brussels

Dr F O Ödberg
Dierlijke Genetica en Veeteelt
Rijksuniversiteit Gent
Fakulteit van de Diergeneeskunde
Heidestraat 19
B-9220 Merelbeke

CANADA

Dr H J Swatland
Department of Animal and Poultry
Science
University of Guelph
Guelph
Ontario N1G 2W1

DENMARK

Dr E Andresen
Department of Animal Genetics
The Royal Veterinary and Agricultural
University
Bulowsvej 13
DK-1870 Copenhagen V

Dr P Barton-Gade
Meat Research Institute
Maglegardsvej 2
DK-4000 Roskilde

Dr P Jensen
Royal Veterinary and Agricultural
University
23 Rolighedsvej
DK-1958 Copenhagen V

Dr P F Jorgensen
Department of Veterinary Physiology and
Biochemistry
The Royal Veterinary and Agricultural
University
Bülowsvej 13
DK-1870 Copenhagen V

Prof J B Ludvigsen
National Institute of Animal Science
Department of Physiology
Rolighedsvej 25
DK-1958 Copenhagen V

494

Ms E Olsen	Meat Research Institute Maglegardsvej 2 DK-4000 Roskilde
Prof H Staun	National Institute of Animal Science Rolighedsvej 25 DK-1958 Copenhagen V
Dr T Vestergaard	National Institute of Animal Science Rolighedsvej 25 DK-1958 Copenhagen V

FRANCE

Dr P Mormède	Psychobiologie des Comportements Adaptifs-INRA U 259 - INSERM Domaine de Carreire Rue Camille St Saens 33077 Bordeaux Cedex
Dr G Kozak-Reiss	Faculté de Medecine Paris Sud Departement de Physiologie Humaine CCML 133 Ave de la Résistance F-92350 Le Plessis Robinson
Dr G Monin	Station de Recherches sur la Viande INRA - CRZV de Theix 63122 Ceyrat
Dr P Sellier	Station de Génétique Quantitative et Appliquée INRA 78350 Jouy en Josas

GERMANY

Mr K Fischer	Federal Institute of Meat Research E C Baumann Strasse 20 8650 Kulmbach
Dr K O Honikel	Federal Institute of Meat Research E C Baumann Strasse 20 8650 Kulmbach
Prof E Kallweit	Institut für Tierzucht und Tierverhalten FAL Mariensee 3057 Neustadt 1
Prof F Schmitten	Institut fur Tierzuchtwissenschaft Endenicher Allee 15 5300 Bonn 1

Pror D Seidler

Fachhochschule Lippe
Studienschwerpunkt Fleischtechnologie
Liebigstr 87
D - 4920 Lemgo 1

IRELAND

Dr N Ahern

Veterinary College of Ireland
Ballsbridge
Dublin 4

Dr P Allen

Meat Research Department
The Agricultural Institute
Grange/Dunsinea Research Centre
Castleknock
Dublin 15

Prof J J A Heffron

Department of Biochemistry
University College Cork
Prospect Row
Cork

Dr D E Hood

Meat Research Department
The Agricultural Institute
Grange/Dunsinea Research Centre
Castleknock
Dublin 15

Dr P McGloughlin

Animal Breeding and Genetics
Department
The Agricultural Institute
Grange/Dunsinea Research Centre
Castleknock
Dublin 15

Prof V McLoughlin

Department of Physiology
Trinity College
Dublin 2

Dr J F O'Grady

President EAAP Pig Production Commission
The Agricultural Institute
Grange/Dunsinea Research Centre
Dunsany
Co Meath

Mr G O'Hagan

Department of Agriculture
Kildare Street
Dublin 2

Dr P V Tarrant

Meat Research Department
The Agricultural Institute
Grange/Dunsinea Research Centre
Castleknock
Dublin 15

ITALY

Prof E Cosentino

Instituto di Produzione Animale
Facolta di Agraria
Portici
Naples

Prof V Russo

Instituto Allevamenti Zootecnici
Facolta di Produzione Animale
Villa Levi - Coviolo
Reggio Emilia

Dr P Santoro

Instituto Allevamenti Zootecnici
Facolta di Produzione Animale
Villa Levi - Coviolo
Reggio Emilia

NETHERLANDS

Dr G Eikelenboom

Instituut voor Veeteeltkundig
Onderzoek
'Schoonoord'
P O Box 501
NL 3700 AM
Zeist

Dr P G van der Wal

Research Institute for Animal
Production
'Schoonoord'
P O Box 501
NL 3700 AM
Zeist

SWEDEN

Dr K Lundström

Department of Animal Breeding and
Genetics
Swedish University of Animal Sciences
S-75007 Uppsala

Dr G Malmfors

Department of Animal Breeding and
Genetics
Swedish University of Animal Sciences
S-75007 Uppsala

UNITED KINGDOM

Dr A Archibald	AFRC Animal Breeding Research Organisation West Mains Road Edinburgh EH9 3JQ
Dr N Gregory	AFRC Meat Research Institute Langford Bristol BS18 7DY
Ms J Guise	Department of Medicine Royal Veterinary College Field Station Hawk's Head Lane North Mymms Hatfield Herts AL9 7TA
Dr A J Kempster	Planning and Development Group Meat and Livestock Commission P O Box 44 Queensway House Bletchley MK2 2EF
Dr D Lister	The Animal and Grassland Research Institute Church Lane Shinfield Reading RG2 9AQ
Dr B W Moss	Department of Agriculture and Food Chemistry Queen's University of Belfast Belfast BT9 5PX
Ms O Southwood	AFRC Animal Breeding Research Organisation West Mains Road Edinburgh EH9 3JQ
Dr P Warriss	AFRC Meat Research Institute Langford Bristol BS 18 7DY
Dr A J Webb	AFRC Animal Breeding Research Organisation West Mains Road Edinburgh EH9 3JQ
Mr D Wright	Department of Agriculture Dundonald House Newtownards Road Belfast

USA

Prof R G Kauffman

University of Wisconsin-Madison
Muscle Biology Laboratory
College of Agricultural and Life
Sciences
1805 Linden Drive
Madison
Wisconsin 53706

YUGOSLAVIA

Dr M Perović

Yugoslav Institute of Meat Technology
11000 Beograd .
Kaćanskog 13

Dr K Benčević

Poslouna Zajednica za Stočarotvo
41000 Zagreb
Amruševa 8